PLANNING AFTER PETROLEUM

The past decade has been one of the most volatile periods in global petroleum markets in living memory, and future oil supply security and price levels remain highly uncertain. This poses many questions for the professional activities of planners and urbanists because contemporary cities are highly dependent on petroleum as a transport fuel. How will oil dependent cities respond, and adapt to, the changing pattern of petroleum supplies? What key strategies should planners and policy makers implement in petroleum vulnerable cities to address the challenges of moving beyond oil? How might a shift away from petroleum provide opportunities to improve or remake cities for the economic, social and environmental imperatives of twenty-first-century sustainability?

Such questions are the focus of contributors to this book with perspectives ranging across the planning challenge: overarching petroleum futures, governance, transition and climate change questions, the role of various urban transport nodes and household responses, ways of measuring oil vulnerability, and the effects on telecommunications, ports and other urban infrastructure. This comprehensive volume – with contributions from and focusing on cities in Australia, the UK, the US, France, Germany, the Netherlands and South Korea – provides key insights to enable cities to plan for the age beyond petroleum.

Jago Dodson is Professor of Urban Policy and Director of the Centre for Urban Research, RMIT University (Melbourne, Australia). His work has addressed theoretical and applied problems in housing, transport, urban planning, infrastructure, energy and urban governance. He has advised governments on urban policy and is active in scholarly and public debates about Australian cities.

Neil Sipe is Professor of Planning in the School of Geography, Planning and Environmental Management at the University of Queensland (Brisbane, Australia). His research interests include transport and land-use planning, natural resource management and international comparisons of planning systems.

Anitra Nelson is Associate Professor at the Centre for Urban Research, RMIT University (Melbourne, Australia). She edited *Steering Sustainability in an Urbanizing World: Policy, Practice and Performance* (2007), co-edited *Sustainability Citizenship in Cities: Theory and Practice* (2016, Earthscan/Routledge) and is writing *Small is Necessary: Shared Living on a Shared Planet* (2017).

"When future generations look back on today's struggle to move off oil as the lifeblood of global society, they will wonder why it took so long for people to see the writing on the wall and find a better way to power the engines of human endeavor. This volume makes an important contribution to that writing on the wall and presents promising tools needed to deal with our energy problems. If contemporary economic and political leaders can learn from the thoughtful approaches in this book, the inevitable post-carbon future that awaits will bring a brighter day for human civilization."

Anthony Perl, Professor of Urban Studies & Political Science, Simon Fraser University, Canada

PLANNING AFTER PETROLEUM

Preparing cities for the age beyond oil

Edited by
Jago Dodson, Neil Sipe and Anitra Nelson

Routledge
Taylor & Francis Group
LONDON AND NEW YORK

earthscan
from Routledge

First published 2017
by Routledge
2 Park Square, Milton Park, Abingdon, Oxon, OX14 4RN

and by Routledge
711 Third Avenue, New York, NY 10017

Routledge is an imprint of the Taylor & Francis Group, an informa business

Library of Congress Cataloging-in-Publication Data
Names: Dodson, Jago, editor. | Sipe, Neil G., editor. | Nelson, Anitra, editor.
Title: Planning after petroleum : preparing cities for the age beyond oil / edited by Jago Dodson, Neil Sipe and Anitra Nelson.
Description: New York, NY : Routledge, 2016. | Includes bibliographical references and index.
Identifiers: LCCN 2016012575| ISBN 9780415504577 (hardback) | ISBN 9780415504584 (pbk.)
Subjects: LCSH: Cities and towns—Energy consumption. | Urban transportation. | Sustainable urban development. | Energy policy. | Urban policy. | Petroleum industry and trade.
Classification: LCC HD9502.A2 P588 2016 | DDC 333.7909173/2—dc23
LC record available at https://lccn.loc.gov/2016012575

ISBN: 978-0-415-50457-7 (hbk)
ISBN: 978-0-415-50458-4 (pbk)
ISBN: 978-1-315-65071-5 (ebk)

Typeset in Bembo
by Apex CoVantage, LLC

CONTENTS

FIGURES

TABLES

ACKNOWLEDGEMENTS

This book grew from research that had culminated in a special issue of the *Australian Planner* 47(4), *Cities and Oil Vulnerability*, which was published in 2010 (see Chapter 1 in this book for details). Therefore, the editors and relevant contributors of this book would like to thank the publisher of the journal, Taylor and Francis, for permitting a range of extracts and images drawn from that special issue, which Jago Dodson and Neil Sipe edited. The articles from which material has been drawn and revised include: "Dark Clouds on the Urban Horizon: Petroleum and Australian Planning" by Jago Dodson (226–31); "Planning Public Transport Networks in the Post-Petroleum Era" by (the late) Paul Mees and John Stone (263–71), which has been revised by Stone with the kind permission of Erica Cervini, who is the copyright holder of Paul Mees' creations; "Rethinking Oil Depletion: What Role Can Cycling Really Play in Dispersed Cities?" by Mathew Burke and Jennifer Bonham (272–83); "The Hope for Oil Crisis; Children, Oil Vulnerability and Independent Mobility" by Scott Sharpe and Paul Tranter (284–92); and "Emerging Australian Planning Practice and Oil Vulnerability Responses" by Jago Dodson and Neil Sipe (293–301).

Samuel Alexander would like to thank Josh Floyd, Matt Mushalik and Jonathan Rutherford for helpful comments on an earlier version of Chapter 3, "The Paradox of Oil: The Cheaper It Is, the More It Costs," with the assurance that errors are the responsibility of the author. Chapter 3 in this book is a revised and updated version of Chapter 8 of Samuel Alexander's collected essays, *Prosperous Descent: Crisis as Opportunity in an Age of Limits*, which was published in 2015 by the Simplicity Institute (Melbourne, Australia).

The authors would like to acknowledge Australian Research Council projects DE120102428 and DP150100299 for Chapter 6, "Post-petroleum Urban Justice."

Jason Byrne gratefully acknowledges the assistance received from: Kie-Wook Kwon, Director General, Water Management Policy Department, Seoul Metropolitan Government, Korea; Assistant Professor Alexander Robinson from the University of Southern California School of Architecture; and Katherine Burgess from the Landscape Architecture Foundation in sourcing images and information for the Korean case study in Chapter 13, "Greenspace After Petroleum: From Freeways to Greenways."

Tiebei Li, Neil Sipe and Jago Dodson, the authors of Chapter 15, "Motor Vehicle Fleets in Oil Vulnerable Suburbs: A Prospect of Technology Innovations," would like to acknowledge the Australian Research Council project DP1095562.

Cheryl Desha and Angela Reeve, the authors of Chapter 16, "Energy for Cities," acknowledge that the content in this chapter references previous work and publications by The Natural Edge Project team, of which the authors are members. The authors acknowledge Dr. Charlie Hargroves, Mr. David Sparks and Ms. Fiona McKeague for their contributions to this research area. Chapter 16 highlights the outcomes and achievements of several initiatives of Townsville City Council. The authors acknowledge the council for its leadership and for sharing its experiences, in particular Mr. Greg Bruce, Mr. Mark Robinson and Mr. Dylan Furnell for their dedication to making a difference, and for their assistance with the case study.

We are grateful for permissions gained from a range of publishers to use other material where the source has been fully specified in the text according to their requests.

Finally, the editors Jago Dodson, Neil Sipe and Anitra Nelson would like to acknowledge the assistance of numerous supporters that enabled this book to come to fruition.

First, we recognize the contributions of our wonderful authors who stuck with the project over a number of years. Their efforts provide new and insightful contributions to understanding the problem of petroleum constraint and oil vulnerability.

Second, we would like to thank the former Queensland Minister of Environment and Sustainability, Mr. Andrew McNamara, and the Queensland Department of Environment and Heritage Protection (formerly the Queensland Environmental Protection Agency) for the small grant that supported background and preparation work for this book.

Third, we would like to thank editors and editorial staff at Routledge and Earthscan – especially Nicole Solano, Nicki Dennis and Judith Newlin – for their patient support during the long preparation of this book.

Fourth, Jago Dodson and Neil Sipe would like to thank the many and various colleagues in academia and beyond who encouraged and supported their research program in this area since 2005, especially those who disseminated their results and reproduced their findings in various papers, reports and PowerPoint presentations.

Fifth, the editors – and the many contributors who worked with him – recognize the important contribution of Paul Mees, who passed away during the preparation of the book. Much of Paul's work drew on the recognition of the finite nature of global oil reserves and the need for cities to move quickly to more sustainable forms of transport and land-use arrangements.

Last, but in no way least, Jago Dodson and Neil Sipe would like to thank colleague and co-editor Anitra Nelson, who came onto the project relatively late in the piece, but without whose contribution, both conceptually and practically, the book would not have been completed.

ABBREVIATIONS

A$	Australian dollars
ABS	Australian Bureau of Statistics
ADM	adult dependent mobility
AEV	alternative energy vehicle
ALP	Australian Labor Party
ATAG	Air Transport Action Group
ATF	aviation turbine fuel
CAFE	Corporate Average Fuel Economy
CBD	central business district
CD	collection district
CIM	children's independent mobility
COAG	Council of Australian Governments
COP21	twenty-first session of the Conference of Parties
EPCA	Energy Policy and Conservation Act of 1975
EROI	energy return on investment
EU	European Union
EV	electric vehicle
GDP	gross domestic product
GFC	global financial crisis
GHG	greenhouse gas
Gt	gigatonne
GW	gigawatt
IA	Infrastructure Australia
IATA	International Air Transport Association
ICAO	International Civil Aviation Organization
IEA	International Energy Agency
IEF	International Energy Forum
IMF	International Monetary Fund
IPCC	Intergovernmental Panel on Climate Change

ITPOES	Industry Taskforce on Peak Oil and Energy Security
JTW	journey to work
kg	kilogram
km	kilometer
km/h	kilometers per hour
KVA	1,000-volt amps
kW(s)	kilowatt/s
kWh	kilowatt hour
L/100km	liter/s per 100 kilometers
mbd	million barrels per day (of oil)
MBM	market-based mechanism
MJ	megajoule/s
MMBW	Melbourne and Metropolitan Board of Works
mpg	miles per gallon
mph	miles per hour
Mtoe	million tonne/s of oil equivalent
NBN	National Broadband Network
n.d.	no date
NHT	National Highway Trust
NOC	national oil company
NSW	New South Wales
NYC	New York City
OAM	Order of Australia Medal
OECD	Organisation for Economic Co-operation and Development
OPEC	Organization of the Petroleum Exporting Countries
p.a.	per annum
PM	particulate matter
PV	photovoltaic
RNP	required navigational performance
RPK	revenue passenger kilometers
SEQ	South East Queensland
SEQIP	South East Queensland Infrastructure Plan
SEQRP	South East Queensland Regional Plan
UK	United Kingdom
UNFCCC	United Nations Framework Convention on Climate Change
US	United States (of America)
US$	US dollars
VAMPIRE	vulnerability assessment for mortgage, petroleum, and inflation risks and expenditure
VFE	vehicle fuel efficiency
VIPER	vulnerability index for petroleum energy rises
WSB	walking school bus
WWII	World War II

NOTES ON CONTRIBUTORS

Samuel Alexander is Research Fellow (Melbourne Sustainable Society Institute) and Co-Director of the Simplicity Institute. He lectures in the Office for Environmental Programs at the University of Melbourne (Australia). Recent books include *Prosperous Descent: Crisis as Opportunity in an Age of Limits* (2015) and *Sufficiency Economy: Enough, for Everyone, Forever* (2015).

Tooran Alizadeh lectures in the School of Environment (Urban and Environmental Planning) at Griffith University (Nathan, Australia) and is an experienced international professional practitioner. Research interests include spatial planning, strategic planning, post-disaster planning and, most recently, the planning and policy implications of telecommunication (specifically high-speed broadband), urban digital strategies and smart cities.

Douglas Baker is Professor of Property and Planning in the School of Civil Engineering and Built Environment at Queensland University of Technology (Brisbane, Australia). His research interests include regional air transport, airport planning and land-use planning.

Brendan F.D. Barrett is Research Fellow at RMIT University (Melbourne, Australia), where he is Research Coordinator of the United Nations Global Compact Cities Programme. His research interests cover urban issues, energy transitions, climate policy and sustainability science. He is a visiting expert at the United Nations University (Tokyo, Japan).

Leslie Belton-Chevallier is Sociologist of Mobility at Université Paris-Est (Paris, France). Her main research themes focus on interactions between – and different forms of – mobility, such as daily, residential, real and virtual mobility within households.

Jennifer Bonham is Senior Lecturer in the School of Social Sciences, University of Adelaide (South Australia). She has a background in human geography and specializes in urbanization and cultural practices of travel. She is co-editor of the book *Cycling Futures* and a member of the Australian Cycling Conference Reference Group.

Matthew Burke is an Australian Research Council Future Fellow in the Urban Research Program, Griffith University. Research interests include transport and land-use planning, travel behavior and transport policy. Matthew's current projects are exploring value capture funding and financing for public transport, light rail in Australian cities and the transport impacts of employment decentralization.

Jason Byrne is Associate Professor in Urban and Environmental Planning at Griffith University (Gold Coast, Australia), where he has lectured since 2006. Research interests include urban nature, parks, green space and environmental justice. Among more than 80 scholarly publications, he co-edited the award-winning *Australian Environmental Planning: Challenges and Future Prospects* (2014).

Cheryl Desha is Senior Lecturer in Sustainable Development and Discipline Leader (Environmental Sciences) in the Science and Engineering Faculty, Queensland University of Technology (Brisbane, Australia). She is also a principal researcher in The Natural Edge Project, a non-profit partnership on research, education and strategy for innovation for sustainable prosperity.

Jago Dodson is Professor of Urban Policy and Director of the Centre for Urban Research, RMIT University (Melbourne, Australia). He has extensive experience addressing theoretical and applied problems in housing, transport, urban planning, infrastructure, energy and urban governance. He advises agencies on urban policy and is prominent in scholarly and public debates about Australian cities.

Ralph Horne is Professor of Geography and Deputy Pro-Vice Chancellor, Research and Innovation, College of Design and Social Context, RMIT University (Melbourne, Australia). His research concerns social and policy change for sustainable design and development, particularly in housing and for households, and environmental, social and policy contexts of production and consumption in urban environments.

Md. Kamruzzaman is Senior Lecturer in Planning at Queensland University of Technology (Brisbane, Australia). His research interests include transit-oriented development, transport disadvantage, travel behavior, and the application of geographic information systems (GIS) and remote sensing in planning.

Lisa de Kleyn is a PhD candidate at RMIT University (Melbourne, Australia), researching the management and use of the Toolangi State Forest (Victoria) from an environmental justice perspective. She has many years of experience as an environmental sustainability consultant for companies, non-governmental organizations and government bodies, supporting environmental projects and programs.

Tiebei Li is a quantitative urban geographer working in the Centre for Urban Research at RMIT University (Melbourne, Australia). His research focuses on developing a geographical understanding of the dynamics of urban and transport systems and their interactions with social and environmental systems in both Australian and international contexts.

Tony Matthews is Lecturer in Urban and Environmental Planning at Griffith University (Nathan, Australia). He is a Chartered Member of the Royal Town Planning Institute. His research focuses on the nexus between spatial planning and climate change adaptation, resource constraints, green infrastructure, urban regeneration and community-led public art.

Paul Mees, OAM, was Associate Professor in Transport Planning at RMIT University (Melbourne, Australia) until his death in 2013. His pioneering research into the relationship between density and public transport performance and insights into efficiencies of well-planned public transport networks stand among his many valuable academic legacies.

Benjamin Motte-Baumvol is Associate Professor at the University of Burgundy (Dijon, France) and completed his PhD in geography. His research focuses on spatial inequalities and daily mobility in France and Brazil.

Anitra Nelson is Associate Professor at the Centre for Urban Research, RMIT University (Melbourne, Australia). She edited *Steering Sustainability in an Urbanizing World: Policy, Practice and Performance* (2007, Ashgate), co-edited *Sustainability Citizenship in Cities: Theory and Practice* (2016, Earthscan/Routledge), and is writing *Small Is Necessary: Shared Living on a Shared Planet* (2017, Pluto Press).

Angela Reeve is Associate Lecturer in the Science and Engineering Faculty, Queensland University of Technology (Brisbane, Australia). An environmental engineer, she specializes in mainstreaming sustainability within the built environment, and is a senior researcher in The Natural Edge Project on research, education and strategy for innovation for sustainable prosperity.

Katelyn Samson is Research Associate for the Australian Research Council Discovery "Enabling Social Innovation for Local Climate Adaptability" project in the Centre for Urban Research at RMIT University (Melbourne, Australia). A social scientist working with the social dimensions of environmental and natural resource governance issues, she lectures while completing her PhD.

Scott Sharpe is a cultural geographer at University of New South Wales, Canberra (Australia). His research interests in cultural geography, and social and cultural theory, center broadly on relationships between thinking and politics, resulting in projects on the academy, affect, humor, aesthetics, movement and habit, racism and anti-racism, and anti-globalization movements.

Neil Sipe is Professor of Planning in the School of Geography, Planning and Environmental Management at the University of Queensland (Brisbane, Australia). His research interests include transport and land-use planning, natural resource management and international comparisons of planning systems.

Wendy Steele is Associate Professor in Urban Planning and Sustainability in the Centre for Urban Research and School of Global, Urban and Social Studies at RMIT University (Melbourne, Australia), a recent Australian Research Council DECRA Fellow and international editorial board member of *Urban Policy and Research*. She wrote *Wild Cities* (Routledge, 2016).

Nicholas Stevens is Researcher and Lecturer at the University of the Sunshine Coast (Maroochydore, Australia). He is a landscape architect and urban planner with research interests in airport and regional planning, and complex systems understandings of land-use planning and urban design.

John Stone is Senior Lecturer in the Urban Planning Program in the Faculty of Architecture, Building and Planning at the University of Melbourne (Australia). His research explores the political, institutional and technical factors that support change to more sustainable urban transport, and techniques to support practical international policy learning.

Paul Tranter, Associate Professor of Geography at the University of New South Wales (Canberra, Australia), has made a pioneering contribution to research on child-friendly environments, active transport, and healthy and sustainable cities – demonstrating that urban environments supporting child-friendly transport modes (walking, cycling and public transport) make our cities more livable, healthy and resilient for residents.

John Whitelegg, Visiting Professor of Sustainable Transport at Liverpool John Moores University and Research Associate of the Stockholm Environment Institute, is editor of the journal *World Transport Policy and Practice* and has written 10 books, including *Quality of Life and Public Management* (2012, Routledge) and *Mobility* (2015, Straw Barnes Press).

FOREWORD

Brendan Gleeson

This comprehensive and timely volume urges us to consider the prospects for *Planning after Petroleum*. The book's chapters consider the new planning imperatives, frameworks and methodologies that will be necessary for *Preparing Cities for the Age Beyond Oil*. There is a lot packed into these urgent injunctions, but perhaps the biggest proposition that the volume contains is that of transformative historical change. *Planning after Petroleum*, as its title suggests, is a work premised on the looming transition to a world where oil is simply less central to human affairs. After a century or more of petroleum-fueled industrialization and urbanization, this is indubitably a change in the very way in which our species does its business. Its impacts cannot be understated and deserve the kind of thorough prospective analysis that this book offers.

It is easy to become somewhat glazed-eyed about this: How many times in recent decades have we heard it announced that "humanity is at a crossroads"? The record shows that too many of these pronouncements were premature, baseless or inflated. Yet, as Terry Eagleton (2015: 96) has pointed out:

> One should not rejoice at the mere prospect of open-endedness. The Third Reich rejected closure, intent as it was on enduring indefinitely. No historical system has been more mutable than capitalism.

Part of this mutability is the market's wondrous capacity to break through barriers to realization via complex, adaptive (though, also, often violent and wasteful) shifts in the use of resources – natural and human. In this sense, capitalism seemed to work around early declarations of "peak oil" over the past decade through technical innovation, as well as rawer methods, including the revalorization of natural residuals (the Canadian tar sands industry comes to mind). There is, however, mounting and irrefutable evidence that even these marvelously adaptive capacities will not prevent an end to the age of oil from occurring in the near future. This book's early chapters explain the prognosis clearly and convincingly.

So, we do not face the end of history – always a ludicrous claim – but the termination of a recent and profoundly important socio-technical phase: the oil age. However, many (including those on the left) will continue to depict this view as Chicken Little thinking – yet another tedious instance of a general failed tendency to predict ruptures, especially transformative ones. Yet, this understandable concept fatigue belies the reality that human history, especially since the rise of capitalism, has been patterned by episodes of rapid, unforeseen change that interrupted what seemed like periods of settled social and technical arrangements. In Western countries, the sudden, in many places precipitous, decline of rail travel in the wake of post-WWII motorization is an instance of such rapid socio-technical change. The decline of the oil age, however, is likely to be much more disruptive, given the centrality of this commodity not only to mobility but also to the entire productive and consumptive circuits of contemporary capitalism. The transition to a post-oil future is made even more complex by the ascension of the urban age and the continued a rapid urbanization of the human species. "Oil" and "urban" have been seen as indissolubly linked features of late capitalism and we do not yet have a general model of city shaping that transcends this relation. New urban frameworks are likely to be forced on us by the changes considered in this book.

Unforeseen and unplanned change is frequently disruptive and costly. The progressive political mind has long argued, and lamented, that this is a hard-wired feature of capitalism, captured in Schumpeter's oft-quoted depiction of "creative destruction." Progressives have argued historically for various versions of a socially and environmentally logical alternative dispensation. To absent that difficult and elusive possibility, the progressive project has focused on the task of imposing the values and techniques of planning and conservation on a political economy that systematically rejects both. This book continues in that latter tradition and presents some extremely insightful and valuable assessments of the many planning tasks that should attend, and preferably precede, the imminent end of the oil epoch. Its chapters mine many important and often neglected themes, including the health and social correlates of urban vulnerability in an age where the "stormy oil horizon" is now clearly in sight. It is especially pleasing to see children considered in this analytical mix given the long inattention to their needs that has regrettably plagued planning thought and practice.

The end of oil coincides with other disruptive changes, notably and most worryingly the heating of the global climate and the continued and seemingly unstoppable deterioration of the global biota. That these now clearly foreseeable changes are likely to be experienced as devastating shocks is further testament to the incorrigible tendency of capitalism not to recognize – let alone rein in – its many destructive forces. None of this, of course, should deter efforts, like this book, to plead the causes of foresight and preparation in an age of global endangerment. We must strive harder than ever to be a disruptive force ourselves, seeking to disturb and overturn a species sensibility that is well described by Augustine's phrase, *perversa securitas*.

I conclude with an observation that might not be shared by all of this book's contributors – that the end of the oil age may well coincide with a more general dissolution of capitalism, a view being retailed with some force recently on the left (see Streeck 2014). In our hour of species peril, those hopeful seek a new world but only from the ruin that seems sure to come, ordained by an order stubbornly clinging to its destructive prowess. Retrenchment of capitalism will not present a straight path back to ecological moderation or social peace. Apart from sparking heaven only knows how many more gruesome wars and migrations, capitalism's death agonies will likely generate many desperate quests for salvation through resource exploitation. These misadventures are prefigured in the contemporary lust for Arctic exploitation, the fracking rush in the New World, and the enthusiasm for newly unlocked carbon, such as Canada's oil sands. Prometheus, it seems, will reign until dethroned by collapse. All cries for abdication will go unheard. The task seems no longer to impose a better future but to prepare for an imposed future.

Capitalism may well have entered its terminal phase, but this does not mean the end of humanity or even modernity. Our species will have to take to the road of history again. Looking ahead to prospect, Eagleton (2015: 115) writes, "Hope . . . is what survives the general ruin," and further:

> Though there will be no utopia, in the sense of a world purged of discord and dissatisfaction, it is sober realism to believe that our condition could be mightily improved. It is not that all will be well, but that all might be well enough.
>
> *(Eagleton 2015: 133)*

In this light, *Planning after Petroleum* can be seen as an investment in the cause of hope, by looking unflinchingly into the maw of disruptive and frightening change, but also seeing beyond to the new human possibilities that will surely flourish in the age beyond oil. It is a commendable project.

Brendan Gleeson
Director of the Melbourne Sustainable Society Institute,
The University of Melbourne, Australia

References

Eagleton, T. (2015) *Hope Without Optimism*, New Haven/London: Yale University.
Streeck, W. (2014) "How Will Capitalism End?" *New Left Review* 87: 35–64.

Introduction

1

INVESTIGATING CITIES AFTER OIL

Planning for systemic urban oil vulnerability

Jago Dodson, Neil Sipe and Anitra Nelson

This book is founded in the recognition that the past decade has been one of the most volatile periods for petroleum markets since WWII. Planners and urbanists need to understand the implications of this new era of volatility for car dependent, petroleum reliant cities. Oil prices grew rapidly from low levels in the early 2000s to near-record post-WWII highs, in real terms, in the late 2000s. The price of oil then fell with the onset of the global financial crisis (GFC) and, as of early 2016, was near levels similar to the early 2000s, in real terms.

The future is always uncertain, but shifts in both the price of oil and dynamics of petroleum production – and, to a degree, consumption – are arguably the most complex since WWII, when petroleum became systematically entrenched as a fuel of globalizing modernity. Although the oil shocks of the 1970s took oil-consuming nations by surprise, their underpinnings were a simple combination of supply restrictions driven by geopolitical tensions. The contemporary era is much more economically and geopolitically complex, with increasing strain between oil producing nations participating in ongoing conflicts within a Middle East that remains wracked with the effects of Western intervention in the 2000s and local revolts against repressive regimes. Some Middle Eastern suppliers face stark national structural economic fragility around their own long-run dependence on petroleum exports and competition from upstart producers, such as the tight oil drillers in the US.

The large consuming nations remain dependent on readily available petroleum but have offered distinct signals that they are prepared to actively shift their reliance away from petroleum imports to domestic supplies or away from petroleum altogether. This includes using deliberate measures to reduce petroleum energy dependence as well as wider global initiatives to curb carbon emissions from the burning of fossil fuels. A growing body of scientific, industry and financial sector analysis anticipates a future in which fossil fuels will no longer provide for global energy consumption. This prospect is very significant for nations whose economic viability is dependent on energy exports.

Woven through this complexity is the struggle within the advanced and emerging economies of the Global North and Global South to pass beyond the low-growth era heralded by the GFC. So far, few countries have found the mix of policy settings that can achieve this; with China's economic trajectory uncertain and with falls in global stock markets in late 2015 and early 2016, the prospect of a collective global geopolitical or economic solution to the low-growth era looks unclear over the medium term.

Although recent reductions in oil prices can offer respite to a struggling global economy, cost reduction needs to be accompanied by moderation in consumption to achieve carbon emissions objectives.

The global petroleum-industrial complex, supporting our urban modernity, looks uncomfortably positioned in an interstitial void. In the foreword to this book, Brendan Gleeson's commentary homes in on the moment of transition between the petroleum era and a more sustainable energy epoch, emerging and yet only embryonically realized. As the gigantic production apparatus propelled by the momentum of a fossil economy fails under the weight of its own supply and demand contradictions, misty visions arise of a new territory – a post-carbon economy – some years off. The world faces an oil crisis, but one that is different to the shocks of the 1970s or the progressive escalation of challenges during the 2000s. The present crisis appears to be a drawn-out crisis of transition between one economic, industrial and urban formation and the not yet established, post-carbon complex. Little can be discerned with certainty.

Many observers of petroleum production speculated that a "peak" would occur in the mid-2010s. Such appraisals appear to have been borne out within the Energy Information Administration (EIA) data for conventional liquid petroleum; new liquids are increasingly being drawn from non-conventional offshore, gas, shale and synthetic sources. Many observers of petroleum depletion did not predict a precipitous collapse in global economies due to increasingly scarce conventional oil but, rather, a volatile period of price surges and declines as uncertainty about supply spurred new investment. Part of this volatility is due to the dynamic of demand destruction – the transfer to alternative fuels and technologies not yet scalable to the level of oil-based systems. Volatility looks to be the overarching characteristic of the petroleum sector in supply, demand and price, for the short term and the medium term.

This problematic context poses many questions for the professional activities of planners and urbanists. How will oil dependent cities respond, and adapt to, the changing structure of petroleum supplies? What key strategies should planners and policy makers implement in petroleum vulnerable cities to address the challenges of moving beyond oil? How might a shift away from petroleum, as a viable form of energy, provide opportunities to improve or remake cities for the economic, social and environmental imperatives of twenty-first-century sustainability? Such questions form the starting point for this book.

Indices of oil vulnerability: VIPER and VAMPIRE

This text has its origins in a stream of work on oil vulnerability begun in early 2005 by its primary contributors, Jago Dodson and Neil Sipe. That work investigated the likely impacts of accelerating oil prices on Australia's predominantly suburban and automobile dependent major cities. It started by developing new methods for examining oil vulnerability at the sub-metropolitan level. This resulted in the "vulnerability index for petroleum expenditure and risks" (VIPER) and the "vulnerability assessment for mortgage, petroleum and inflation risks and expenditure" (VAMPIRE). These indices were presented at conferences and published in key urban journals, and reproduced, by invitation, in showcase volumes. The associated work generated a high level of public and professional interest leading to references in key policy reports in Australia.

In the late 2000s Dodson and Sipe were asked to edit a special issue of the prestigious *Australian Planner* journal. This issue examined the problem of oil vulnerability from a multidisciplinary perspective. That effort, with the addition of further international perspectives from the UK and France, led to the present text. Drawing on such various works, this book has provided an opportunity to explore in further detail the problem of oil vulnerability and the questions it raises for cities.

The book format has allowed for a wider array of perspectives, covering topics not previously considered in oil vulnerability scholarship, such as walking ("Walking the City," Chapter 7 by John Whitelegg); access to parks ("Greenspace After Petroleum: From Freeways to Greenways," Chapter 13 by Jason Byrne); broadband digital connectivity ("The Role of Telecommunication in Post-petroleum Planning," Chapter 17 by Tooran Alizadeh); and airports ("Peak Oil: Challenges and Changes for the Air Transport

Industry," Chapter 18 by Douglas Baker, Nicholas Stevens and Md. Kamruzzaman). We've also expanded our scope beyond metropolitan centers of the US, the UK and Australia to incorporate challenges, research and initiatives in European and Asian cities, such as Paris ("Outer Suburbs, Car Dependence and Residential Choice in France," Chapter 12 by Benjamin Motte-Baumvol and Leslie Belton-Chevallier); Seoul (Chapter 13); Freiburg-im-Breisgau (Chapter 7); the Dutch Sustainable Technological Development program ("Energy for Cities," Chapter 16 by Cheryl Desha and Angela Reeve); and an international survey of energy transition efforts in eighteen cities ("Local Energy Plans for Transitions to a Low Carbon Future," Chapter 14 by Brendan F.D. Barrett and Ralph Horne).

Oil vulnerability

This book approaches oil vulnerability from the standpoint that petroleum conditions globally shape supply and prices at the national level. In turn, the costs of automobile travel within cities are conditioned by the price of petrol at the pump. This price, in turn, has an array of direct and indirect effects. The primary effect of higher oil prices is to increase the cost of automobile travel both in absolute terms and relative to the cost of other modes, such as walking, cycling or public transport.

Simple micro-economic theory suggests that a rise in the oil price will lead to a reduction in the quantity of fuel demanded by motorists, but conversely will signal to producers to pump more. If prices remain high, they will spur substitution effects as consumers look for more cost-effective means of traveling, whether by mode or by vehicle efficiency. Systemic and sustained price rises, such as those implied by notions of "peak oil," will likely lead to systemic shifts in petroleum demand.

Such an environment is relatively new to urban planning, which has produced urban systems – especially in the US and Australia – that are highly dependent on petroleum for the viability of their transport and land-use systems. Such arrangements involve large capital and institutional structures to support them and are difficult to adjust or adapt within short time frames. Adaptive success depends on long planning horizons. With the global oil production system in crisis and the future uncertain, it is time that urban planners begin to prepare cities for the period after petroleum. This compilation aims to contribute to the intellectual demands of this task.

The structure

This work is structured in three parts. Part I reviews the strategic horizons for energy economics, and for societies and cities that are dependent on petroleum for their functional viability. This topical focus begins with Chapter 2, "A Stormy Petroleum Horizon: Cities and Planning Beyond Oil," by Jago Dodson, who assesses the trajectory of petroleum prices from the early 2000s. Dodson argues that official analyses became increasingly concerned about oil price rises in the late 2000s, but the GFC distracted them from this focus by posing an immediate and systemic problem for global economies. Dodson reviews the notion of peak oil to argue that theories of peak oil are not disproven by recent shifts in petroleum production and prices observed in petroleum markets. Rather, Dodson argues, the potential for supply disruption remains strong within the context of an increasingly complex global supply and consumption dynamic. The main regions of the globe where the effects of this increasing volatility will be felt are the large car dependent cities of the US, Canada and Australia. They are the areas, Dodson concludes, that should be the focus of efforts to reduce oil vulnerability.

This theme of global petroleum economics is developed further by Samuel Alexander in Chapter 3, "The Paradox of Oil: The Cheaper It Is, the More It Costs." Alexander's premise is that the underlying dynamics of petroleum production and supply in global markets faces systemic contradictions that will likely lead to long-term price escalation or supply shocks. Alexander focuses on recent developments in oil markets to argue that low oil prices are as much a challenge as high prices because of the erratic signals that they send to markets where the development of new production operates on multiyear, typically decadal, investment cycles. This leads to retardation of midterm to long-term production with ongoing prospects of capital destruction where new production investment is stranded by short-term price falls. The only choice for industrial (urban) society, Alexander argues, is to play a different game, "beyond oil." But transitioning to such a new world would require bravery and decisiveness, factors plainly lacking in contemporary political and policy spheres.

In Chapter 4, "Institutional Planning Responses to a Confluence of Oil Vulnerability and Climate Change," Tony Matthews and Jago Dodson deploy the notion of a "transformative stressor," which has the potential to drive systemic changes within institutional and governance structures. These contributors see climate change and oil vulnerability as confluent stressors whose effects will likely occur in tandem. Matthews and Dodson explore the ramifications of the intersection of these transformative stressors within spatial and land-use planning governance and identify key areas for policy change to respond to associated stresses.

Chapter 5, "Energy Security and Oil Vulnerability Responses," by Jago Dodson and Neil Sipe, investigates how national and subnational policy institutions in both the US and Australia are responding to the stresses caused by changing oil market dynamics, through national energy strategies, state and regional policies, and metropolitan or local planning. They reveal that, although energy security is clearly a problem for national policy agendas, the substantive actions taken by governments in the US, and especially in Australia, are inadequate to deal with the economic adjustment implied by underlying dynamics of oil production and consumption (as outlined in earlier chapters). Dodson and Sipe further demonstrate that, despite some notable exceptions, state and local policy to reduce oil vulnerability is rare. Consequently, most cities are poorly positioned to respond to any rapid escalation of oil prices.

Part I is rounded off by a reflective and normative discussion in Chapter 6, "Post-petroleum Urban Justice," by Wendy Steele, Lisa de Kleyn and Katelyn Samson. They center their discussion on two questions: "Who is served by the transition to a post-petroleum city?" and "What is the role of planning in this transition?" They identify key assumptions, or foundations, for the practice of urban climate justice and the relationship between energy equity and oil justice in cities. They critique the role of planners and highlight, rather, how local movements of citizens can participate in redesigning and reinvigorating planning in the oil constrained city. In particular, they argue that collective local community action can re-envisage the city through alternative stories and practices in order to create – on the ground – new post-petroleum possibilities for cities.

Part II focuses on transport and land use – the key features of cities that are likely to be most directly and obviously affected by petroleum production and supply constraints. This part begins with John Whitelegg's Chapter 7, an appraisal of the neglect of walking within transport and urban policy around the globe, a failing that poorly serves the task of adapting cities to a petroleum constrained future. Whitelegg reviews the benefits of walking, and assesses the factors that contribute to higher and lower levels of use of this most fundamental travel mode. He argues that both the prosperity and the inclusiveness of cities depend on raising the significance and practice of walking (and cycling), irrespective of oil vulnerability. Nevertheless, prudence suggests the case is even stronger when questions of petroleum costs (and climate) are considered. Whitelegg prescribes speed limits and zero automobile death targets, along with the eradication of motorist subsidies, as key means to improve the status of walking in cities right away.

In Chapter 8, Jennifer Bonham and Matthew Burke investigate "Cycling Potential in Dispersed Cities." They recognize that cycling is receiving more attention as a transport strategy, but that this attention is rarely linked to oil vulnerability. Bonham and Burke argue that cycling has moved well past its post-WWII trough as a travel mode, even in car dependent cities of Australia and the US. They contend that cycling can play an even greater role as a transport mode if supported by measures to make it easier and safer for cyclists. Bonham and Burke foresee a scenario in which reduced automobile demand, due to higher fuel prices, is paired with modest transport networks and land-use changes to increase cycling's mode share. In particular, under a sustained period of oil crisis, cycling is likely to become the key mode for trips of up to 5 km, including trips feeding public transport.

The focus in this part of the book on active travel takes a novel turn in Chapter 9, "Children's Active Transport: An Upside of Oil Vulnerability?" by Scott Sharpe and Paul Tranter. They argue that the petroleum transport era has generated many costs from children's perspective: reductions in active travel compared to historical patterns, with accompanying negative effects on cognitive development, emotional wellbeing, social participation and physical health. In turn, Sharpe and Tranter suggest that a post-petroleum world will offer the opportunity for children to embrace more traditional modes, such as walking and cycling within local environments less affected by the dangers and risks of the automobile. In particular, they argue that the institution of the school, currently based around the automobile as the key mobility mode, will need to adapt and respond to petroleum constraint by offering new ways of organizing children's education in urban settings.

The next four chapters of the book move beyond the questions of local scale of active transport to those of more systemic urban transport. In Chapter 10, "Public Transport Networks in the Post-petroleum Era," John Stone and the late Paul Mees observe that, while many commentators have suggested that dispersed car dependent cities will require widespread changes in land use to cope with the effects of constrained petroleum supply, this may not be necessary. Rather, current models exist for the coordination of public transport networks in dispersed settings that can offer high levels of mobility and operational efficiency. Stone and Mees argue that such coordination requires sophisticated organizational and governance capacity, which is lacking in most cities, resulting in transport networks that are poorly positioned for a petroleum constrained future. They describe the key changes needed to ensure that public transport systems are optimized for the era after oil.

In Chapter 11, "Oil and Mortgage Vulnerability in Australian Cities," Jago Dodson and Neil Sipe describe the development of their VAMPIRE Index and, via a case study, show how it can be operationalized. The study reveals that Australian cities have large areas of oil vulnerable suburbia with a regressive dimension; that is, lower-income households tend to be located in car dependent areas where they face high relative travel costs and limited access to alternative travel modes, particularly public transport. Dodson and Sipe argue that policy effort is urgently needed to transform these dispersed suburban zones by improving public transport to prepare them for a constrained petroleum environment.

The focus then shifts to parallel developments within France, in Chapter 12 by Benjamin Motte-Baumvol and Leslie Belton-Chevallier. The authors' investigations reveal how suburban residential developments produce car dependence within Paris as households seek to achieve home ownership. Furthermore, as many households in new residential estates become much more car dependent than they were in more central locations, householders adapt to and try to overcome transport poverty through a range of economizing and mutual aid practices, using neighborhood and familial supports. The authors argue in favor of policy to reduce oil vulnerability in these zones, which they contend has been underappreciated by policy makers who downplay the risks of petroleum constraint.

The final contribution in Part II is Chapter 13 by Jason Byrne. He explores how households fare under petroleum constraint in terms of access to "greenspace," that is, vegetated public and private areas such as parks, a variety of recreational areas and verges of waterways and roads, community gardens, and even cemeteries, rock walls, rooftop gardens and drainage corridors. Byrne argues that, under petroleum constraint, greenspaces will become vitally important because of the ecosystem services they contribute, the access to exercise and leisure opportunities that they present and the rejuvenation of urban environments they permit. Urban greenspaces were developed prior to the automobile and mechanized public transport. Petroleum constraint will shift demand for greenspace away from remote, car accessible sites to more local places served by transit. Byrne concludes that reduced motor vehicle use would offer opportunities to reclaim road space for greenspace – an unexpected advantage from this potentially disruptive shift.

Part III examines an array of planning questions, in a wider spatial and infrastructure sense, covering topics such as energy transitions, measurement of oil vulnerability, energy use in cities, telecommunications as a means of relieving oil vulnerability and the challenges posed for air travel and airports.

In Chapter 14, Barrett and Horne review concepts and plans for energy transitions, including diminishing petroleum consumption. They focus on the capabilities and capacities of local governments within the context of increasingly fragmented and splintered governance of infrastructure networks, including urban projects and energy systems. They investigate how local authorities are responding to the imperatives of an energy transition and the sorts of plans that are emerging from this effort, focusing on the results of a survey of 18 global cities within the UN Global Compact Cities Program. They find a shift in energy priorities from a direct concern with energy per se to a wider focus on framing energy concerns within "climate change."

In Chapter 15, "Motor Vehicle Fleets in Oil Vulnerable Suburbs: A Prospect of Technology Innovations," Tiebei Li, Neil Sipe and Jago Dodson discuss techniques used to assess the spatial distribution of oil vulnerability in cities. They consider how fuel-efficient vehicles might be taken up spatially – particularly useful for addressing questions of the transitional phase. Their work shows how moves to increase the fuel efficiency of motor vehicles in response to oil vulnerability will be socially conditioned. The most oil vulnerable populations are likely to be those least able to afford technologies abetting a transition away from oil, resulting in long time frames for change for those who most need it. If left to market forces, technology-based energy transitions will have a strong regressive dimension.

In Chapter 16, Cheryl Desha and Angela Reeve investigate a range of questions for city-wide systems where fossil fuels are no longer the dominant energy sources. Deploying the notion of "decoupling" – the separation of economic prosperity from fossil fuel use – Desha and Reeve use the Dutch Sustainable Technological Development program of the 1990s as an example of an attempt to manage the transition of a complex system to a more sustainable model. They draw on the efforts by Townsville City Council, in Australia's north, which has shown leadership in energy management to drive shifts in resident and institutional behavior.

In Chapter 17, Tooran Alizadeh investigates the potential of high-speed broadband Internet to substitute for physical travel and enable the restructuring of urban spatial and economic relationships in response to petroleum constraint and oil vulnerability. Using the case of Australia's National Broadband Network, Alizadeh identifies key policy issues that impact on the ability of digital communications technologies to support post-petroleum transitions within the wider context of reducing general urban energy demand. One of the challenges that Alizadeh identifies is the poor coordination of digital governance that hinders efforts to link new broadband networks to the wider social and economic imperatives demanded by energy consumption and climate change.

The final topical chapter in the book is by Douglas Baker, Nicholas Stevens and Md. Kamruzzaman. In Chapter 18, the authors assess the implications of petroleum constraint for airports and the cities that rely on them. Baker et al. argue that profit margins in the contemporary aviation sector are already very thin, with little capacity to maneuver in response to higher fuel prices, whether due to petroleum depletion or climate change regulation. They argue that low prices for oil herald uncertainty rather than security. They assess the complex intersection of energy, air transport demand and operations, and the prospects for alternatives to resolve the contradictions posed. They find that the air transport industry has recognized its significant reliance on high octane liquid fuels and is rapidly developing a range of measures – technological, efficiency, infrastructure and market-based – to meet the challenges of a fossil fuel constrained future.

We sought contributions to this book that would provide a multidimensional perspective on oil vulnerability, drawing colleagues from a range of disciplines, including planning, social and cultural geography, transport, human geography, economics, sociology and engineering. Although Part II in particular includes dedicated transport content, we have not construed this book as a transport text. Similarly, even though the petroleum phenomenon can be portrayed in terms of economic narratives and analyses – which various contributors have explored, especially in Part I – our foci have been rather more social and governance (policy) related. In short, we sought to compile a range of perspectives that offer insights into the challenges of petroleum constraint and the possible means to reduce vulnerability to its adverse effects. We might have included analyses of many other worthy questions, but they will have to wait their chance to appear in future books on this topic.

PART I

Energy horizons

2

A STORMY PETROLEUM HORIZON

Cities and planning beyond oil

Jago Dodson

Petroleum is the fuel of globalized modernity. Around one-third of total global energy consumption – especially within transport systems – depends on oil. Cities are where the majority of humanity now lives, and petroleum fuels provide the essential basis for urban travel, most prominently among the spatially dispersed car dependent urban regions of North America and Australia. Since the early 2000s, urban planners have begun to develop a closer understanding of petroleum security issues and to track shifts in global petroleum supplies so they can better craft measures to address their implications for cities. An informed and energy literate planning profession is essential to future public debates over energy use in cities, whether in the car dependent global urban regions or in those urban areas less reliant on petroleum.

This chapter charts the global trajectory of petroleum supply and addresses the urban security and vulnerability questions that this pattern poses. It examines the oil price volatility of the 2000s and 2010s, and the debate that has accompanied those market shifts, and then reviews the problem of petroleum depletion and debates over its probability and timing. The chapter concludes by sketching the initial implications of petroleum depletion for cities and planning from which other authors in this book will expand and expound.

A volatile flow

Cities face an array of social, economic and environmental forces that both spur their ongoing transformation and add to urban stresses. Rapid urbanization, population growth, infrastructure and resource demands all place pressure on urban systems. A major strategic challenge for urban regions – especially among developed nations of the Global North, in North America and Australia – is the vulnerability of cities to the systemic and particular effects of future constraints on global petroleum supplies. The potential for disruptions to oil supplies has been appreciated since the oil shocks of the 1970s. This issue recaptured public and policy attention in the mid-2000s as global oil prices departed from the low and stable levels of the preceding two decades to head sharply upward.

In late 2003 the price of petroleum on global markets was around US$15 per barrel but, by late 2004, had increased to nearly US$50 per barrel. The subsequent six years witnessed extreme volatility in petroleum markets, with prices rising to a peak of US$140 per barrel by late 2008. Oil prices slipped back to US$30 in early 2009 with the onset of the global financial crisis but, by mid-2010, had returned to their pre-crisis levels of approximately US$80 per barrel. Since then they have ranged between US$60 and US$100, although with a sharp decline toward US$30 in 2014–15.

Despite a recent weakening in oil prices, the future trajectory of oil prices remains hugely uncertain – an unruly mix of economic, geopolitical and technical factors have unsettled both petroleum demand and reconfigured the capacity to satisfy it. It seems unlikely that the world will return to the persistently low and stable price patterns seen from the late 1980s to the mid-2000s. A more likely pattern is medium-term and long-run volatility as the relationship between consumers and producers is restructured at the global scale. The reasons for the structural realignments in the global petroleum sector merit close attention from planners charged with the complex task of piloting cities through the troubled petroleum seas of the next few decades. This book offers insights that can support this critical task.

Underlying factors during the late 2000s

Why have oil prices departed so dramatically from their historical script into a prolonged bout of upward and downward volatility? To understand these patterns, we need to examine global economic conditions in the late 2000s and early 2010s. Commentators and scholars have identified a range of factors as contributing to higher oil prices in the late 2000s, which included accelerating global demand, geopolitical risks and production limits. A commonly cited causal factor underpinning high oil prices in the 2000s was the global economic expansion over the decade leading up to 2009, which intersected with static or only gradually increasing oil production. China's economy, for example, grew by at least 8.3 percent per annum (p.a.) in GDP terms from the start of the decade (IMF 2010).

A number of official estimates had anticipated much greater future demand for petroleum as a result of continuing global economic growth. The International Energy Agency (IEA), which is arguably the most authoritative international adviser on energy matters, anticipated that global petroleum demand would increase from around 85 million barrels per day (mbd) in 2008 to over 105 mbd by 2030 (IEA 2009: 38). The International Energy Forum (IEF) – an increasingly prominent venue for dialogue, debate and negotiation between producing and consuming nations – suggested a global production ceiling of 95–100 mbd would not meet demand (IEF 2009: 9). In the late 2000s, price rises were anticipated where global petroleum production failed to satisfy current or rising levels of demand for oil.

The world's ability to match petroleum output to expanding global oil demand in the late 2000s was under deepening doubt. Most official assessments predicted that production could be expanded if certain conditions were achieved. These included consistent flows of financing for oil production, continuing exploration and discovery of oil resources, improved extraction technologies, and assured political and military security of oil supplies. The investment task was viewed as monumental. The IEA (2008: 39) projected that US$26 trillion worth of production investment was needed by 2030 for anticipated global oil demand. Other assessments were even greater, with Simmons (quoted in Izundu 2008) suggesting US$50 trillion would be necessary. Much of this investment was expected simply to maintain existing levels of production as current production infrastructure aged.

Geopolitical tensions in key petroleum producing regions also added to higher oil prices. Expanding global demand implied the need for rapidly developing nations to gain access to scarce oil resources. The IEA (2007) estimated that, to achieve their projected 2030 oil demand of 23.2 mbd, China and India needed access to a further 12.5 mbd of oil production, which could potentially push them into competition for the 29.2 mbd used by the US and Europe. The Middle East is the most obvious oil region where geopolitical problems have been manifest (Fattouh 2007) through military invasions and armed conflict. Other oil-bearing regions, such as Central Asia and West Africa, have also witnessed tensions that have amplified concerns about the political security of global oil supplies (Klare 2008; Kleveman 2003).

China has been careful not to intrude too assertively into the Middle East, lest it threaten other interests. Instead, China has pursued its oil interests elsewhere, such as in Africa, within a wider mineral resource strategy (Carmody and Owusu 2007). Given the concentration of global oil reserves within the Middle East, however, the future need to secure oil supplies may force China to develop a greater presence there.

The concentration of oil supplies into a shrinking set of producing nations also bolstered oil prices in the late 2000s. Global oil reserves are unevenly distributed and their exploitation has varied sharply. Once the largest oil producer, the US has fallen in importance compared to the Middle East. As other reserves decline, the share of global oil production provided by members of the Organization of the Petroleum Exporting Countries (OPEC) – dominated by Middle Eastern nations – is expected to increase from 44 percent in 2008 to 52 percent by 2030, giving this group considerable influence on global supplies. Some 28 percent of global oil supply in 2030 is projected to come from just six Middle Eastern OPEC countries: Iran, Iraq, Kuwait, Qatar, Saudi Arabia and the United Arab Emirates (IEA 2009: 83–84), most of which are politically fragile and willing to use oil to political advantage.

The balance of institutional control of oil production is also slipping from the hands of private corporations into the grasp of national oil companies (NOCs). The NOCs' strategic roles are, in turn, adapting from delivering oil to global commodity markets in favor of a strategy of "resource nationalism" that views oil as a lever for national advantage. As early as 1972, Adelman (1972: 69) described national oil companies as "OPEC's tax collectors" – "taxing people in consuming countries – both rich and poor – and transferring the lion's share of the proceeds to the governments of the oil-producing countries." Likewise, Hamilton (2008) has detected "scarcity rents" within the oil hikes of the late 2000s.

A final complicating factor in the petroleum supply environment, which exploded dramatically into the global consciousness in mid-2010 with the BP Deep Horizon oil well catastrophe, is the risk and danger associated with expanded oil exploration in difficult, demanding and technically complex environments. The need to expand global oil reserves has forced exploration into increasingly remote and often hostile environments, such as the Alaskan arctic or the southern US continental shelf. Australia has recent experience of this problem, as witnessed by the ten-week spill of oil from the West Atlas rig into the Timor Sea. The recent pattern of spills occurring at remote or deep ocean extraction points (such as a wellhead failure), rather than from within the distribution system (such as due to a tanker stranding), testifies to the challenges and risk of global petroleum thirst. The levying of higher recovery and remediation costs on oil companies by governments and insurance companies will inevitably increase the cost of petroleum in world markets and constrain the capacity of future exploration to generate new extraction.

Emerging factors in the early 2010s

The petroleum environment in the mid-2010s, however, is superficially different to that observed half a decade earlier. Many of the underlying dynamics in the petroleum production sector remain, although with a markedly changed global economic and geopolitical environment to that of the late 2000s, which has altered demand and led to a realignment of the strategic outlook.

The largest shift in the global petroleum environment has been in the pronounced weakness in global economic conditions during much of the present decade due to the lingering consequences of the global financial crisis (GFC) in North America and Europe. Economic growth in these regions has been muted by relatively high unemployment and investment uncertainty around longer-term prospects. In the US, for example, unemployment went above 5 percent in March 2008, peaked at 10 percent in October 2010, and only returned to 5 percent seven years later, in October 2015 (Bureau of Labor Statistics 2015). The economic weakness implied by high unemployment rates in much of the Global North has dampened both domestic and business sector demand for petroleum leading to relative flat prices.

Weak economic conditions in the early 2010s combined with the rollout of investment in new petroleum production that had been planned in the high-demand and high-price conditions of the previous decade. This included reinvigoration of existing oil fields and further exploitation of non-conventional oil reserves, such as offshore and remote basins. In addition, new forms of non-conventional petroleum began to be exploited. The most significant of these non-conventional forms of petroleum for the global petroleum market has been "tight oil," which is a form of crude oil contained within oil shale rock. The term "tight" refers to the geological structure within which the oil is embedded, meaning that greater force is required to extract it when compared to conventional oil reservoirs. The US possesses large tight oil reserves and, with oil prices around US$100 in the late 2000s, the exploitation of these reserves became economically viable. Production increased rapidly in the late 2000s, such that the US grew its domestic oil production by 70 percent from 2007 to 2014 (BP 2015). This growth had the effect of accelerating declines both in US oil imports and due to demand loss from the GFC. Crude petroleum imports to the US peaked at 3.7 gigabarrels in 2005 declining to 2.7 gigabarrels by 2015 (EIA 2015).

Negative import demand growth in the US, which consumes one-fifth of the world's petroleum, has been a major contributor to overall global demand weakness, particularly in the Global North. This has been accompanied by weaker demand in Europe during the 2010s. In 2014, petroleum demand in Germany, France, the UK, Spain and Italy was lower than the immediate pre-GFC level (BP 2015). Total global petroleum demand has grown during the post-2007 period, however this growth has been biased toward the less developed world and the Global South. Thus, total OECD petroleum consumption declined from 50 mbd in 2007 to 45 mbd by 2014, while non-OECD demand grew from 35 to 47 mbd. The removal of much of the Global South's demand from the growth in global petroleum demand, in a period following very high price hikes, has negatively affected investment. The ramp-up of prices and projections around future demand and prices had the effect of stimulating production investment. Major production capacity investment is developed over long time frames. Generally, investment initiated in 2009 or 2010 would not begin production until after 2012. Thus new production capacity has accompanied weak economic conditions, leading to a decline in global oil prices.

A further array of changes in long-term expectations around global fossil fuel use have influenced the international petroleum market since 2010. Reports on the effects of fossil fuel consumption on global climate change became more emphatic during the early 2010s. This appreciation strengthened recognition of the need to end fossil fuel use to avert catastrophic climate change. A new seriousness accompanied major global initiatives to avoid climate change, such as the twenty-first session of the Conference of Parties (COP21) of the United Nations Framework Convention on Climate Change (UNFCCC). The conference, held in Paris in 2015, culminated in a global agreement on carbon emissions reduction. Among the objectives of COP21 is the cessation of fossil fuel use by the end of the twenty-first century. In the years leading up to COP21, the political legitimacy of fossil fuels, particularly of coal, had come into increasing doubt. Major divestment campaigns that sought to reduce fossil fuel investment have been established and have had a growing influence on corporate investment strategies, as the business sector has begun to adjust its long-term view of the prospects of future growth in fossil energy use.

Furthermore, technological advances began to influence the fossil fuel economy more tangibly in the early 2010s. Renewable energy generation experienced rapid improvement in efficiency and production cost during the early 2010s, with projections of unit cost parity with conventional coal power by the end of the 2010s, in turn forcing reappraisal of historic generation investment strategies in favor of renewables. Many investment banks have begun to avoid coal-based investments, in part because of the reputational risk associated with a climate-harming energy form but also based on the demand risks associated with fossil fuels.

Although petroleum differs from coal in terms of its sectoral use and emissions profile, this fuel has also been targeted in the anti-fossil fuel campaigns. The rapid development of fully electric passenger vehicles, which do not require petroleum, has aided the increasing wariness around long-term oil demand. Most prominent among these has been the corporation US Tesla, which has sold some 90,000 electric cars into the North American market in a few years since 2012. This is a very small proportion of total North American vehicle sales but, with technology improvements and production cost declines – especially in conjunction with emerging renewable energy battery storage technologies – the prospect of wholesale manufacture of non-fossil fuel automobiles now appears to have greater potential than at any previous moment.

Irrespective of electric mobility, conventional fuel economy improvements have also contributed to a moderation in petroleum demand. For example, US passenger vehicle fuel economy improved by barely 2 miles per gallon (mpg) between 1986 and 2005 (27.6–30.1 mpg) but, by 2013, saw a nearly 6 mpg improvement (to 36 mpg) (BTS 2015). The prospects of continued fuel economy gains and long-term shifts to electric vehicles in North America and other major automobile-using regions has led to increased pessimism about the long-run demand for petroleum fuels. This has had flow-on effects to other sectors, such as government revenue. As fuel efficiency has increased and total petroleum consumption per vehicle has declined, so have excise revenues, leading to reduced funding available for road construction in jurisdictions where taxes and roads are linked.

A final significant factor underpinning the changed petroleum market context in the mid-2010s has been a series of ongoing geopolitical realignments. These include continued instability in the Middle East, increasing Russian adventurism and the expansion of China's presence within the Asia-Pacific. Of these, the most important factor for oil prices has been the relationship between the Middle East powers of Saudi Arabia and Iran. Saudi Arabia is the world's largest oil producer, with an historic influence over oil prices through its ability to expand or contract global production. Saudi Arabia's strategy appears to have shifted, during the early 2010s, to one of defending market share against new producers who have been encouraged by high oil prices and depend on those high prices for their viability. This has meant maintaining production output despite price falls.

A peak debate

A thread throughout both the high oil price period of the late 2000s and the relatively lower price period of the early 2010s centers on a cluster of questions around long-term petroleum supplies. A major scholarly, industry and policy debate has surrounded the continuity of global conventional (light sweet crude) petroleum output as the world's finite set of exploitable current reserves approach exhaustion. Many petroleum geologists have noted that there will be an inevitable moment when rates of conventional global oil production can no longer be expanded, due to production constraints or to geological and reserve limitations, and production levels plateau or decline. An oil production peak, due to a plateau or decline of output in the context of growing demand, implies higher oil prices if either viable alternatives or radical demand destruction measures are not adopted.

Peak oil has proven a controversial topic. Warnings about a peak in global oil production in the late 1990s and early 2000s (Campbell and Laherrere 1998) were dismissed by many petroleum industry observers (Clarke 2007; Huber and Mills 2005; Yergin 2006). Yet the last decade has seen growing recognition among informed opinion that the world will face increasingly constrained supply conditions in coming years and into the long term. The IEA's official reports have been cautious on the question of production constraints and peak oil. For much of the 2000s the agency rejected concerns about a medium-term production peak. But, in 2008, the agency's head of economic forecasting indicated that a peak in oil production by 2020 is probable (Fatih Birol, quoted in Monbiot 2008). This concern appears to have been partly reflected in the IEA's projection of future petroleum sources.

A chart in the agency's 2008 *World Energy Outlook* (Figure 15.1 in IEA 2008: 250) indicated rapid production declines in currently producing oil reserves from around 2008, which would need to be made up from as yet undeveloped and undiscovered fields. The chart suggested conventional crude oil production would plateau from around 2009, and then begin to decline from around 2014. For peak oil observers, such assessments suggest that absent "non-conventional" oils – tar sands, oil shale and biofuels – and natural gas liquids, global supplies of conventional oil have either peaked or will do so by mid-decade. That extraction of non-conventional oils, such as tar sands and oil shale, is often highly costly and energy intensive adds to the challenge of a peak in conventional supplies.

Subsequent IEA *World Energy Outlook* reports replicated the format of the 2008 figure with updated production data. By 2011, the IEA (Figure 3.16 in 2011: 123) was estimating a plateau in production around 2006, with a peak around 2015, a trend anticipated by a similar figure in the 2012 *World Energy Outlook* (Figure 3.15 in IEA 2012: 103). Later reports have offered similar insights. The 2015 IEA *World Energy Outlook* (Table 3.5 in IEA 2015: 134) implied a peak in total conventional crude oil production by 2025 and projected a 0.1 percent compound annual decline on 2014 production levels by 2040, including yet to be developed and yet to be found oil, as well as enhanced recovery. Although this data included projections for tight oil, heavy bitumen and natural gas and liquids contributing an additional 0.5 percent compound annual supply growth by 2040, these latter forms of supply are considerably costlier than conventional crude (IEA 2008: 250).

Wider anxieties

The IEA has not been the only agency concerned by petroleum challenges. In 2008 a US Energy Information Administration projection suggested that global oil production will peak in 2012 and decline to create a shortfall relative to projected demand of 43 mbd by 2028 (Sweetnam 2009). The chief of the Shell Oil company warned, in 2008, that the international community risks a dangerous geopolitical scramble for remaining oil reserves (van der Veer 2008).

In the late 2000s, private industry began to actively engage with the question of petroleum depletion. The Volvo (2008) automotive company was one of the first major corporations to acknowledge petroleum depletion problems, expecting a global oil production peak within a decade. A consortium of major UK companies – including the Arup consulting firm, the Stagecoach bus company and Virgin Airlines – argued that in the short term oil prices on world markets would be significantly higher than historic averages, with potentially significant price volatility, including high peaks and the possibility of supply disruptions (ITPOES 2008, 2010). Meanwhile, a senior Macquarie Bank oil and gas analyst evaluated the problem of petroleum depletion to conclude that "capacity has pretty much peaked in the sense that declines equal new resources" (Ian Reid, cited in Sheppard 2009).

Probably the most significant and imperative business statement on the problem of petroleum security, peak oil and wider energy constraints came from Lloyds – "the world's leading specialist insurance market" and "often the first to insure new, unusual or complex risks" (Froggatt and Lahn 2010: verso). Lloyds' 360 Risk Insight program and Chatham House June 2010 white paper on sustainable energy (Froggatt and Lahn 2010: 4) stated:

> Companies which are able to plan for and take advantage of this new energy reality will increase both their resilience and competitiveness. Failure to do so could lead to expensive and potentially

> catastrophic consequences . . . An oil supply crunch in the medium term is likely to be due to a combination of insufficient investment in upstream oil and efficiency over the last two decades and rebounding demand following the global recession. This would create a price spike prompting drastic national measures to cut oil dependency.

This is perhaps the most significant statement on the petroleum challenge by any business representatives to date. That such alarm is expressed by one of the world's greatest insurance corporations in such plain terms suggests that concerns about petroleum depletion have passed from the domain of speculation into the realm of probability. If, as Lloyds suggests, businesses face "catastrophic consequences" from an energy crunch, then governments and cities would likely face similar threats. Like businesses, planners, policy makers and elected representatives should be preparing for the impact on governments and cities.

Formal scholarship has tested concerns about petroleum depletion and identified the potential for a medium-term oil shock (Hirsch 2008; Jakobsson et al. 2009). Jakobsson et al. (2009: 9) undertook a comprehensive analysis of official petroleum production forecasts and concluded that "an imminent peak in production cannot be ruled out." Hirsch (2008) noted that, in a best-case scenario, a stable plateau in oil production followed by an annual 2–5 percent decline could be expected, and a worst-case scenario would involve greater deficits. A review of oil reserves by Owen et al. (2010: 1) concluded:

> While there is [*sic*] certainly vast amounts of fossil fuel resources left in the ground, the volume of oil that can be commercially exploited at prices the global economy has become accustomed to is limited and will soon decline.

Nashawi et al. (2010) analyzed global oil production patterns to predict a global peak in 2014 and an OPEC production peak by 2026.

Some governments have begun to respond to strengthening evidence of declining petroleum security, often focusing on national security and infrastructure policy, although such response is often patchy and uneven. The US Government Accountability Office (2007) has observed the potential for a decline in global petroleum production and proposed that the US should prepare a national strategy to deal with the problem. The US military, which is the world's largest institutional petroleum consumer, has affirmed these petroleum anxieties and noted that a "severe energy crunch is inevitable without a massive expansion of production and refining capacity" (Joint Forces Command 2010: 28). The US military now fears further calls upon its services for interventions in new future oil-driven conflicts. The UK government was unwilling to recognize petroleum depletion risks for nearly a decade after UK oil production peaked in 1999. In early 2010, however, the UK's Energy Minister was reported to have belatedly changed this stance, convening a (not so well-kept) secret meeting of key industry leaders to address oil security issues (Macalister 2010). This laggard approach contrasts with the commitment by the Swedish government (2009) to reduce that country's dependence on oil by 50 percent by 2020.

Perhaps the most substantive, if opaque, signal that petroleum security is now a major international governmental concern is the shifting political and institutional relations between producing and consuming nations. The IEF meeting in early 2010 – gathering together 68 national energy ministers, including Australia's – sought to establish a formal global venue for "consumer-producer dialogue" focusing on measures to mitigate energy market volatility, and the security implications of this problem. The meeting in part responded to an EIF background report, which suggested that global production of conventional liquid fuels would peak at just over 90 mbd by 2020, with shortfalls needing to be drawn from costly, technically complex, non-conventional sources such as tar sands and biofuels (IEF 2009). The IEF's subsequent communiqué signaled its intent to manage political relationships between suppliers and consumers to both avoid tensions arising and ensure a smooth and stable environment for production investment.

Over the course of the 2010s, the IEF has seen a strengthening of its coordinating role between oil producing and consuming nations with joint analyses conducted in partnership between the IEA and OPEC with further national, intergovernmental and private sector petroleum institutional actors. This remarkable global ambition of the EIF, and its program of international engagement, reveals the strengthening hold exerted on global political attention by petroleum issues and the febrile global petroleum environment.

At the time of writing (late in 2015), the projections of the past decade deserve some appraisal. There is relatively broad recognition that a peak in conventional crude oil production is inevitable, given continuing demand expansion, but the date for this is highly contested. Those who argued that an extended "bumpy plateau" in crude oil production would occur (Sorrell et al. 2010; Warrilow 2015), rather than a sharp peak, have so far proven the more accurate assessors. Supply assessments were, perhaps, overly pessimistic in relation to the elasticity of supply to price signals, particularly around the combined energy sector and governmental determination to increase US tight oil production. Assessments that the world would see high oil prices over the very long term have, perhaps, been contradicted by the post-2013 decline in the price of crude. But this is not a consequence of underlying petroleum geology but rather the effect of weak economic demand; the maneuvering of major producing nations in relation to their market share and that of others; and shifting, perhaps even declining, global appetite for fossil fuels. The destruction of demand resulting from the GFC was unforeseen by most petroleum commentators and, thus, should not be seen as invalidating projections of petroleum supply capacity. The world remains dependent on petroleum for energy, particularly in transport. Just as oil prices can shift downward in relation to market positioning by producing countries, they can also shift upward should the dynamics of geo-economics push in that direction. Dependence on declining long-run global supplies of petroleum remains a vulnerability for many cities around the world.

The urban planning challenge for North American and Australian cities

Modern cities are concentrated sites of petroleum consumption. This consumption occurs primarily through transport, although in some regions, particularly in the US, oil for heating is also an important component of urban petroleum demand. Two major regions, North America and Australia, are particularly exposed to the effects of higher petroleum prices due to their extensive car dependence. In the US, an average of 86 percent of work travel is undertaken by automobile, a similar figure to that found in Australia. Many US cities, such as Los Angeles, Atlanta, Miami and Phoenix, have journey-to-work mode shares for the automobile that are higher than 90 percent (Mees 2009). Among US cities, New York City has the lowest use of cars for work journeys (67.7 percent), slightly less than Sydney, Australia's least car dependent city (71.2 percent). Among Canadian cities, Ottawa's work journeys were the least car dependent, relying on automobiles for 68.1 percent of trips (Mees 2009).

Two major factors underpin the high levels of car reliance in North American and Australian cities. First, the possession of large supplies of petroleum during the twentieth century meant that the US could cheaply deploy automobiles and roads as urban and inter-urban transport systems. Second, strategic government decisions were taken to develop the US around the automobile after WWII, particularly through the interstate highway network, including shifting finance from transit to roads.

This deliberate shift to petroleum and automobile priority, and then dependence, had two flow-on effects. Road- and automobile-based transport networks tend to disperse land uses and, thus, many US cities developed in a highly dispersed land-use pattern in the latter half of the twentieth century. With transit investment languishing, and the car a favored mode in government policy, it is no surprise that automobile dependence became endemic. Similar polices, albeit less extreme, were followed in Canada and Australia from the 1960s onward. Although many observers argue that density is now the fundamental factor determining transport demand patterns, there is a strong argument that density is a consequence of oil prices and transport policy settings in relatively lightly regulated land-use markets, rather than a determinant of policy (see Stone and Mees, Chapter 10 in this book). In any case, reorganizing land uses at mass scale for a petroleum constrained energy environment will be almost impossible, given the spatial fixity of most of the built environment.

Given the high levels of car dependence for such a basic trip as the journey to work, it is no surprise that North American and Australian cities are among the most energy dependent in the world. Newman and Kenworthy (Figure 1.14 in 2015: 25) have demonstrated that per capita urban travel in US cities is among the most highly energy intensive globally, with around 50,000 MJ energy consumed p.a. By contrast, Australian and Canadian cities are slightly less energy dependent per capita, consuming around 30,000 MJ p.a. European and Asian cities, by contrast, are much less energy dependent, consuming around 5–20 MJ per capita p.a. (Newman and Kenworthy 2015).

The US, Canada and Australia share similar characteristics, which make them vulnerable to constrained petroleum supplies. The US is the second largest energy consumer internationally, using 2,298 million tonnes of oil equivalent (Mtoe) energy annually – some 673 Mtoe fewer than the world's largest consumer, China, but achieves this feat with a population nearly three-quarters smaller than China's (BP 2015: 40). The US is one of the world's largest petroleum-consuming nations, accounting for one-fifth of total oil consumption (BP 2015: 41). Canada exhibits similar levels of energy use per capita, while Australia is the greatest OECD energy consumer per capita (Garnaut Climate Change Review 2008).

The US, Canada and Australia each possess large energy reserves. The US has extensive coal, oil and gas deposits, while Canada has abundant shale oil reserves. Australia possesses huge energy resources in coal and gas but only modest petroleum reserves, holding just 0.3 percent of the world's oil. Conventional oil production in the US, Canada and Australia has peaked, although recent developments such as high global oil prices have spurred renewed production levels of unconventional oil. But the US resurgence is likely to be transitory. Projections from the IEA suggest that tight oil will peak in the US shortly after 2020 and decline thereafter. Although tight oil becomes less viable as prices decline, it can be relatively easily mothballed and restarted in response to increasing price conditions. Thus, the US is likely to be able to extend its periods of relative independence in oil supply, but not indefinitely.

Canada has extensive shale oil supplies that can be mined and converted to petroleum. However, this conversion process is highly energy intensive and depends on high energy prices globally to maintain viability. The energy intensity of shale oil production also means that the carbon emissions from Canadian supplies is much higher than for conventional oil. With carbon reduction becoming a significant global imperative, the potential for Canadian shale oil to become stranded by either regulation or carbon pricing appears very possible. Urban transport systems in the US, Canada and Australia are dependent on petroleum for their viability. Petroleum futures, volatile as they appear now, look set to remain that way for the foreseeable future. There is just as strong a prospect of oil prices returning to levels higher than the US$140 per barrel seen in 2008 as there is of prices remaining at sub-US$40 levels.

The future trajectory and timing of global petroleum price changes is highly uncertain and difficult to predict. The portents presented in industry analyses, scholarship and government reports are steeped in uncertainty, and the gyrations of petroleum markets are hard to anticipate over the medium and long terms. This uncertainty, and the profound significance of declining – or carbon constrained – petroleum supplies for human societies in the future, suggest that urban policy makers and planners must take the petroleum security and dependence challenge very seriously.

With US, Canadian and Australian cities so clearly exposed to the effects of depleting global oil supplies or carbon pricing, our urban planners should turn their attention now to mitigating oil vulnerability and adapting car dependent cities to an oil constrained world. This chapter has set out the international dimensions of this problem. It is hoped that the remaining scholarship assembled in this collection will expand on the planning thoughts and urban deeds needed to achieve the adjustment to less oil vulnerable cities as the petroleum horizon darkens.

References

Adelman, M.A. (1972) "Is the Oil Shortage Real? Oil Companies as OPEC Tax-Collectors," *Foreign Policy* 9(Winter): 69–107.

BP (2015) *Statistical Review of World Energy*, London: BP Plc.

BTS (2015) "Table 4–23: Average Fuel Efficiency of U.S. Light Duty Vehicles," *Bureau of Transportation Statistics*, accessed 18 January 2016 — www.rita.dot.gov/bts/sites/rita.dot.gov.bts/files/publications/national_transportation_statistics/html/table_04_23.html

Bureau of Labor Statistics (2015) "Seasonal Unemployment Rate," accessed 18 January 2016 — http://data.bls.gov/timeseries/LNS14000000

Campbell, C. and J.H. Laherrere (1998) "The End of Cheap Oil," *Scientific American* 278(3): 78–83.

Carmody, P.R. and F.Y. Owusu (2007) "Competing Hegemons? Chinese Versus American Geo-Economic Strategies in Africa," *Political Geography* 26(5): 504–24.

Clarke, D. (2007) *The Battle for Barrels: Peak Oil Myths and World Oil Futures*, London: Profile Books.

EIA (2015) "Petroleum and Other Liquids: US Imports by Country of Origin," Energy Information Administration, Washington, DC, accessed 18 January 2016 — www.eia.gov/dnav/pet/pet_move_impcus_a2_nus_epc0_im0_mbbl_a.htm

Fattouh, B. (2007) *How Secure Are Middle East Oil Supplies?* Oxford: Oxford Institute for Energy Studies, University of Oxford.

Froggatt, F. and G. Lahn (2010) *Sustainable Energy Security: Strategic Risks and Opportunities for Business*, London: Lloyds and Chatham House.

Garnaut Climate Change Review (2008) *Garnaut Climate Change Review – Final Report*, Melbourne: Garnaut Review of Climate Change.

Government Accountability Office (2007) *Crude Oil: Uncertainty about Future Oil Supply Makes It Important to Develop a Strategy for Addressing a Peak and Decline in Oil Production*, Washington, DC: US Government.

Hamilton, J. (2008) "Understanding Crude Oil Prices," *National Bureau of Economic Research Working Papers*, Working Paper No. w14492 (November).

Hirsch, R.L. (2008) "Mitigation of Maximum World Oil Production: Shortage Scenarios," *Energy Policy* 36(2): 881–89.

Huber, P. and M.P. Mills (2005) *The Bottomless Well: The Twilight of Fuel, the Virtue of Waste, and Why We Will Never Run Out of Energy*, New York: Basic Books.

IEA (2015) *World Energy Outlook 2015*, Paris: International Energy Agency and Organisation for Economic Co-operation and Development.

——— (2012) *World Energy Outlook 2012*, Paris: International Energy Agency and Organisation for Economic Co-operation and Development.

——— (2011) *World Energy Outlook 2011*, Paris: International Energy Agency and Organisation for Economic Co-operation and Development.

——— (2009) *World Energy Outlook 2009*, Paris: International Energy Agency and Organisation for Economic Co-operation and Development.

——— (2008) *World Energy Outlook 2008*, Paris: International Energy Agency and Organisation for Economic Co-operation and Development.

——— (2007) *World Energy Outlook 2007: Focus on China and India*, Paris: International Energy Agency and Organisation for Economic Co-operation and Development.

IEF (2009) *Unpacking Uncertainty: Investment Issues in the Petroleum Sector*, Riyadh: International Energy Forum.

IMF (2010) "World Economic Outlook Database, April 2010," International Monetary Fund, accessed 21 January 2016 — www.imf.org/external/pubs/ft/weo/2007/01/data/download.aspx

ITPOES (2010) *The Oil Crunch: A Wake-Up Call for the UK Economy – ITPOES Second Report*, London: UK Industry Taskforce on Peak Oil & Energy Security.

——— (2008) *The Oil Crunch: Securing the UK's Energy Future – First ITPOES Report*, London: UK Industry Taskforce on Peak Oil and Energy Security.

Izundu, U. (2008) "OTC Speakers Highlight Offshore Industry's Future," *Oil and Gas Journal* 12 May: 20.

Jakobsson, K., Soderbergh, B., Hook, M. and K. Aleklet (2009) "How Reasonable Are Oil Production Scenarios from Public Agencies?" *Energy Policy* 37(11): 4809–18.

Joint Forces Command (2010) *The Joint Operating Environment*, Washington, DC: US Department of Defense.

Klare, M. (2008) *Rising Powers, Shrinking Planet: The New Geopolitics of Energy*, New York: Metropolitan Books.

Kleveman, L. (2003) *The New Great Game: Blood and Oil in Central Asia*, New York: Atlantic Monthly Press.

Macalister, T. (2010) "Energy Minister Will Hold Summit to Calm Rising Fears Over Peak Oil," *The Guardian*, 22 March, accessed 21 January 2016 — www.theguardian.com/business/2010/mar/21/peak-oil-summit

Mees, P. (2009) "How Dense Are We? Another Look at Urban Density and Transport Patterns in Australia, Canada and the USA," *Road and Transport Research* 18(4), accessed 21 January 2016 — www.ppt.asn.au/pubdocs/Mees Paul-HowDenseAreWe.pdf

Monbiot, G. (2008) "When Will the Oil Run Out?" *Guardian*, 15 December.

Nashawi, I.S., Malallah, A. and M. Al-Bisharah (2010) "Forecasting World Crude Oil Production Using Multicyclic Hubbert Model," *Energy Fuels* 24(3): 1788–800.

Newman, P. and J. Kenworthy (2015) *The End of Automobile Dependence: How Cities Are Moving Beyond Car-Based Planning*, Washington, DC: Island Press.

Owen, N.A., Inderwildi, O.R. and D.A. King (2010) "The Status of Conventional World Oil Reserves – Hype or Cause for Concern?" *Energy Policy* 38(8): 4743–49.

Sheppard, D. (2009) "When Will We Hit Peak Oil?" *Reuters*, 16 September.

Sorrell, S., Speirs, J., Bentley, R., Brandt, A. and R. Miller (2010) "Global Oil Depletion: A Review of the Evidence," *Energy Policy* 38(9): 5290–95.

Swedish Government (2009) *A Sustainable Energy and Climate Policy for the Environment, Competitiveness and Long-Term Stability*, Stockholm: Regeringskansliet.

Sweetnam, P. (2009) "Meeting the World's Demand for Liquid Fuels: A Roundtable Discussion," *Proceedings of a New Climate for Energy: EIA 2009 Energy Conference*, 7 April, Washington, DC: Energy Information Administration, accessed 30 January 2016 — www.eia.gov/conference/2009/session3/Sweetnam.pdf

Van der Veer, J. (2008) "Two Energy Futures," 21 January, accessed 21 January 2016 — www.project-syndicate.org/commentary/two-energy-futures?barrier=true

Volvo (2008) *Future Fuels for Commercial Vehicles*, Goteborg: AB Volvo.

Warrilow, D. (2015) "A Bumpy Road to the Top: Statistically Defining a Peak in Oil Production," *Energy Policy* 82: 81–84.

Yergin, D. (2006) *Why the "Peak Oil" Theory Falls Down – Myths, Legends, and the Future of Oil Resources*, Cambridge, MA: Cambridge Energy Research Associates.

3

THE PARADOX OF OIL

The cheaper it is, the more it costs

Samuel Alexander

The timing of the sudden drop in the price of oil since June 2014 took energy and financial analysts by surprise. After averaging around US$110 per barrel since 2011 (IEA 2013: 6) – suggesting a "new normal" – by August 2015 the price of oil had fallen to around US$50 per barrel. Although the timing of this price drop was not forecast by analysts with any precision, there were economic, geological and geopolitical dynamics at play that made this kind of price volatility not so surprising.

In "The New Economics of Oil," published a few months prior to the fall in price, I (Alexander 2014: 9) explained that

> as oil production slows or stagnates, oil prices may continue to increase until they reach an economic breaking point, crashing or destabilizing economies, which would lead to a crash in oil prices; the low oil prices would then facilitate economic recovery, which puts more demand pressure on oil, leading prices to rise till economic breaking point, and so on and so forth.

Further, "this cycle of bust-recovery-bust is what we may face in coming years and decades."

Furthermore, I outlined why expensive oil can incentivize greater investment in its production while disincentivizing its consumption, a dynamic that can increase oil production faster than demand and, thereby, generate short-term oil gluts that can also lead to price volatility. These supply and demand dynamics go a long way to explaining the current state of oil markets. While the exact timing of the fall in prices that started mid-2014 may have come as a surprise, the phenomenon is quite comprehensible when one recognizes the intimate connection between energy (especially oil) and economics. The ever-present influence of geopolitics is shaping oil markets too.

What is so frustrating about much oil commentary today is the tendency for analysts to focus on the immediate or short-term situation, often from a purely financial or economic perspective, neglecting the larger social, political and environmental contexts in which oil markets unfold. When those larger contexts are given due attention, it becomes clear that oil is a commodity that defies reductive analysis and that can only be understood through a multidimensional, interdisciplinary lens.

In this chapter I outline and analyze various explanations for why the price of oil fell so dramatically from mid-2014 and present some considered but tentative hypotheses about what we can expect from the oil markets in coming years. I also hope to challenge the naive conclusion – drawn all-too-hastily in the mainstream media – that the drop in price somehow debunks the analytical framework of the "peak oil school" (Sakya 2015). Although it may sound counterintuitive, cheap oil is actually a complicated function or symptom of peak oil dynamics and, far from solving oil problems, the drop in price merely creates new problems of equal or greater weight, in ways that will be explained. Those who claim that the effects of cheap oil are clearly positive are at best being simplistic and are at worst just plain wrong (*Economist* 2014b).

The main conclusion I defend in this chapter is that so-called cheap oil (averaging US$50 per barrel) is just as problematic as expensive oil (say, more than US$100 per barrel), but for very different social, political, economic and environmental reasons. Just as expensive oil suffocates industrial economies that are dependent on cheap energy inputs, cheap oil merely propagates and further entrenches the existing order of global capitalism that is, however, in the process of growing itself to death (Turner 2014). The fall in prices undermines the oil industry by scaring off capital investment due to declining energy returns on investment (Murphy 2014) in an age when the costs of establishing and drilling new fields is relentlessly on the rise (Kopits 2014). Therefore, cheap oil is likely to retard midterm to long-term production, setting the scene for a mid-range supply crunch that will soon enough push prices back up (Kent and Faucon 2015; Mushalik 2015b; Zumbrun 2015).

Accordingly, we should not be fooled by this current period of depressed prices. As the world continues to replace easy, conventional oil with ever more marginal unconventional oils – such as deep water, shale oil and tar sands, and alternative biofuels (see Baker et al., Chapter 18 in this book) – resource depletion will be putting upward pressure on production costs forever. So, despite currently depressed prices, it remains true to say that we live in an age of expensive oil in geological context: the low-hanging fruit is gone. The only way that oil will remain cheap over the long term is if our economies are doing so poorly from a conventional growth perspective that we cannot afford for oil to be any more expensive, making oil demand weak and keeping prices deflated (Meijer 2014a).

Looking at the current situation from a different angle, cheap oil makes renewable energy alternatives less cost competitive, which will have disastrous ramifications on climate change mitigation by disincentivizing the necessary transition beyond fossil fuels at a critical time. This ecological issue is typically overlooked by those oil analysts who are blinded by the apparent, short-term economic benefits of cheaper oil. Herein lies the paradox of oil: the cheaper it is (economically), the more it costs (environmentally).

For these types of reasons, I argue that there is no "optimal" price for oil in much the same way as there is no "optimal" price for heroin. This analogy between oil and heroin may appear like a polemical exaggeration, but I hope to show that it is, in fact, worryingly apt. When heroin is expensive, addicts cannot afford what they desperately need, or feel they need, and suffer accordingly. Expenditure on more worthwhile things is cut back in order to fund the increasingly expensive and debilitating addiction. But, when heroin is cheap and readily available, the negative effects of addiction become even more pronounced through overconsumption, and the addiction only deepens as hopes of rehabilitation fade. Oil acts as industrial civilization's own form of heroin and, whether it is cheap or expensive, addicts today are in as much trouble as ever.

The new economics of oil

Before focusing on the specific issue of the recent fall in prices, I will briefly describe the fundamental changes that have taken place over the last decade with respect to the relationship between oil demand, geology and economic activity (for more detail, see Alexander 2014). Only by understanding these changes can we begin to gain insight into the diverse forces that shape oil markets today.

Throughout most of the twentieth century oil supply was able to meet increasing demand without much trouble. Leaving aside the geopolitical oil crises of 1973 and 1979, cheap oil in the range of US$20–25 was readily available (WTRG Economics 2011). Industrial economies came to rely on cheap energy inputs and structured their production and consumption accordingly, assuming energy costs would remain marginal and that economic growth trajectories could be maintained indefinitely. Around 2005, however, conventional crude oil production stagnated (Miller and Sorrell 2014: 6) and the theory of peak oil began to be taken seriously by more people and institutions (Munroe 2010).

Peak oil refers to that point when the rate of oil extraction is at an all-time high. This point arrives not because oil is running out but because the low-hanging fruit of easy-to-produce oil has already been discovered, leaving only the more marginal oil reserves. Then producers have to drill more and in less ideal places to keep up stable or even declining supplies (Likvern 2012), meaning diminishing marginal returns. Eventually producers cannot maintain supply rates, the flow of oil stops growing or peaks, and eventually begins to fall, despite the fact there is still lots of oil left.

This is not merely a geological phenomenon. In ways outlined later in this chapter, geology and economics (and geopolitics) become intertwined, forming a complex interrelationship, with various factors giving shape to the rise, peak and decline of oil supply. The primary concern of the peak oil school is that the peak arrives while demand for oil keeps growing. According to basic economic principles, a stagnating supply coupled with increasing demand would lead to a spike in oil prices, and this would place a huge financial burden on oil dependent economies, with destabilizing effects.

As conventional oil supply began to stagnate in 2005 while global demand continued to increase, the price of oil began a steady incline, moving from its historic average of US$20–25 per barrel (where it sat even in the late twentieth century) to more than US$100 by 2008. This basic dynamic played out as the peak oil school predicted (Heinberg 2003, 2011), even if the interplay between geology, economics, technology, culture and geopolitics proved more complicated and nuanced than petroleum geologists and other analysts anticipated. Today conventional crude oil remains on what is often called a "corrugated" or "undulating" plateau (Jackson and Smith 2014), a phenomenon that has been acknowledged by mainstream institutions, including the International Energy Agency (BBC 2013; IEA 2010: 6). In other words, conventional crude oil seems to have peaked. Any gains from now on, if they occur, will be negligible.

Nevertheless, as the rate of conventional crude oil production stopped growing, the consequent rise in the price of oil made various unconventional oils more economically viable, facilitating their production and incentivizing the development of new or more refined technologies (including "fracking" techniques). What this meant was that global supply of oil was able to keep up with a growing global demand, delaying a peak in overall liquid fuels.

But meeting this growing demand came at a huge financial cost, and the intimate relationship between energy and economics became clearer. No longer could oil be considered a marginal cost of negligible economic significance to the processes of production. After a century of cheap energy inputs, industrial economies (especially the oil importers) found their dependence on oil to be an increasingly debilitating financial burden (Ayres 2014; Ayres and Warr 2009; Murphy and Hall 2011a; Tverberg 2015).

It is worth being clear about the extent of this financial burden. By 2012 the global economy was consuming around 90 million barrels of oil every day and, when trying to maintain those levels of consumption, the difference between oil at US$25 per barrel and oil at more than US$100 per barrel becomes hugely significant. To be precise, it constitutes an extra cost to the global economy of around US$7.2 billion per day, or US$3.6 trillion per year – money that would otherwise have been spent in the broader economy. If we look specifically at the US – the world's largest oil consumer – the rise in the price of oil from US$25 to more than US$100 meant that the US was spending an extra US$600 million every day on oil imports, money that was not just being sucked into the energy sector but being sucked out of the national economy all together (Alexander 2014).

In light of these figures, it is not difficult to understand why ten of the last eleven recessions in the US have been associated with high oil prices (Hamilton 2011), or why the implosion of the global economy in 2008 correlated so closely with oil prices spiking at US$147 per barrel (Hamilton 2012; Murphy and Hall 2011b). When oil gets more expensive, everything dependent on oil – such as transport, mechanized labor, industrial food production and plastics – gets more expensive. This pricing dynamic siphons discretionary expenditure and investment away from the rest of the economy or out of the national economy altogether, causing debt defaults, economic stagnation, recessions, or even longer-term depressions (Tverberg 2012). While it would be one-dimensional to argue that expensive oil was the only cause of the global financial crisis – and the ongoing economic stagnation – it would be just as blind to deny the defining role expensive oil played both in the global financial crisis and state of the deflated global economy today (Ayres 2014).

Two principal factors influencing the fall in the price of oil

Against this background, the two principal factors influencing the fall in the price of oil can be inferred with a degree of confidence. The first is a demand-side factor; the second, a supply-side factor. These are not mutually exclusive and, in fact, have fed off each other to exacerbate their individual effects, explaining why the fall in price has been so dramatic.

The demand-side factor influencing the price drop was the deflated global economy (Hamilton 2014; Mearns 2014), in large part owing to several years of expensive oil (averaging more than US$100 since 2011) which has had a suffocating effect on expected growth trajectories. In mid-2015 the economies of countries in the European Union and Japan remain very weak; China's growth is slowing; and the Russian economy is sinking quickly, all of which reduces oil demand, and expected demand (Meijer 2014b). When economic growth is strong, oil demand is high; when economies are weak, stagnant, or in recession, oil demand is weak.

When oil demand is weak while supply is maintained, however, basic economic principles dictate that the price of oil will fall, and this is precisely what we have seen. Another way to frame this demand-side point is to say that when oil is expensive, it becomes increasingly unaffordable, especially when wages stagnate, and this unaffordability induces "demand destruction," which puts less pressure on oil supply chains. It could even be said that there is not so much a glut of cheap oil so much as there is a glut of consumers that cannot afford expensive oil (Mushalik 2015c). Consequently, the reduced pressure on the oil markets manifests in reduced prices. All this is perfectly comprehensible, even if the exact timing of the effects could never be predicted with any precision. Economics is not a hard science.

The second principal factor influencing the depression in prices starting mid-2014 can also be understood in relation to the prolonged period of expensive oil in recent years, but this time from the supply-side. Historically, the vast reserves of unconventional oil around the world (especially in tar sands of Canada and Venezuela and shale oil plays in the US) have been underexploited, because the capital expenditures needed to extract oil from them have been so great that it would have been uneconomic to do so. But, once conventional oil began to plateau around 2005, putting supply pressure on global oil markets, this induced the steady rise in the price of oil. As oil reached beyond US$100 and seemed to stabilize, it suddenly appeared as if much more of the unconventional oils could be produced for a profit. This naturally provoked something of an investment frenzy, especially in the US and Canada, resulting in the significant uptick in US oil production and the steady rise in Canadian tar sands production. Several years of "manic drilling" (*Economist* 2014b) resulted in a short-term glut in oil supply, and whenever there is a glut in supply, prices inevitably fall. (Why the so-called glut is likely to be short term is addressed later in this chapter). A boost in Libyan oil production also magnified the temporary oversupply (Patterson 2014).

It is worth highlighting the important interactions here between the demand-side and the supply-side dynamics. As we have seen, expensive oil places a burden on oil dependent economies, making it difficult to maintain expected or desired growth trajectories and inducing demand destruction. But, just as oil demand was weakening due to poor economic performance, the very same phenomenon of expensive oil was bringing new supply chains to the market. If these supply and demand dynamics were at play in isolation, they would have produced a drop in the price of oil. When they occur together (when demand is being destroyed by expensive oil just as expensive oil is incentivizing increased production), it should come as no surprise that at some point the markets would react, and we have seen precisely that occur (Berman 2015).

Why cheap oil is a mixed blessing and, ultimately, a curse

As noted in the introduction, the sudden and drastic fall in the price of oil has been widely interpreted as "good news" for economies. In a superficial sense, this is quite an understandable reaction. While many people seem resistant to the thesis that expensive oil inhibits economic growth, more people seem willing to accept the flip side of the same coin, that cheap oil is good for economic growth. In an age of deep economic uncertainty and widespread economic instability, anything that is perceived to be good for growth is generally regarded as something worth celebrating. Unfortunately, the implications of cheap oil are far more complicated and by no means so positive.

It is certainly true that cheap oil makes conventional economic growth easier than if oil was expensive, so if returning to historic growth trajectories is considered the ultimate goal, then the celebration of the falling price of oil is justified, so far as it goes. But the following analysis unpacks the situation in more detail and fleshes out some of the intricacies in this situation in order to show why cheap oil is likely to cause as many problems as it solves.

The first thing to note is that, irrespective of the current market price of oil, the energy return on investment (EROI) of oil is in terminal decline (Murphy 2014). We must not forget that it is "net energy" that is the important measure of energy supply, not total barrels extracted and consumed. Due to declining EROI, it is possible that oil production can increase in gross supply while net energy from oil can be flat or in decline, which might happen as high-EROI conventional oil is replaced with low-EROI unconventional oil. Indeed, this would disguise the peak in useful energy supply from oil. Could we be at that point now, even though total liquid fuels still seem to be creeping upward? It is hard to be sure, but it is important that we take account of the subtle phenomenon of peak oil, because it is on its way, if it is not already upon us.

The increasing financial costs of production are easier to quantify. Oil's declining EROI translates as increasing costs of production, especially in new oil fields. According to a recent analysis (Kopits 2014: 43), capital expenditure in the large oil firms has been rising at 11 percent per annum (p.a.) since 1999. When the price of oil was hovering above the US$100 mark, it made economic sense to invest and produce many of these unconventional oils because, despite the increasing costs of production, it seemed that profits could still be made. But now that oil has dropped to around US$50 per barrel, a large proportion of this new production no longer seems profitable. Highlighting the case, the January 2015 *Monetary Policy Report* of the Bank of Canada (2015: 3) suggested:

> Based on recent estimates of production costs, roughly one-third of current production could be uneconomical if prices stay around US$60, notably high-cost production in the United States, Canada, Brazil and Mexico. More than two-thirds of the expected increase in the world oil supply would similarly be uneconomical. A decline in private and public investment in high-cost projects

> could significantly reduce future growth in the oil supply, and the members of the Organization of the Petroleum Exporting Countries (OPEC) would have limited spare capacity to replace a significant decrease in the non-OPEC supply.

In much the same vein, in the *Financial Times*, Raval (2014) concluded that Canadian oil sands had a break-even price of US$80 per barrel compared with US shale plays (and other areas of tight oil) at around US$76, and that Brazil's deep water fields required US$75 to break even with Mexican projects more around US$70. If these estimates are even roughly accurate, the recent price drop to around US$50 per barrel means that all these technically recoverable oil resources may become vulnerable to their own high (and increasing) production costs. Needless to say, profit-seeking businesses will not produce oil that costs more than US$70 if they can only sell it for US$50 (Carroll and Klump 2013).

Of course, significant portions of the costs of production occur in the early stages of setting up a field for drilling, which means that most of the current projects already have "sunk costs." Because of this, most of those kinds of projects were not going to stop producing in the short term. But at US$50 per barrel, many (if not most) new unconventional oil projects may not be profitable, and investors and oil companies alike were already beginning to show signs of caution or withdrawal. Notably, in November 2014, there was a 40 percent drop in new oil and gas permits in the US, being attributed to the lower price of oil (Hays 2014). Similarly, since October 2014, rigs in the US fell by 34 percent (Inman 2015a, 2015b; Mushalik 2015d) according to the Baker Hughes index.

As of August 2015, shale rigs in the US had dropped by an astounding 58 percent over the previous year (Hamilton 2015). Likely consequences were reported in the *Economist* (2014b) in early December 2014:

> A rash of bankruptcies is likely. That, in turn, would bespatter shale oil's reputation among investors. Even survivors may find the markets closed for some time, forcing them to rein in their expenditure to match the cash they generate from selling oil. Since shale-oil wells are short-lived (output can fall by 60–70% in the first year), any slowdown in investment will quickly translate into falling production.

A different article in the same issue of the *Economist* (2014a) was even more explicit:

> Wood Mackenzie, a research consultancy, estimates that the "break-even price" of American projects is clustered around $65–70, suggesting many are vulnerable (these calculations exclude some sunk costs, such as building roads). If the oil price stays at $70, it estimates investment will be cut by 20% and production growth for America could slow to 10% a year. At $60, investment could drop by as much as half and production growth grind to a halt.

This is hugely significant, especially when it is recognized that the growth in US shale oil and Canadian tar sands in recent years has been essentially the only thing that has disguised peaking production of liquid fuels in the rest of the world (Mushalik 2013). In fact, the current oil situation, which some are claiming debunks peak oil, may in fact be announcing its arrival. Conventional oil production is already on a corrugated plateau that almost certainly represents the highest "peak" it will ever reach, but it could be that the current supply and demand dynamics mark the onset of peak liquid fuels. As oil analyst, Ron Patterson (2015), notes:

> Peak oil will be the point in time when more oil is produced than has ever been produced in the history of the world, or ever will be in the future of the world. It is far more likely that this period will be thought of as a time of an oil glut rather than a time of an oil shortage.

Could it be that 2015–17 will be characterized by an oil glut that marks the peak in liquid fuels? If cheap oil is in the process of jeopardizing future production or if the "shale boom" peters out in the next year or two (Heinberg 2013; Hughes 2013, 2014; Mushalik 2014; Zittel et al. 2013), such a near-term peak could indeed eventuate. In fact, US shale oil has been in decline since April 2015 (Hamilton 2015), suggesting the end of the boom.

Furthermore, a 2014 study conducted by Goldman Sachs (Adams 2014) concluded that the lower oil prices meant that US$1 trillion of oil investment funds were at risk of being withdrawn from projects, and that this would reduce production by 7.5 million barrels of oil per day over the coming decade. Following that study, prices fell further. Even before the price slump, in fact, the biggest oil companies were shelving expansion plans and shredding operations with profit margins too tight to justify (Gilbert and Scheck 2014; Tverberg 2014). Maintaining current production looks like a Herculean task.

Nevertheless, the likely consequence of falling production and investment would be a tightening of global oil supply, thus increasing the price of oil, especially if demand increased at the same time. This upward pressure, of course, could bring some of the high-cost producers back online, although investors would be more cautious and funds harder to come by, for fear of another price collapse. Furthermore, if the economy cannot accommodate a return to expensive oil, there might be a subsequent price slump and further drop in production. This is the kind of volatility that we can expect in coming years and decades. It is too simplistic to suggest that lower prices have meant that our oil troubles are over. They are merely challenging oil dependent economies in new ways, primarily by threatening to render huge amounts of existing production "uneconomic." At the World Economic Forum in 2015, the chief economist of the IEA, Fatih Birol (quoted in Mushalik 2015a), described the oil situation as follows:

> In 2015 we expect oil and gas upstream investments to decline US$100 billion or 15%. And the big chunk of it will come from the high cost areas. And this will have implications, not perhaps immediately but for 2016–17. And if this comes together with a stronger demand this will have strong implications for the price and the markets.

Another issue that needs to be borne in mind is the economic instability that arises in oil-exporting nations when the price of oil drops so suddenly and deeply. Significant exporters such as Russia, Iran, Iraq and Venezuela are highly dependent on high oil prices to balance their fiscal budgets (Herszenhorn 2014). In October 2014 the International Monetary Fund assessed what ideal oil price governments needed to balance their budgets, and drew some disturbing conclusions: Russia needed oil at $101 per barrel; Iran needed $136; Venezuela and Nigeria needed US$120 (see Viscusci et al. 2014). When we recognize that oil and gas have been making up 50 percent of Russia's federal budget, it becomes clear that a drop in price from US$110 (June 2014) to US$50 (August 2015) more than halves that oil revenue stream. If cheap oil means extreme economic hardship for exporters, this could well provoke social discontent and political instability. Geological and economic issues quickly become infused with geopolitics.

Further geopolitical complexities and speculations concerned the fall in the price of oil. For instance, it provoked some speculation that Saudi Arabia desired these lower prices because cheap oil would undermine competition in global oil markets, especially the US shale plays and Canadian oil sands, both of which required higher prices to maintain existing production over the long term (Critchlow 2015; Meijer 2014a; Solomon and Said 2014). For decades Saudi Arabia had been the so-called swing producer that increased or decreased production as necessary to stabilize the price of oil where it could. By 2014, however, Saudi Arabia made it clear that it had no intention of reducing production to push the price of oil higher (Defterios 2014). "If we cut production then there will be spare capacity and producers will not invest, or postpone projects," stated the secretary general of OPEC, Abdalla Salem el-Badri, quoted in Critchlow (2015): "The market will rebound back higher than the $147 we saw in 2008."

Perhaps most importantly, with cheap oil, Saudi Arabia is able to punish or put pressure on some of its geopolitical enemies and those of the US, including Iran and Russia – two oil exporters much harder hit by US$50 oil than the wealthier Saudi Arabia (Mazzetti et al. 2015). As of early 2015, the Russian economy seemed particularly weak and unstable, and there was some speculation that the US has colluded with Saudi Arabia to flood the markets for this very purpose (Elliot 2014; Topf 2015; Whitney 2014), even if this hurt US shale producers. In fact, some analysts argue, with plausibility, that in our neoliberal era oil markets provide a means for the US government and the broader "Transnational Elite" to insidiously wage economic war, especially against Russia (Fotopoulos 2014).

It is very difficult to know how far these geopolitical influences have been shaping the oil markets – and space does not permit a more elaborate analysis – but there certainly seems to be more than plain "supply and demand" issues at play. It seems that the geology is fundamental, which then enters a dialectical relationship with the economics, leaving the very real geopolitical tensions and strategies to play out against that background.

The environmental costs of "cheap" oil

The analysis so far has focused primarily on the central role energy plays in economic processes, touching also on a couple of important geopolitical issues. The environmental impacts of oil consumption are too often left out of this picture. Not only does oil consumption facilitate the depletion of natural resources and the devastation of biodiversity as a result of ever-expanding, globalized economies, perhaps most importantly we now know that the consumption of oil and other fossil fuels contributes directly to climate change (Hansen and Kharecha 2008; IPCC 2013).

Analysts tend to try to deal with these issues in isolation, exemplified most strikingly by Fatih Birol. On the one hand, his position as chief economist of the IEA demands that he does all he can to ensure that enough affordable oil is supplied to global markets in such a way that facilitates stable economic growth. In fact, that is essentially the reason the IEA was formed, in the wake of the 1973 oil crisis. On the other hand, Birol sees the world continuing its addiction to fossil fuels in such a way that is locking humanity into decades of high carbon living. The implications of this on climate change will be disastrous, as the chief economist acknowledges. He seems to be torn apart by the contradiction of trying to facilitate conventional (oil dependent) growth economics in the grim context of climate change.

At least Fatih Birol is forthright enough to acknowledge the intractable problem posed by this situation, even if he still tries to address the problems in isolation. Environmentalists often fail to understand how destabilizing it would be, from a conventional economic perspective, to swiftly and significantly reduce oil consumption. Economists, however, are often too quick to celebrate the economic benefits of cheap oil, neglecting to mention the fact that cheap oil will incentivize increased fossil fuel consumption at a time when the world's climate scientists are crying out that we must swiftly move away from fossil fuels (IPCC 2013).

According to carbon budget analyses, such as Carbon Tracker (2013), 60–80 percent of known fossil fuels reserves must be left in the ground if the world is to have a good chance of keeping temperature rises less than 2 degrees above pre-industrial levels. Climate scientists Kevin Anderson and Alice Bows (2011) have shown that, in order to keep within a fair share of the carbon budget, the wealthier nations need to decarbonize their economies by 8–10 percent p.a. over coming decades. Guess what would happen to the price of oil if oil availability was reduced by 8–10 percent p.a. as a climate change response strategy? Even reducing availability at 3–4 percent p.a. would produce a price spike to unprecedented highs and probably crash many economies, just like in 2008.

But cheap oil only makes continued oil consumption more affordable, while at the same time making renewable energy alternatives less price competitive. In this light, cheap oil is a catastrophe for climate-response strategies. As the *Financial Times* reports: "falling oil prices threaten to make economies more carbon-intensive and less energy efficient" (Wolf 2014; Zumbrun 2015). The reality is that if a peak and decline in liquid fuels is not imposed upon us for geological-cum-economic reasons, we should nevertheless be embracing it voluntarily for reasons of climate mitigation. Of course, I do not claim that this climate-response strategy is likely. I only claim that the challenge of climate change clearly shows that the question of how to deal with a peak and decline of oil supply is more relevant today than ever before.

These points ultimately highlight the incoherence of talking of "cheap oil." The only reason it can be considered "cheap" is because the environmental costs of oil consumption are externalized (i.e. not included in the market price of oil). If the costs of climate change, biodiversity loss, pollution and resource depletion were built into the price of oil, there is no way it would be cheap. And what of the social and economic costs that will be borne by future generations? In decades to come, as climate change wreaks havoc on global food systems and increases the severity and regularity of extreme weather events, how will today's language of "cheap oil" be received? I suspect that US$50 oil will be seen for what it is – something that came at far too great a cost.

This is the paradox of oil: The cheaper it is, the more it costs.

Conclusion

What lies ahead for oil markets? Nobody really knows. There are too many economic, geopolitical, technological and social variables at play for any certainty. Black swans could lie around every bend in the river. The unpredictable actions of OPEC have significant implications. Will they cut production after their next meeting? It is anybody's guess. There is also the ever-present possibility of ongoing geopolitical disruptions, as evidenced especially by the instability in several oil-rich nations, such as Iraq, Iran and Russia. If a new war broke out in any of these areas, or if Russia's economic decline intensifies, markets would be thrown into further turmoil. This could push prices back up very quickly, but high prices could assist the producers of non-conventional oil that need high prices to make any money. However, the point I have been laboring is that those high prices could again squeeze the life out of oil dependent economies and place further obstacles in the way of economic recovery. Another global financial crisis would only overturn the oil markets again, as happened in 2008, leading to a pricing collapse. There is also the threat of the "carbon bubble" bursting, if shareholders in fossil fuel companies begin to worry that their shares could become "stranded assets" should nations or the international community decide to take climate change seriously (Alexander et al. 2014).

In light of all this, my view is that we should expect continued price volatility. That may sound like I am hedging my bets, but the fact is that the oil situation is so unstable that cycles of bust-recovery-bust are the most likely future we face. It could well be that we will see a further fall in the price of oil, although claims of US$20 oil (Worstall 2015) seem very unlikely to be realized. If oil ever fell to this price it would necessarily be short-lived, for reasons already outlined: Many unconventional producers need prices of more than US$70 per barrel simply to break even. Indeed, early 2015 the IEA reported that "a price rebound . . . seems inevitable" (Kent and Faucon 2015).

Where to? Again, no one can be sure, but the high production costs of unconventional oil suggest that a reasonable "floor" for oil prices in the midterm to long term (within a couple of years) may be in a range of more than US$80 – higher if the shale boom goes bust or if war or political instability enters the scene more significantly. But, remember, part of my argument is that there is no safe operating space for oil prices. The "floor" of oil prices is likely to be too high for the economy, just as the "ceiling" is likely to be too low for the climate. We live in paradoxical times indeed.

What we do know is that the EROI of oil is in terminal decline, and it is this geological reality that means there will forever be upward pressure on the price of oil, and that is forever going to put pressure on oil dependent, growth-oriented economies. As Murphy and Hall (2011a: 52) argue: "Increasing the oil supply to support economic growth will require high oil prices that will undermine that economic growth." This is the world we now live in.

In the introduction, I stated that there is no optimal price for oil. In an age of increasing capital expenditure on new oil fields, due to declining EROI, oil needs to be sufficiently expensive for oil supply to keep up with demand. But when oil is too expensive, economies that rely on cheap energy inputs cannot function and demand dries up, reducing the price of oil. Some analysts argue that there is a "narrow ledge" (Nelder and Macdonald 2011), where the price of oil is high enough to procure the necessary investments and production but not so high as to inhibit so-called healthy growth of the economy. That may have been the case in recent history, but my suspicion is that this narrow ledge has crumbled away. There is no longer an optimal price that falls within such a ledge. Oil is now either too cheap to procure ongoing investments and production or too expensive for oil dependent economies to function well – and, perhaps, even both too cheap to meet demand and too expensive for growth. When these issues are placed in the context of climate change and the need to transition beyond fossil fuels, it becomes clear that there is no such thing as cheap oil.

In short, industrial civilization now finds itself between a rock and a hard place; or, to change the metaphor, we now find ourselves in "checkmate," with nowhere to move. Our only option is to start playing a different game "beyond oil" – a choice we should have made many years, if not decades, ago. Unfortunately, building a post-petroleum civilization would require a bravery and boldness that we have hitherto lacked (Alexander 2015; Trainer 2010). Can we yet muster the courage? The challenge, admittedly, is to find ground between naive optimism and despair.

References

Adams, C. (2014) "Oil Price Fall Threatens $1tn of Projects," *Financial Times*, 15 December 2014, accessed 10 January 2015 — www.ft.com/intl/cms/s/0/b3d67518–845f-11e4-bae9–00144feabdc0.html#axzz3RxSpUHzf

Alexander, S. (2015) *Sufficiency Economy: Enough, for Everyone, Forever*, Melbourne: Simplicity Institute.

——— (2014) "The New Economics of Oil," *MSSI Issues Paper 2*, March, 1–15.

Alexander, S., Nicholson, K. and J. Wiseman (2014) "Fossil Free: The Development and Significance of the Fossil Fuel Divestment Movement," *MSSI Issues Paper 4*, September, 1–16.

Anderson, K. and A. Bows (2011) "Beyond 'Dangerous' Climate Change: Emission Scenarios for a New World," *Philosophical Transactions of the Royal Society* 369: 2–44.

Ayres, R. (2014) *The Bubble Economy: Is Sustainable Growth Possible?* Cambridge, MA: MIT Press.

Ayres, R. and B. Warr (2009) *The Economic Growth Engine: How Energy and Work Drive Material Prosperity*, Cheltenham: Edward Elgar.

Bank of Canada (2015) *Monetary Policy Report*, accessed 17 February 2015 — www.bankofcanada.ca/wp-content/uploads/2014/07/mpr-2015–01–21.pdf

BBC (2013) "Interview with Fatih Birol, Chief Economist of IEA," HARDtalk, British Broadcasting Corporation, 10 January.

Berman, A. (2015) "No Oil Price Rebound Yet: An Explanation in Two Charts," *The Petroleum Truth Report*, 11 February, accessed 17 February 2015 — www.resilience.org/stories/2015–02–11/no-oil-price-rebound-yet-an-explanation-in-two-charts

Carbon Tracker (2013) "Unburnable Carbon 2013: Wasted Capital and Stranded Assets," *Carbon Tracker Initiative*, accessed 10 January 2015 — www.carbontracker.org/report/wasted-capital-and-stranded-assets/

Carroll, J. and E. Klump (2013) "Oil's $5 Trillion Permian Boom Threatened by $70 Oil," *Bloomberg Business*, 26 October, accessed 15 September 2015 — www.bloomberg.com/news/2013–10–24/oil-s-5-trillion-permian-boom-threatened-by-70-crude.html

Critchlow, A. (2015) "Saudi Arabia Increases Oil Output to Crush US Shale Frackers," *Telegraph*, 27 January, accessed 15 February 2015 — www.telegraph.co.uk/finance/newsbysector/energy/11372058/Saudi-Arabia-increases-oil-output-to-crush-US-shale-frackers.html

Defterios, J. (2014) "Saudi Arabia: We'll Never Cut Oil Production," *CNN Money*, 22 December, accessed 10 January 2015 — http://money.cnn.com/2014/12/22/news/economy/saudi-arabia-oil-production/

Economist (2014a) "In a Bind: Will Falling Oil Prices Curb America's Shale Boom," *Economist*, 6 December, accessed 25 January 2015 — www.economist.com/news/finance-and-economics/21635505-will-falling-oil-prices-curb-americas-shale-boom-bind

——— (2014b) "Sheiks v Shale," *Economist*, 6 December, accessed 25 January — www.economist.com/news/leaders/21635472-economics-oil-have-changed-some-businesses-will-go-bust-market-will-be

Elliot, L. (2014) "Stakes Are High as US Plays the Oil Card Against Iran and Russia," *The Guardian*, 10 November, accessed 10 January 2016 — www.theguardian.com/business/economics-blog/2014/nov/09/us-iran-russia-oil-prices-shale

Fotopoulos, T. (2014) "Oil, Economic Warfare, and Self-Reliance," *Inclusive Democracy Weekly Column*, 27 October, accessed 15 February 2015 — www.inclusivedemocracy.org/journal/vol10/vol10_no1-2_Oil_economic_warfare_and_self-reliance.html

Gilbert, D. and J. Scheck (2014) "Big Oil Feels the Need to Get Smaller: Exxon, Shell, Chevron Pare Back as Rising Production Costs Squeeze Earnings," *Wall Street Journal*, 2 November, accessed 10 January 2015 — www.wsj.com/articles/big-oil-feels-the-need-to-get-smaller-1414973307

Hamilton, J. (2015) "US Tight Oil Production Decline," *Econbrowser*, 30 August, accessed 30 August 2015 — http://econbrowser.com/archives/2015/08/u-s-tight-oil-production-decline

——— (2014) "Oil Prices as an Indicator of Global Economic Conditions," *Econbrowser*, 14 December, accessed 25 January 2015 — http://econbrowser.com/archives/2014/12/oil-prices-as-an-indicator-of-global-economic-conditions

——— (2012) *Oil Prices, Exhaustible Resources, and Economic Growth*, National Bureau of Economic Research Working Paper 17759, accessed 10 September 2013 — www.nber.org/papers/w17759.pdf

——— (2011) *Historical Oil Shocks*, National Bureau of Economic Research Working Paper 16790, accessed 10 September 2013 — www.nber.org/papers/w16790.pdf

Hansen, J. and P. Kharecha (2008) "The Implications of 'Peak Oil' for Atmospheric CO_2 and Climate," *Global Biochemical Cycles* 22: GB3012, 1–10.

Hays, K. (2014) "Exclusive: New U.S. Oil and Gas Well November Permits Tumble by Nearly 40%," *Reuters*, 2 December, accessed 10 January 2015 — www.reuters.com/article/2014/12/02/us-usa-oil-permits-idUSKCN0JG2C120141202

Heinberg, R. (2013) *Snake Oil: How Fracking's False Promise of Plenty Imperils Our Future*, Santa Rosa, CA: Post Carbon Institute.

——— (2011) *The End of Growth: Adapting to Our New Economic Reality*, Gabriola Island, British Columbia: New Society.

——— (2003) *The Party's Over: Oil, War, and the Fate of Industrial Civilization*, Gabriola Island, British Columbia: New Society.

Herszenhorn, D. (2014) "Fall in Oil Prices Poses a Problem for Russia, Iraq, and Others," *New York Times*, 15 October, accessed 10 January 2015 — www.nytimes.com/2014/10/16/world/europe/fall-in-oil-prices-poses-a-problem-for-russia-iraq-and-others.html?_r=1

Hughes, J.D. (2014) *Drilling Deeper: A Reality Check on US Government Forecasts for a Lasting Tight Oil and Shale Gas Boom*, Santa Rosa, CA: Post Carbon Institute, accessed 10 January 2015 — www.postcarbon.org/publications/drillingdeeper/

——— (2013) *Drill, Baby, Drill: Can Unconventional Fuels Usher in a New Era of Energy Abundance*, Santa Rosa, CA: Post Carbon Institute, accessed 10 September 2013 — www.postcarbon.org/reports/DBD-report-FINAL.pdf

IEA (2013) *World Energy Outlook 2013: Executive Summary*, Paris: International Energy Agency, accessed 10 February 2014 — www.iea.org/publications/freepublications/publication/WEO2013_Executive_Summary_English.pdf

——— (2010) *Global Energy Outlook 2010: Executive Summary*, Paris: International Energy Agency, accessed 10 September 2013 — www.worldenergyoutlook.org/media/weowebsite/2010/WEO2010_es_english.pdf

Inman, M. (2015a) "Early Signs of the Bakken Oil Slow Down," *Beacon*, 26 January, accessed 20 February 2015 — www.beaconreader.com/mason-inman/early-signs-of-the-bakken-oil-slow-down?ref=profile

——— (2015b) "The Investment Gap," *Beacon*, 12 January, accessed 20 February 2015 — www.beaconreader.com/mason-inman/the-investment-gap

IPCC (2013) *Climate Change 2013: The Physical Science Basis*, Fifth Assessment Report, (WG1) Geneva: Intergovernmental Panel on Climate Change, accessed 15 September 2015 — www.ipcc.ch/report/ar5/wg1/#.Uk6k-CjqMRw

Jackson, P.M. and L.K. Smith (2014) "Exploring the Undulating Plateau: The Future of Global Oil Supply," *Philosophical Transactions of the Royal Society* 372(2006): 1–20.

Kent, S. and B. Faucon (2015) "Oil-Price Rebound Predicted: IEA Adds to Chorus of Voices Saying Glut with Abate," *Wall Street Journal*, 9 February, accessed 10 February 2015 — www.wsj.com/articles/demand-for-opec-crude-will-rise-this-year-says-group-1423482563

Kopits, S. (2014) "Oil and Economic Growth: A Supply-Constrained View," presentation to Center on Global Energy Policy, Columbia University, 11 February, accessed 15 January 2015 — http://energypolicy.columbia.edu/sites/default/files/energy/Kopits%20-%20Oil%20and%20Economic%20Growth%20%28SIPA,%202014%29%20-%20Presentation%20Version%5B1%5D.pdf

Likvern, R. (2012) "Is Shale Oil Production from Bakken Headed for a Run with 'the Red Queen'," *Oil Drum*, 25 September, accessed 20 February 2015 — www.theoildrum.com/node/9506

Mazzetti, M., Schmitt, E. and D. Kirkpatrick (2015) "Saudi Oil Is Seen as Lever to Pry Russian Support for Syria's Assad," *New York Times*, 3 February, accessed 10 February 2015 — www.nytimes.com/2015/02/04/world/middleeast/saudi-arabia-is-said-to-use-oil-to-lure-russia-away-from-syrias-assad.html?ref=todayspaper

Mearns, E. (2014) "The 2014 Oil Price Crash Explained," *Energy Matters*, 24 November, accessed 10 January 2015 — http://euanmearns.com/the-2014-oil-price-crash-explained/

Meijer, R. (2014a) "Cheap Oil a Boon for the Economy? Think Again," *Automatic Earth*, 29 November, accessed 10 January 2015 — www.theautomaticearth.com/2014/11/cheap-oil-a-boon-for-the-economy-think-again/

——— (2014b) "The Price of Oil Exposes the True State of the Economy," *Automatic Earth*, 27 November, accessed 10 January 2015 — www.theautomaticearth.com/2014/11/the-price-of-oil-exposes-the-true-state-of-the-economy/

Miller, R. and S. Sorrell (2014) "The Future of Oil Supply," *Philosophical Transactions of the Royal Society A* 372: 1–27.

Munroe, R. (2010) "Energy Security: An Annotated Military/Security Bibliography (2010 update)," *Resilience*, 28 September, accessed 10 January 2015 — www.resilience.org/stories/2010–09–28/energy-security-annotated-militarysecurity-bibliography-2010-update

Murphy, D. (2014) "The Implications of the Declining Energy Return on Investment of Oil Production," *Philosophical Transactions of the Royal Society A* 372, 20130126: 1–19.

Murphy, D. and C. Hall (2011a) "Adjusting to the New Energy Realities of the Second Half of the Age of Oil," *Ecological Modelling* 223: 67–71.

——— (2011b) "Energy Return on Investment, Peak Oil, and the End of Economic Growth," *Annals of the New York Academy of Sciences* 1219: 52–72.

Mushalik, M. (2015a) "Free Oil! Next Stop Free Oil Crunch," *Crude Oil Peak*, 15 February, accessed 20 February 2015 — http://crudeoilpeak.info/free-oil-next-stop-free-oil-crunch

——— (2015b) "NorthConex Road Tunnel Contract Signed Only Days after US$150 Oil Price Warnings in Davos," *Crude Oil Peak*, 10 February, accessed 15 February 2015 — http://crudeoilpeak.info/northconnex-road-tunnel-contract-signed-only-days-after-us-150–200-oil-price-warnings-in-davos

——— (2015c) "Peak Affordable Oil' *Crude Oil Peak*," 2 February, accessed 5 February 2015 — http://crudeoilpeak.info/peak-affordable-oil

——— (2015d) "US Drilling Count," *Crude Oil Peak*, 13 February, accessed on 17 February 2015 — http://crudeoilpeak.info/us-drilling-rig-count

——— (2014) "IEA Report Implies US Crude Production May Start to Peak 2016," *Crude Oil Peak*, 14 August, accessed 10 January 2015 — http://crudeoilpeak.info/iea-report-implies-us-crude-production-may-start-to-peak-2016

——— (2013) "US Shale Hides Crude Oil Peak in Rest of World," *Crude Oil Peak*, 11 September, accessed 15 October 2013 — http://crudeoilpeak.info/us-shale-oil-hides-crude-oil-peak-in-rest-of-world

Nelder, C. and G. Macdonald (2011) "There Will Be Oil, But at What Price?" *Harvard Business Review*, 4 October, accessed 10 January 2015 — https://hbr.org/2011/10/there-will-be-oil-but-can-you

Patterson, R. (2015) "Texas RRC Oil and Gas Production Data," *Peak Oil Barrel*, 20 January, accessed 25 January 2015 — http://peakoilbarrel.com/texas-rrc-oil-gas-production-data/

——— (2014) "OPEC October MOMR and Other News," *Peak Oil Barrel*, 10 October, accessed 10 January 2015 — http://peakoilbarrel.com/opec-october-momr-news/

Raval, A. (2014) "Oil Price Plunge Means Survival of Fittest: Crude Oil at $70 Puts at Least 1.5m b/d of Projects 2016 at Risk," *Financial Times*, 10 December, accessed 10 January 2015 — www.ft.com/intl/cms/s/0/51cc00ba-7f85–11e4–86ee-00144feabdc0.html#axzz3RxSpUHzf

Sakya, P. (2015) "BP PLC, Shell PLC and Petrofac PLC: Why Peak Oil Theory Was Wrong," *Yahoo Finance*, accessed 25 January 2015 — https://uk.finance.yahoo.com/news/bp-plc-shell-plc-petrofac-093705013.html

Solomon, J. and S. Said (2014) "Behind OPEC Decision, a Saudi Fear of US Shale," *Wall Street Journal*, 22 December, accessed 10 January 2015 — http://online.wsj.com/public/resources/documents/pageone122214.pdf

Topf, A. (2015) "Did the Saudis and the US Collude in Dropping Oil Prices?" *On Line Opinion*, 5 January, accessed 20 February 2015 — www.onlineopinion.com.au/view.asp?article=16981&page=0

Trainer, T. (2010) *The Transition to a Sustainable and Just World*, Sydney: Envirobook.

Turner, G. (2014) "Is Global Collapse Imminent? An Updated Comparison of *The Limits to Growth* with Historical Data," *MSSI Research Paper* 4, August.

Tverberg, G. (2015) "A New Theory of Energy and the Economy: Part I – Generating Economic Growth," *Our Finite World*, 21 January, accessed 20 February 2015 — http://ourfiniteworld.com/2015/01/21/a-new-theory-of-energy-and-the-economy-part-1-generating-economic-growth/

——— (2014) "Beginning of the End? Oil Companies Cut Back on Spending," *Our Finite World*, 25 February, accessed 20 February 2015 — http://ourfiniteworld.com/2014/02/25/beginning-of-the-end-oil-companies-cut-back-on-spending/

——— (2012) "Oil Supply Limits and the Continuing Financial Crisis," *Energy* 37(1): 27–34.

Viscusci, G., Patel, T. and S. Kennedy (2014) "Oil at $40 Possible as Market Transforms Caracas to Iran," *Bloomberg Business*, 1 December, accessed 10 January 2015 — www.bloomberg.com/news/articles/2014–11–30/oil-at-40-possible-as-market-transforms-caracas-to-iran

Whitney, M. (2014) "The Oil Coup," *Counterpunch*, 16 December, accessed 10 January 2015 — www.counterpunch.org/2014/12/16/the-oil-coup/

Wolf, M. (2014) "Two Cheers for the Sharp Falls in Oil Prices," *Financial Times*, 2 December, accessed 10 January 2015 — www.ft.com/intl/cms/s/0/18a2df62–7949–11e4–9567–00144feabdc0.html#axzz3RxSpUHzf

Worstall, T. (2015) "Citigroup: Oil's Heading to $20 and OPEC's Days are Over," *Forbes*, 10 February, accessed 15 February 2015 — www.forbes.com/sites/timworstall/2015/02/10/citigroup-oils-heading-to-20-and-opecs-days-are-over/

WTRG Economics (2011) "Oil Price History and Analysis," at WTRG Economics (site), accessed 15 September 2015 — www.wtrg.com/prices.htm

Zittel, W., Zerhusen, J., Zerta, M. and N. Arnold (2013) *Fossil and Nuclear Fuels – The Supply Outlook*, March, Berlin: Energy Watch Group.

Zumbrun, J. (2015) "Oil's Plunge Could Help Send Its Price Back Up," *Wall Street Journal*, 22 February, accessed 16 September 2015 — www.wsj.com/articles/oils-plunge-could-help-send-its-price-back-up-1424632746

4

INSTITUTIONAL PLANNING RESPONSES TO A CONFLUENCE OF OIL VULNERABILITY AND CLIMATE CHANGE

Tony Matthews and Jago Dodson

There is a growing awareness that human development has reached a point where our collective activities shape the natural world more than we are shaped by it. We are entering a new epoch, termed the Anthropocene, where human activities are fast becoming the dominant global force (Davis 2010). The power of nature to condition our world is being overtaken by the power of human impacts. Our relentless drive for economic, social, industrial and technological advancement now places critical strain on our environment.

This shift in balance is the result of a dysfunctional relationship between what we add to the environment and what we take from it. This is perhaps most vividly demonstrated by the profoundly negative interrelationship between fossil fuel combustion and the accelerating pace of greenhouse gas accumulation in the planet's atmosphere. We extract fossil fuels from nature and burn them to power our constructed world. In doing so we risk ourselves and our environment, as ever more greenhouse gases build up in the atmosphere where they provide an intensifying basis for devastating climate change.

Oil is the most significant of all fossil fuels shaping and enabling human civilization and advancement due to the mobility it has offered. Now, in the early twenty-first century, we face a stark reality as global oil supplies peak and the intensification of climate change becomes more clearly evident. This century will be the stage upon which the next major phase in human civilization occurs. A key part of this change will concern how we respond to the challenges of oil vulnerability and climate change, especially in cities.

This chapter addresses the issue of such a response from a specific perspective – conceptually depicting oil vulnerability and climate change as "transformative stressors" (Matthews 2012) capable of driving radical changes to institutional governance structures. Rather than considering the phenomena individually, this chapter proposes that a confluence of transformative stressors may occur, meaning that impacts associated with oil vulnerability and climate change begin to occur in tandem. Impacts associated with each may be profound, with implications across scales, sectors and time. Both phenomena are likely to generate profound economic, social, environmental, spatial and institutional impacts.

We examine responses to a confluence of these impacts in terms of institutional transformation, a social process that leads to new forms of governance better designed to respond to transformative stressors. We address the intersection of both transformative stressors within spatial and land-use planning governance. We identify and critically discuss five areas where changes to planning governance may be targeted in an effort to simultaneously respond to the impacts of both transformative stressors: green infrastructure, urban agriculture, urban consolidation, renewable energy and urban transport.

A confluence of transformative stressors

In the social sciences, the term "institution" refers to social constructions that shape social practices by creating and imposing forms of governance (Bell 2002; Kingston and Caballero 2009). Institutions guide social interaction through various forms of governance that are designed to provide stable social structures to facilitate human interaction (Connor and Dovers 2004; Hall and Taylor 1996). Institutional governance can be expressed formally or informally. Examples include laws, statutes, policies, traditions and customs. Spatial and land-use planning systems can be understood as social institutions given that planning provides collective forms of governance intended to coordinate spatial development activities toward desired outcomes (Alexander 2005; Matthews 2013). Institutional planning governance focuses on regulating spatial and land-use development in an effort to balance specific development aims with broader social needs, such as environmental protection, infrastructural provision and the preservation or creation of amenity (Faludi 2000).

Institutional governance frameworks are often faced with "change imperatives." A change imperative is a particular phenomena or stressor that cannot be managed adequately through existing governance. Broader external circumstances and phenomena often provide prompts for institutions to change (McFaul 1995). The process of modifying existing institutional governance or creating new forms in response to a change imperative may be referred to as "institutional transformation." Institutions generally respond to change imperatives in one of three ways. One option is to consciously assess the nature, character and extent of impact associated with the change imperative and to respond to it through a process of institutional transformation. A second option is for affected institutions to either consciously or unconsciously resist or ignore the change imperative. A third option may be a mix of the two.

The role of institutional actors in enabling or blocking institutional and systemic transformation is widely debated in the social sciences (Cortell and Peterson 1999; Hogan 2006; Young 2010). Rather than respond to the imperative itself, institutional actors are often considered to be more influential in conditioning institutional responses to change imperatives instead (Matthews 2012). In other words, a change imperative may manifest and create a compelling impetus for institutional transformation, yet institutional actors may try to ignore or resist the problem. Consequently, institutional transformation may be slow, unsuitable or entirely absent.

By transformative stressor we mean "a chronic large-scale phenomenon which triggers a process of institutional change whereby institutions seek to re-orientate, reorganize and restructure their activities in order to better manage the social, economic and environmental impacts created by the transformative dynamic" (Matthews 2012: 1093). Our conceptual model of "transformative stressors" recognizes that institutional actors have power over transformative processes but that their power is not absolute. The model proposes that certain change imperatives – referred to here as transformative stressors – are sufficiently powerful to compel institutional transformation, even if some actors seek to block the process. Their severity and longevity, combined with an escalation of the associated stressor impacts, can quickly become too acute to be ignored. Institutional actors remain important but their capacity to resist institutional transformation may be greatly diminished, or made impotent or irrelevant, by the manifestation of a transformative stressor.

The impacts associated with transformative stressors are acute, severe, intense and critical. Transformative stressors can quickly manifest as an institutional change imperative. They may be chronic in environmental, economic or social terms, or in any combination of these. They may occur across any spatial scale, from the global to the local or any in between. They can also impact across systemic scales, allowing a single transformative stressor to impact on more than one institution. Their temporal impact is expected to occur most visibly in the medium to long term. As such, a transformative stressor may manifest at any time, although the most profound and damaging impacts may be years or decades away. The length of potential impacts is not a justification for short-term institutional inaction. Indeed, short-term inaction will likely serve to prolong the challenge of managing a transformative stressor. Allowing its impacts to intensify in the short term may ultimately increase the level of response required later.

Climate change and oil vulnerability are both examples of transformative stressors. Each phenomenon has the potential to severely challenge institutional governance frameworks across scales. Both are likely to have global, national, regional and local impacts. Each possesses the potential to compel institutional transformation through the severity, longevity and escalation of their impacts. Both are likely to generate profound negative impacts economically, socially, environmentally, spatially and institutionally.

Institutional transformation leading to new forms of governance better designed to respond to transformative stressors associated with oil vulnerability and climate change will be necessary and vital. Increasingly, the discipline and practice of planning will be required to respond to these stressors. Responses will need to focus heavily on the creation of new institutional planning governance frameworks purposefully designed to manage the spatial and sectoral impacts associated with the emergence and intensification of both change imperatives.

Oil vulnerability refers to the social and economic effects of a sustained increase in the price of petroleum fuels (Dodson and Sipe 2008). Petroleum price increases can be understood as a transformative stressor because stable supplies of relatively cheap petroleum underpin the function and viability of contemporary human civilization. Oil fuels road, water and air travel; generates electricity; and provides agricultural fertilizer and many chemicals used in production and manufacturing. A reliable, affordable supply of oil has enabled the rapid expansion of industry, transport, trade and agriculture across the globe, especially since the mid-twentieth century (Campbell 2005). Oil vulnerability is likely to intensify during the twenty-first century as a consequence of constrained oil supplies. This might be due to climate mitigation actions or to events such as "peak oil." Peak oil describes the moment where global oil production peaks and declining supply becomes insufficient to meet demand, leading to significant price rises.

When peak oil will occur, or indeed whether it already has, is a much debated question. Some assessments suggest that the point of peak oil has already occurred, while others suggest it may be some years away. Irrespective of a specific date, there appears to be general consensus that the peak will be an event of the early decades of the twenty-first century (Newman 2006). Oil vulnerability due to dependence on relatively cheap petroleum fuels will likely bring economic problems, including the need to reappraise the nature of both public and private transport, requirements for alternative energy mixes to meet local and national power demands, as well as changes to urban form and function to ensure the spatial and energy efficiency of cities. Responses to these pressures will need to be sustained, perhaps over many decades. As such, the escalating implications of oil vulnerability are likely to be transformative and new institutional governance frameworks necessary to effectively respond.

Climate change also corresponds to the characteristics of a transformative stressor (Matthews 2012). Contemporary climate science demonstrates clearly that the climate change phenomenon is large-scale and its impacts will escalate, with potential to generate many different forms of stressors at global, national, regional and local levels (CSIRO 2012; IPCC 2012, 2014). Predicted impacts include damage to physical infrastructure, extreme weather events, physical harm to natural and artificial environments, escalating economic costs, biodiversity losses and resource reductions (IPCC 2012, 2014; Stern 2006). Climate change impacts are predicted to be chronic and acutely felt across human and natural systems. They are likely to negatively affect and stress societies in many ways.

If left unaddressed, climate change will create substantial cost impacts, which are likely to eventually challenge cost tolerability at both the macro-economic and micro-economic scales (Stern 2006). In addition, the institutional management of climate change will be a long-term and intergenerational challenge. Due to historical emissions remaining in the atmosphere (CSIRO 2012; IPCC 2014), climate science indicates that impacts will continue to occur even if anthropogenic greenhouse gas emissions stop abruptly. For instance, global sea levels are predicted to rise 27–71 cm during the twenty-first century due to greenhouse gases already locked into the atmosphere through human activity (AMS 2012).

The specter of impacts associated with either climate change or constrained petroleum supplies becoming part of daily life is unsettling. Economic calamity, social upheaval and rapid environmental degradation are all possibilities associated with the emergence of either transformative stressor. Of more concern is the potential for a confluence of both stressors, with impacts associated with oil vulnerability and climate change occurring in tandem. The reliance of modern civilization on plentiful and affordable oil supply, along with stable climactic conditions, would be profoundly disrupted if both transformative stressors were to unfold in parallel. Institutional governance frameworks would be heavily challenged if social, economic and environmental stability became the exception rather than the norm in many countries.

Built environments, in particular, could face extensive challenges associated with impacts of oil vulnerability and climate change. These are likely to be especially acute in cities, where the majority of the global population now lives. It is probable that spatial and land-use planning systems will need to become key response agencies as transformative stressors spread throughout urban environments. Substantial changes to institutional planning governance frameworks will be required. This begs the question of how planning governance could be changed to accommodate the shifting conditions of this new paradigm and what specific responses warrant close attention. The next section addresses this question.

Planning responses to a confluence of transformative stressors

Impacts associated with oil vulnerability and climate change will be most acutely felt in cities, where the majority of the world's population now lives (Dodson and Sipe 2010; Matthews 2011). Cities are heavily dependent on fossil fuels, particularly oil, to provide electricity, transport, heating and cooling, food supply, and materials needed for urban construction and maintenance such as road surfacing materials, plastics and cabling. The combustion of oil to facilitate urban life directly contributes to climate change as development demands lead to escalating greenhouse gases and further encroachment of urban environments into rural areas.

By relying heavily on oil, cities intensify greenhouse gas emissions and increase the prospect of more severe climate change impacts. Many cities are expected to become vulnerable to a range of climate change impacts over the coming century. Depending on geographic location, these changes may include urban heat island effects, increased incidence of flooding, declining water supplies, heat waves, sea level rise and damage to critical infrastructure (Matthews 2011).

While there is a causal relationship between oil use and climate change in an urban context, constraints in petroleum supplies that create oil vulnerability are unlikely to offset climate change impacts. Rather, many cities may have to respond to a confluence of transformative stressors, as the social and economic impacts of oil vulnerability occur in tandem with the social, economic and environmental impacts of climate change. Responding to these twin threats should be a major concern for spatial and land-use planning systems charged with ensuring the ongoing viability of cities worldwide. If a confluence of transformative stressors occurs, how might institutional planning governance respond? What policy areas offer most potential for meaningful response? Is it possible to target governance to respond to impacts associated with both transformative stressors simultaneously? These questions are considered through an examination of five areas where changes to planning governance could be targeted in an effort to respond simultaneously to a confluence of these urban transformative stressors.

Green infrastructure

"Green infrastructure" typically refers to an interconnected, strategically planned and managed network of multifunctional green spaces that are designed to provide a range of ecological, social, and economic benefits (Benedict and McMahon 2006; Kambites and Owen 2006; Wright 2011). Examples of green infrastructure include green roofs, permeable vegetated drainage systems, urban forests, green alleys and streets, biophylic buildings, open green spaces and urban wetlands (Byrne and Yang 2009; Douglas 2011). Green infrastructure appears to offer some potential as a planning-led response to some impacts associated with both climate change and oil vulnerability.

A large-scale switch to green roofing in cities could potentially mitigate urban heat island effects, reduce the volume and intensity of storm water runoff during intense rainfall events, and reduce heat absorption by buildings, in turn lowering energy demand for thermal cooling (Byrne and Yang 2009; Matthews 2011). These benefits occur when roof-based vegetation reflects solar radiation, absorbs and retains some rainfall on-site and increases local humidity, providing a cooling effect to surrounding areas. Such features offer associated benefits by reducing power demand from air conditioning units and wastewater treatment plants.

As another example, permeable vegetative surfacing could be implemented as a default infrastructural standard in outdoor car parks, residential streets, alleyways and urban plazas. They are designed to allow surface water to pass readily through and infiltrate into the ground. They can be utilized to slow the conveyance of excess surface water, thus reducing water carriage volumes, improving flood risk management, better protecting groundwater resources and reducing energy demand associated with wastewater treatment (Matthews 2011; Scholz and Grabowiecki 2007). Again, there is potential for combined benefit in terms of simultaneously responding to stresses associated with oil vulnerability and climate change.

Planning systems are important agents in establishing policy and regulatory standards to direct provision of specific infrastructure and design goals. As such, providing green infrastructure is potentially within the remit of institutional planning governance in many jurisdictions. Notwithstanding, it can be difficult for planning systems to encourage development sectors and the general public to take up new infrastructure and design standards. However, the potential extent and impact of stresses associated with oil vulnerability and climate change provide an institutional context for planning and a rational context for stakeholders to turn to green infrastructure as a form of combined response. New standards potentially offer a variety of positive returns, so they may become more widely sought in an era in which escalating climate change impacts oil prices. Transformations in institutional planning governance should aim to facilitate this transition and incentivize the take-up of green infrastructure if necessary. Even if initially resisted by planners and stakeholders alike, it is possible that the multiple benefits associated with green infrastructure will provide rational and economic incentives for medium- to long-term change.

Urban agriculture

Adverse impacts associated with oil vulnerability and climate change may negatively affect global food supplies (Burton et al. 2013; Newman 2006). Global agriculture is heavily dependent on oil to provide fertilizers, fuel agricultural machinery, facilitate the transport of produce over long distances and power food processing activities. Global agriculture is also very reliant on stable climactic conditions.

This reliance is especially acute in modern times, as agriculture is highly industrialized and entire regions can be given over to the cultivation of a single food type. Oil vulnerability and climate change may profoundly challenge such agricultural practices.

Rapidly inflating oil prices may mean that it is no longer economically viable to preserve and transport food over long distances, while intensifying extreme weather events may lead to widespread crop failures. Therefore, the food security of cities might be undermined by a confluence of oil vulnerability and climate change stressors. This poses a real and serious threat to the ongoing viability of cities and the wellbeing of their residents.

Planning can respond to stresses linked to food security by providing new governance frameworks designed to facilitate greater resilience in urban food systems. Urban agricultural programs are an area of promise, particularly if backed by astute planning governance (Burton et al. 2013; Gosh 2011; Newman 2006). Urban agriculture refers to the urban production of fruit and vegetables, as well as the raising of animals and cultivation of fish with the intent to supply an immediate, local market (Hodgson et al. 2011). Horticultural precincts, community gardens, suburban garden farming, allotments and vertical farms are examples of urban agricultural activities.

Gosh (2011) estimates that an average Australian quarter-acre suburban garden could produce 800–1,100 kg of produce annually. Around 1,000 kg of produce annually would be sufficient to meet a typical household's requirements for fresh vegetables, while providing a small surplus of fruit. During the adverse economic period of the 1930s, Gaynor (2006) estimates that around one-third of all food consumed in Australia was produced in domestic backyards. New planning governance designed to respond to food security threats could aim to facilitate urban agriculture at different scales. New regulations to enable and codify land requirements for backyard agriculture, including more collective forms, could be introduced and the on-site collection of rainwater for irrigation could be supported.

On a larger scale, planning policies could encourage the creation and preservation of horticultural precincts immediately adjacent to cities (Newman 2006). These precincts could be protected through planning policy, ensuring that co-operatives and corporations can invest in necessary infrastructure and technology, safe in the knowledge that the land itself is free from the threat of changing its use or opportunistic rezoning.

Urban consolidation

Urban form is likely to influence the impact and incidence of transformative stresses linked to oil vulnerability and climate change. Cities with low density urban form may be especially vulnerable as long distances between different parts of the city increase fuel demand and require connection to lengthy utility supply networks (Newman 2006). An oil constrained future may mean that residents living in distant urban fringe areas face the problem of rapidly escalating petroleum prices with the possibility of limited recourse to public transport under existing configurations (Dodson and Sipe 2008).

Extended utility supply networks are vulnerable to damage and interruption caused by severe weather associated with climate change (Wilson and Piper 2010). These networks include communications infrastructure, water and energy distribution, waste removal systems and sewerage systems. Consequently, a severe weather event may have wider and more damaging implications for those cities where urban form is predominantly low density and dispersed. Moreover, utility networks often rely heavily on oil as part of the energy mix that provides electricity to power them. A potential response for planning governance is to activate policy levers that can refocus urban form toward more consolidated patterns. Such "urban consolidation" involves increasing densities of built stock in order to increase population and dwellings per unit area (Ewing et al. 2008; Hamin and Gurran 2009; Shaw et al. 2007).

The principle of urban consolidation holds that higher urban densities may assist to reduce transport greenhouse gas emissions and fuel demand, while consolidating the delivery of services and utilities. Increased densities potentially allow for providing greater amounts of open space and green space, which can naturally cool urban areas and absorb greenhouse gases. Higher densities and more compact urban form can bring a mix of land uses into proximity (Shaw et al. 2007). Planners can zone residential, employment, retail and other services closer together. This can potentially reduce petroleum demand and greenhouse gas emissions, and limit the vulnerability of utility supply networks to extreme weather events.

However, such benefits of urban consolidation for the climate and oil vulnerability are contested. A substantial literature attests to the risks of poorly designed consolidation in relation to household water and energy consumption (Dodson 2012). Unless careful design is undertaken, high-rise development may perform relatively poorly in water and energy consumption when contrasted to detached single-storey dwellings, especially if measured per capita. Similarly, Mees (2009, 2010) and Stone and Mees (Chapter 10 in this book) have shown that urban consolidation is not a prerequisite for good quality public transport in either suburban or semi-rural settings.

A further concern with responding to stresses associated with oil vulnerability and climate change through urban consolidation is the fact that building stock and service infrastructure is designed to last for several decades (Biesbroek et al. 2009). The lifespan of existing urban stock may act as a barrier to radical consolidation of cities, especially under constrained energy conditions. However, conflated transformative stressors associated with climate change and oil vulnerability might bring ongoing challenges to low density living, leading to radical reappraisals of institutional planning governance and rapid shifts toward more consolidated urban form. In this scenario, existing stock may be retained through strategic densification. However, it remains possible that impacts of stressors will mean that much existing stock will have to be replaced, notwithstanding the existing financial and social investments therein.

Renewable energy

A shift to decentralized renewable energy production has the potential to offer substantial benefits in regards to stressors generated by climate change and oil vulnerability (Newman et al. 2009). Centralized electricity networks, which are standard in many countries at present, can be vulnerable to a number of stresses associated with oil vulnerability and climate change. These include vulnerability to extreme weather associated with climate change, dependence on oil as a staple fuel for electricity production, large energy demands to transmit electricity over long distances, and cumulative energy losses as electricity is lost from the system during transmission.

Potential benefits of an infrastructural reorientation toward decentralized local renewable energy production may offer possibilities to redress this situation. They include the possibility for stable, clean and cheap electricity production, reduced vulnerability to extreme weather and minimization of energy waste. Decentralized but interconnected renewable energy systems may be viable from regional to building scales. These systems can be powered by a combination of renewable energy types, with specific mixes reflecting local conditions and how well they support particular forms of renewable energy generation (Newman et al. 2009). Recent advances in battery technology are likely to make the small-scale storage of electricity increasingly cost-competitive with grid electricity, thereby facilitating greater uptake of rooftop solar generation and thus contributing to climate mitigation.

Notwithstanding the potential benefits of decentralized renewable energy production, utility companies may resist such moves based on fears of revenue declines and falls in profitability. However, the spectrum of potential stressors that may be wrought by a confluence of oil vulnerability and climate change may transform future economic landscapes, leading to rapid innovation and a shift to new paradigms. If this happens, planning systems may be able to assume a lead role in guiding this transition. Indeed, new strategic partnerships may form between planning systems and utility companies.

Reorientations in planning governance may support the widespread rollout of renewable technologies that cumulatively constitute decentralized energy systems. Relevant technologies include photovoltaic solar panels on roofs, micro wind turbines freestanding in backyards or mounted on gable walls, and geothermal and aerothermal power systems. Planning systems may seek to develop new regulatory frameworks to guide rolling out such technologies and to ensure that new buildings are adequately designed and equipped to meet or exceed minimum thresholds for on-site renewable energy generation. Standards for retrofitting existing building stock may be developed and implemented. Planning governance could be used to establish energy efficiency standards that complement the widespread rollout of renewable energy technologies by ensuring that buildings maximize benefit from minimal electricity use, particularly with respect to thermal heating and cooling demands.

Urban transport

In addition to the intersection of climate change and oil vulnerability impacts on land use and built environments, the urban transport sector is a major potential sphere of change. Road transport has been identified by Unger et al. (2010) as the worst contributing sector to climate change because of the particular composition of vehicle emissions. Although coal-based electricity produces more carbon than road transport fuels, thus adding to climate forcing (heating), its relatively dirty accompanying particulate emissions also contribute to climate mitigation through reduced radiative forcing (cooling). The result is that, in terms of net climate impact, road transport fuels have a higher relative forcing (heating) effect than coal-based fuels (Unger et al. 2010). From a climate change mitigation perspective, therefore, reducing the carbon burden from road transport is a major global imperative. Combined with the problem of oil vulnerability, the imperative to reduce both the climate impact of urban transport and urban transport reliance on petroleum fuels become major tasks for urban planning.

The suburban regions of Australia and North America are highly dependent on motor vehicles for urban transport. In such areas, around 80 percent of travel is undertaken by automobile. In some parts of the US, especially sunbelt cities largely developed in the twentieth century, this level is closer to 95 percent. Therefore, highly car dependent areas are vulnerable to the adverse impacts of higher petroleum fuel costs, whether this is due to either harsh carbon pricing or to higher global fuel prices.

Transitioning to a more fuel sustainable form of urban transportation has been a major policy challenge over the past few decades in Australia and North America. Although much effort has been dedicated to reducing suburban oil dependence, progress has been modest. Much of the effort has focused on providing new public transport infrastructure and modifying transport and land-use relationships, including higher density urban development and "transit-oriented development" (TOD). This can be beneficial, even though individual infrastructure developments, such as light or heavy rail lines, can serve only a limited catchment without wider network connectivity to other lines or bus networks (Mees 2010). The effort to reduce car dependence through positive interventions, such as public transport infrastructure and TOD, are often contradicted by ongoing metropolitan strategic policy stances that continue to supply road capacity in response to traffic demand growth. This appears to be occurring despite the emergence of a "peak" in car travel in much of the developed world of the Global North (Metz 2010; Puentes 2012).

Reducing transport energy demand is recognized as an important component of both climate mitigation and limiting oil vulnerability. There are some important subtleties, however, to the intersection of climate impact from urban transport with wider consumption-based climate impacts and with oil vulnerability. Lenzen et al. (2004), for example, demonstrated with respect to Sydney (Australia) that households in the dispersed relatively low density middle and outer suburban zones poorly serviced by public transport had much higher levels of automotive fuel use, due to their high degree of car dependence, than inner and central households. Yet, across all consumption classes, households located in inner and central areas were the greatest carbon emitters because of the intersection of the urban geography of household wealth, car dependence and general consumption. Middle and outer suburban households tend to be relatively less wealthy and, thus, they have lower overall consumption levels than wealthier inner urban households.

Although automotive energy is an important component of total household energy use, the energy embodied in broader consumption of goods and services – of which the wealthy consume proportionately more – is much greater. Any carbon mitigation effort that focuses solely on car dependence of suburban households is likely to miss the wider target of general carbon mitigation. Rather, reducing oil vulnerability needs to be viewed as a systematic response to a transformative stressor via an array of land-use, transport and institutional responses.

Since the early 2010s, electric vehicles have emerged as a major area of policy interest in transport. Electric vehicles, especially those powered by renewably generated electricity, promise to reduce the carbon intensity and energy vulnerability of urban transport systems. Although fully electric vehicles have been adopted in some jurisdictions, particularly in the US, they remain a fraction of total motor vehicle fleets. Transition to fully electric vehicle fleets will likely be a long process confounded by a number of important intersecting effects.

In Australia, for example, the most car dependent households tend to be those on lower incomes located in the middle and outer suburbs of the major cities. This group is likely to be relatively less able to afford relatively expensive new electric vehicles compared to wealthier households whose residents, in turn, tend to drive less. Consequently, the carbon reduction potential of electric vehicles is likely to be limited in the short and medium terms by such socio-economic factors.

In addition, equilibrium effects are likely with the take-up of electric vehicles. If fuel costs rise markedly, the demand for electric vehicles will grow. However, this will reduce demand for conventional fuel vehicles, meaning that their prices (especially for secondhand vehicles) will drop and, in turn, make them relatively more affordable. An equilibrium in terms of full purchase and operational costs between conventional and electric new and secondhand vehicles would be likely, which would potentially delay the rate of transition to an all-electric fleet, if this happened at all.

A post-carbon, oil-resilient urban transport strategy involves much more policy effort than hands-off reliance on an electric vehicle transition.

Conclusion

Climate change and oil vulnerability are two major transformative stressors looming on the strategic horizon for cities. These twin problems and responses to them are closely related dimensions of unsustainable energy use in cities and the transition to more sustainable urban systems. Recognizing the transformative nature of climate change and oil vulnerability as major urban stressors is essential to the task of responding to them. The responses to climate change and oil vulnerability in cities have two dimensions: on the one hand, technical adaptations, adjustments and refitting of cities, and on the other hand, the governance and policy adjustments, including institutional and political rearrangements needed to guide cities through this period of adjustment.

This chapter has identified in broad outline many of the major climate mitigation and adaptation responses that will be required in cities, and has assessed them in conjunction with responses to oil vulnerability. Previous literature records relatively few attempts to consider climate change and oil vulnerability responses together. By combining discussion of these two major stressors, the chapter has made a contribution to understanding the kinds of responses that may be necessary as both global processes unfurl. More research, however, will be required to specify in greater detail the joint technical responses to climate change and oil vulnerability needed to reduce their impacts, and to identify further contradictions and tensions to avoid in such joint responses.

References

Alexander, E.R. (2005) "Institutional Transformation and Planning: From Institutionalization Theory to Institutional Design," *Planning Theory* 4(3): 209–22.

AMS (2012) "Climate Change: An Information Statement of the American Meteorological Society," adopted by AMS 20 August, Boston: American Meteorological Society, accessed 14 January 2016 — www2.ametsoc.org/ams/index.cfm/about-ams/ams-statements/statements-of-the-ams-in-force/climate-change/

Bell, S. (2002) "Institutionalism: Old and New," in J. Summers (ed.) *Government, Politics, Power and Policy in Australia* (7th edn), Sydney: Pearson Education Australia, 363–80.

Benedict, M.A. and E.T. McMahon (2006) *Green Infrastructure: Linking Landscapes and Communities*, Washington, DC: Island Press.

Biesbroek, G.R., Swart, R.J. and W.G.M. van der Knaap (2009) "The Mitigation-Adaptation Dichotomy and the Role of Spatial Planning," *Habitat International* 33(3): 230–37.

Burton, P., Lyons, K., Richards, C., Amati, M., Rose, N., Des Fours, L., Pires, V. and R. Barclay, (2013) *Urban Food Security, Urban Resilience and Climate Change*, Gold Coast: National Climate Change Adaptation Research Facility.

Byrne, J. and J. Yang (2009) "Can Urban Greenspace Combat Climate Change? Towards a Subtropical Cities Research Agenda," *Australian Planner* 46(4): 36–43.

Campbell, C.J. (2005) "Peak Oil: A Turning Point for Humankind," *Social Contract* 15(3): 180–83.

Connor, R.D. and S.R. Dovers (2004) *Institutional Change for Sustainable Development*, Cheltenham: Edward Elgar.

Cortell, A.P. and S. Peterson (1999) "Altered States: Explaining Domestic Institutional Change," *British Journal of Political Science* 29: 177–203.

CSIRO (2012) *State of the Climate Report 2012*, Melbourne: Commonwealth Scientific and Industrial Research Organisation.

Davis, M. (2010) "Who Will Build the Ark?" *New Left Review* 61: 29–46.

Dodson, J. (2012) "Transforming Australia's Housing Solution: How We Can Better Plan Suburbia to Meet Our Future Challenges," in R. Tomlinson (ed.) *Australia's Unintended Cities: The Impact of Housing on Urban Development*, Clayton: CSIRO, 19–34.

Dodson, J. and N. Sipe (2010) "Emerging Australian Planning Practice and Oil Vulnerability Responses," *Australian Planner* 47(4): 293–301.

——— (2008) *Shocking the Suburbs: Oil Vulnerability in the Australian City*, Sydney: University of New South Wales Press.

Douglas, I. (2011) "The Role of Green Infrastructure in Adapting Cities to Climate Change," in I. Douglas, D. Goode and M. Houck (eds) *Handbook of Urban Ecology*, Florence, KY: Routledge, 583–88.

Ewing, R., Bartholomew, K., Winkelman, S., Walters, J. and D. Chen (2008) *Growing Cooler: The Evidence on Urban Development and Climate Change*, Washington, DC: Urban Land Institute.

Faludi, A. (2000) "The Performance of Spatial Planning," *Planning Practice and Research* 15(4): 299–318.

Gaynor, A. (2006) *Harvest of the Suburbs: An Environmental History of Growing Food in Australian Cities*, Crawley: University of Western Australia Press.

Gosh, S. (2011) *Growing Healthy Local Food: Sustainability Potential and Household Participation in Home Gardens*, Melbourne: State of Australian Cities Research Network.

Hall, P. and R. Taylor (1996) "Political Science and the Three New Institutionalisms," *Political Studies* 44(5): 936–57.

Hamin, E.M. and N. Gurran (2009) "Urban Form and Climate Change: Balancing Adaptation and Mitigation in the U.S. and Australia," *Habitat International* 33(3): 238–45.

Hodgson, K.C., Campbell, M.C. and M. Bailkey (2011) *Urban Agriculture: Growing Healthy, Sustainable Places*, Chicago: American Planning Association.

Hogan, J. (2006) "Remoulding the Critical Junctures Approach," *Canadian Journal of Political Science* 39: 657–79.

IPCC (2014) "Summary for Policymakers," in C.B. Field, V.R. Barros, D.J. Dokken, K.J. Mach, M.D. Mastrandrea, T.E. Bilir, M. Chatterjee, G.-K.K.L. Ebi, Y.O. Esandra, R.C. Genova, B. Girma, E.S. Kissel, A.N. Levy, S. MacCracken, P.R. Mastrandrea and L.L. White (eds) *Climate Change 2014: Impacts, Adaptation, and Vulnerability. Part A: Global and Sectoral Aspects. Contribution of Working Group II to the Fifth Assessment Report of the Intergovernmental Panel on Climate Change.* Cambridge, UK/New York: Cambridge University Press.

——— (2012) "Summary for Policymakers," in C.B. Field, V.R. Barros, T.F. Stocker, D. Qin, D.J. Dokken, K.L. Ebi, M.D. Mastrandrea, K.J. Mach, G.-K. Plattner, S.K. Allen, M. Tignor and P.M. Midgley (eds) *Managing the Risks of Extreme Events and Disasters to Advance Climate Change Adaptation: A Special Report of the Working Groups I and II of the Intergovernmental Panel on Climate Change*, Cambridge/New York: Cambridge University Press.

Kambites, C. and S. Owen (2006) "Renewed Prospects for Green Infrastructure Planning in the UK," *Planning Practice & Research* 21(4): 483–96.

Kingston, C. and G. Caballero (2009) "Comparing Theories of Institutional Change," *Journal of Institutional Economics* 5: 151–80.

Lenzen, M., Dey, C. and B. Foran (2004) "Energy Requirements of Sydney Households," *Ecological Economics* 49: 375–99.

Matthews, T. (2013) "Institutional Perspectives on Operationalising Climate Adaptation through Planning," *Planning Theory and Practice* 14(2): 198–210.

——— (2012) "Responding to Climate Change as a Transformative Stressor through Metro-Regional Planning," *Local Environment: The International Journal of Justice and Sustainability* 17(10): 1089–103.

——— (2011) *Climate Change Adaptation in Urban Systems: Strategies for Planning Regimes*, Brisbane: Urban Research Program, Griffith University.

McFaul, M. (1995) "State Power, Institutional Change and the Politics of Privatisation in Russia," *World Politics* 47: 210–43.

Mees, P. (2010) *Transport for Suburbia: Beyond the Automobile Age*, London: Earthscan.

——— (2009) "How Dense Are We? Another Look at Urban Density and Transport Patterns in Australia, Canada and the USA," *Road and Transport Research* 18(4), accessed 21 January 2016 — www.ppt.asn.au/pubdocs/Mees Paul-HowDenseAreWe.pdf

Metz, D. (2010) "Saturation of Demand for Daily Travel," *Transport Reviews* 30(5): 659–74.

Newman, P. (2006) "Beyond Peak Oil: Will our Cities and Regions Collapse?" *Res Publica* 15(1): 1–16.

Newman, P., Beatley, T. and P. Boyer (2009) "A Vision for Resilient Cities," in P. Newman, T. Beatley and P. Boyer (eds) *Resilient Cities: Responding to Peak Oil and Climate Change*, Washington, DC: Island Press, 55–85.

Puentes, R. (2012) *Have Americans Hit Peak Travel? A Discussion of the Changes in US Driving Habits*, Paris: International Transport Forum of the OECD.

Scholz, M. and P. Grabowiecki (2007) "Review of Permeable Paving Systems," *Building and Environment* 42(11): 3830–36.

Shaw, R., Colley, M. and R. Connell (2007) *Climate Change Adaptation by Design: A Guide for Sustainable Communities*, London: Town and Country Planning.

Stern, N. (2006) *Stern Review on the Economics of Climate Change*, London: HM Treasury.

Unger, N., Bond, T., Wang, J., Koch, D., Menon, S., Shindell, D. and S. Bauer (2010) "Attribution of Climate Forcing to Economic Sectors," *Proceedings of the National Academy of Sciences* 107(8): 3382–87.

Wilson, E. and J. Piper (2010) *Spatial Planning and Climate Change*, London: Routledge.

Wright, H. (2011) "Understanding Green Infrastructure: The Development of a Contested Concept in England," *Local Environment* 16(10): 1003–19.

Young, O.R. (2010) "Institutional Dynamics: Resilience, Vulnerability and Adaptation in Environmental and Resource Regimes," *Global Environmental Change* 20(3): 378–85.

5

ENERGY SECURITY AND OIL VULNERABILITY RESPONSES

Jago Dodson and Neil Sipe

Many of the primary economic and social effects of higher oil prices are likely to be experienced in the vast car dependent suburban zones of the US and Australia. Therefore, national energy security policy in both countries should recognize urban oil vulnerability and adopt an appropriate national security perspective. Conversely, national and state policy should recognize emerging local government policy development on petroleum security issues sprouting at the ground level of major conurbations. Despite ongoing structural changes in the international energy context, national and local policies have often been conservative on energy questions, with local government the main site of policy progress.

This chapter examines policy responses to petroleum depletion and oil vulnerability policy development at the federal and local levels in the US and Australia. The chapter argues that, while local governments have begun to address oil vulnerability questions, there remains a vertical disconnect between national scale policy development and planning at the metropolitan and local scale.

The policy terrain

Policy makers and planners face a number of conceptual, empirical and programmatic challenges in dealing with the issue of declining global petroleum security. These challenges can be divided into three broad categories. First, there is the challenge of comprehending and conceptualizing the problem of petroleum depletion, the vulnerability of economies and societies to the direct adverse effects of higher oil prices and possible shortages, and the wider systemic effects of costlier oil. This task of comprehension and conceptualization has not been easy to initiate or pursue. Much of the literature on petroleum depletion has been generated by sources outside of government, science or the energy industry.

While, in the late 2000s, a consensus seemed to settle around a price crunch, by the end of the 2010s the degree of volatility of oil prices, and the scale and distribution of price impacts, remained uncertain. At the middle of this apparent "crunch decade," global oil prices were hitting lows not seen since 2003. Formulating authoritative and defensible policy analysis from this diverse mix has not always proven easy, especially given the early polarity in debate between the optimistic forecasts of official organs and the doubtful warnings of "depletionists." A complex dynamic in strategies among oil-producing nations developed, intersecting with a shifting global appetite for fossil fuels and set within a distinctly fragile global economic growth context. Discerning the long-term structural shifts in the geological basis for petroleum supply is now a more difficult task than in the late 2000s. Nonetheless, global petroleum reserves are material and finite while demand for petroleum is a human phenomenon. The factors contributing to very low oil prices in the mid-2010s do not necessarily reflect conditions in the underlying resource base and its rate of consumption. There is a risk, then, that policy and practice are not adequately tracking resource reality.

The second factor complicating the development of policies and plans to address the effects of petroleum depletion centers on political questions for societies that depend on oil supplies for their basic functioning and for which the cheap and abundant availability of petroleum has been a critical factor underpinning prosperity. With the cost of transport fuels having stirred social concern in cities since the mid-2000s, the necessary task of openly and transparently airing the potential consequences of higher oil prices seems to be a great challenge for many political representatives. This challenge is confounded by the trajectory of much current policy, which remains based on cheap and abundant oil and is often diametrically opposed to the strategies needed for an expensive and scarce petroleum environment.

Furthermore, despite recent global agreements – such as the twenty-first Conference of the Parties (COP21) for the United Nations Framework Convention on Climate Change (UNFCCC), held in Paris in December 2015 – there is limited dedicated national effort to reduce oil consumption and carbon emissions in the US or Australia. In both regions, there are many interests whose fortunes are founded in the petroleum present, not the post–fossil fuel future. This is especially complicated in the US and Australia, where the effects of high oil prices have spurred major investment in petroleum and gas production, giving an apparent positive economic gloss to wider social consequences of high oil prices.

National strategic reviews have proven very cautious in identifying petroleum security risks, perhaps because of the chance that the dismal conclusions of an unbiased assessment might be rejected by voting publics. The connection between the burning of petroleum as a liquid fossil fuel, and the wider fossil fuel economy, is strong. National and global agreements to limit emissions from the latter have, at least until late 2015, proven very difficult to achieve. The US and Australia have been among the most recalcitrant nations in this regard. The result is a petroleum security policy and metropolitan planning stasis in which obscuration or deferment of problem acknowledgement substitutes for the formulation of a response. The strategy of "denial and delay" so prominent in climate change policy is equally apparent in the area of domestic petroleum security.

The third major challenge impeding the preparation of policies to respond to oil depletion surrounds the identification of depletion impacts and consequent policy and planning instruments capable of, and necessary for, addressing such impacts. The policy and planning field is replete with an array of methodologies and measures for managing cities but with relatively modest certainty as to their efficacy and efficiency across a range of criteria. Furthermore, any planning interventions to address oil depletion would have to sit within a crowded planning frame, in which other strategic considerations such as climate change, housing affordability, environmental protection, economic efficiency, water security and social equity all clamor for attention.

Added to these concerns is the shifting, rescaling and reframing of urban governance among both formal structures of government and the reconfiguration of networks of influential non-government planning actors. The task of devising strategies that can address oil depletion within the policy and planning sphere, while simultaneously resolving the accompanying economic and environmental threats assailing cities, is no small mission. Thus, it is no wonder, although no excuse, that contemporary urban policy has shied away from the task of protecting cities against the economic and social risks associated with much higher oil prices.

The remainder of this chapter reviews efforts within the US and Australia to respond to, and plan for, reducing oil vulnerability. Our discussion addresses a central question: Is the scale of the problem of oil depletion matched by commensurate policy and planning responses? Our assessment cascades from the higher scale of federal policy, through state level assessment and response, down to the grounded terrain of local government action.

National level responses and policy

At the national level, the US and Australia share broadly similar responses to rising oil prices, questions of energy security and oil vulnerability. The trajectory of strategic response might be typified as an evolution of fossil fuel dependence through shifts in supply mix and a moderate focus on fuel efficiency. In general, the response of both federal governments to the changing global petroleum environment has been slow, especially in the case of Australia. In part, this is due to relatively large domestic reserves of fossil energy, such as oil, gas and coal. Relative energy wealth, and economic dependence on it, has contributed to international stances on energy transition and carbon emissions abatement that are considerably more conservative than most other developed nations of the Global North. Consequently, national efforts to either reduce fossil fuel use or reduce emissions from fossil fuels in these countries have tended to be modest. The result is very little effort at the national scale to respond to subnational challenges around oil vulnerability, particularly for cities.

US federal energy strategy

The energy policy of the US – the world's largest petroleum consuming nation – has an extensive influence on global energy markets and geopolitics. Describing the full international breadth of US energy strategy is beyond the scope of this chapter. US international energy strategy has been organized around ensuring continuity of petroleum supply through support for key supplier nations, particularly in less politically secure regions such as Saudi Arabia in the Middle East. Since the oil price rises of the late 2000s, some observers, such as Hughes (2014), have suggested that a shift in US policy has occurred as domestic production increased via improved shale and tight oil recovery.

The main elements of current US domestic petroleum energy policy are encapsulated in President Obama's policy *Blueprint for a Secure Energy Future* (White House 2011). This policy aspired to increase domestic petroleum production, improve consumer access to fuel-efficient vehicles and support innovation in clean energy technology. In practice, the first element of this policy has enabled and supported the exploitation of existing and new oil fields via more advanced production technologies, and has encouraged the expansion of production of shale oil and tight oil. The second strategy concerning fuel-efficient vehicles has been pursued via federal fuel economy standards, for both light and heavy vehicles, which have been introduced in two stages from 2012. The US has operated fuel economy standards since the Energy Policy and Conservation Act of 1975 (EPCA). The recent policy adjustments initially required fuel economy improvements for new light vehicles, first to 6.72 L/100km from 2016, and then to 4.32 L/100km by 2025. Estimates released by the White House have suggested that these improvements would equate to US$1 per gallon in price savings for most vehicles. The final component of the strategy is being implemented via the Department of Energy's research and development program, which focuses on the development of clean energy production.

Beyond the *Blueprint for a Secure Energy Future*, most federal policy in the US remains largely business-as-usual in relation to infrastructure investment. However, serious doubts have been raised about the sustainability of the National Highway Trust (NHT), which hypothecates fuel excise revenues for road construction. Projections prepared by the US Department of Transportation (DOT 2015) anticipate that NHT revenues will fall below financially sustainable levels as of mid-2016. A shortfall in funds would require a change in approach to the historic means of constructing and maintaining federal transport infrastructure in the US. This would offer a moment for fresh contemplation of an appropriate modal balance for both the energy and climate problems faced by the US.

Australian federal energy strategy

Despite considerable public support for energy efficiency and climate abatement in the late 2000s, Australia's national energy policy has been haphazard and often contradictory. Policy around petroleum fuels has often disregarded problems of oil vulnerability. A national Australian Labor Party (ALP) government was elected in 2007, with a robust climate policy that included signing the Kyoto Protocol and introducing an emissions trading scheme. However, overall national energy policy has barely recognized petroleum security issues.

The most significant national statement on energy security and oil vulnerability was an inquiry into Australia's future oil supply and alternative transport fuels by the Australian Senate (2007). This report represented an initial and broadly spread political response to the higher oil prices after 2004 and raised new queries about the sustainability of global oil supplies. The Australian Senate report (2007: 33) concluded that

> the possibility of a peak of conventional oil production before 2030 should be a matter of concern. Exactly when it occurs (which is very uncertain) is not the important point. In view of the enormous changes that will be needed to move to a less oil dependent future, Australia should be planning for it now.

Despite the conclusions of this bipartisan report, a Senate resolution to compel the Australian government to prepare a strategy on petroleum depletion was resoundingly defeated (Senate Hansard 2009).

Since the Senate inquiry, there has been little national government action on petroleum security beyond the development of new national energy strategies. Some of the analysis undertaken by the national government has been dismissive of petroleum security concerns. The *Liquid Fuels Vulnerability Report* (ACIL Tasman 2009), prepared as part of the background to the development of a new energy white paper, downplayed the significance of global petroleum production constraints in favor of ensuring efficiencies in market processes. In contrast, a discussion paper prepared as part of a *National Energy Security Assessment* (DRET 2009: 8, 13) identified a likely scenario by and for 2023, as follows:

> Tight global supply/demand balance returns with ongoing demand growth and mature field decline. Development of more difficult geological and geo-political regions continues . . . In 2023, the key issue expected to reduce liquid fuels security is access to reliable and affordable crude oil.

The 2011 *National Energy Security Assessment* (DRET 2011) mainly endorsed this assessment and rated Australian petroleum security risks as low or moderate through 2035. This assessment was shared in the 2011 *Energy White Paper* (DIS 2011), which presumed that the depth and breadth of international petroleum markets would be sufficient to meet Australia's needs until 2035. That white paper had, in part, reflected the energy blueprint adopted by the US government (White House 2011) – although notably the Australian version recognizes that Australia has no vehicle fuel efficiency standards and proposes none.

Agencies with responsibility for long-term infrastructure planning, including highly petroleum dependent major roads, have adopted a more precautionary approach to national energy security although they have had little direct power to influence energy policy. The agency Infrastructure Australia (IA 2008: 33), which is charged with selecting and funding priority national infrastructure, has acknowledged:

> At best, the world will reach an oil production "plateau" by 2015; at worst, there will be a production peak by 2013, with reserves declining rapidly thereafter . . . Governments can do more to encourage private sector investment in less carbon-intensive energy and transport infrastructure.

To date, disbursements of Infrastructure Australia's A$8 billion urban transport funding have favored rail projects over roads. While not necessarily oriented to resolving petroleum security challenges, this pattern does imply that oil supply analyses influence investment outcomes, even if only very indirectly.

Infrastructure Australia's concerns about petroleum supply have been only partially reflected in the agency's *State of Australian Cities* reports. For instance, in one report (IA 2008), the agency barely discusses the global energy security context even though the high reliance of cities on energy and the implications of resource depletion are noted. While that report discussed research on the oil vulnerability of Australian suburbs by Dodson and Sipe (2008b), it was defined as an issue of living affordability, rather than security or functionality.

The final area where the Australian government has directly engaged in energy regulation was via a carbon pricing scheme that only operated from mid-2012 through mid-2014. That scheme established a carbon trading mechanism, initially with a fixed price for carbon. However, the scheme deliberately excluded road transport fuels, reflecting a similar exemption in the equivalent European carbon trading scheme. Although the Australian government has released analytical papers examining potential measures to require motor vehicle manufacturers to progressively improve the efficiency of new vehicles (e.g. Climate Change Authority 2014), such measures have not been adopted as policy. Australia operates regulations to control the carbon emissions levels of new vehicles, but these are relatively modest and are not oriented to achieve fuel economy.

In summary, the Australian national response to emerging threats to global energy security has been to either ignore or avoid discussion of the problem. Little material could be described as either policies or plans dedicated to ensuring Australian cities are protected from high or volatile petroleum prices. Efforts to pair energy security and energy efficiency to climate change mitigation have stalled at the national scale in line with global inaction.

US subnational policy development

Energy security strategies adopted by state and local governments throughout the US have largely mirrored those of the federal government. The key difference with most state or local government energy strategies is that, unlike their federal counterparts, such strategies rarely consider the security dimensions of petroleum supply. Ensuring national supply of oil is, necessarily, a national concern. State and local governments are much less likely to have the capacity to secure petroleum supply within their own jurisdiction, primarily because the resource is unavailable. Just five states, plus the Gulf of Mexico Federal Marine Territory, produce 80 percent of US domestic supply (EIA 2014).

Energy security and oil vulnerability are taken into limited account in US state plans, through energy efficiency requirements for private motor vehicles, overall transportation planning and, increasingly, climate mitigation policy. The energy efficiency approach reflects both the original EPCA requirements and standards required under the Clean Air Act of 1963, which sought to regulate environmental emissions. The main instrument for the current iteration of the EPCA in relation to petroleum consumption is the Corporate Average Fuel Economy (CAFE) program, which requires motor vehicle manufacturers to meet set fuel economy standards on average across the full range and volume of vehicles sold in any given year. Some states, particularly California, have adopted more stringent state regulation of emissions than the federal requirements, although primarily addressing pollution and climate concerns. Subsequent distortions that these enhancements introduced into the calculation of overall national CAFE regulation have led to a national effort to harmonize state and federal CAFE standards. Currently, most state efforts to manage dependence on motor vehicles are undertaken via general transportation strategies. They typically contain some discussion of fuel efficiency, but only rarely do transport plans actively consider energy consumption as a transport question.

Perhaps the best example of an attempt to contribute an oil vulnerability perspective to a state plan has been the *Florida Energy Resiliency Report* (Florida Regional Councils Association 2013), which canvassed the state's exposure to energy security and climate problems. The report noted that Florida was exposed to possible difficulties from constrained petroleum supply but offered few strategies to deal with the problem. Of the six recommendations emanating from the Florida report, only one proposed substantive effort around increasing energy efficiency of motor vehicles. All other recommendations mainly centered on information provision to consumers or monitoring and evaluation energy demand and supply conditions. The report offered little in the way of a sense of transformation of transport and urban relationships arising from increasing petroleum constraint. The Florida Department of Transportation is currently preparing a new transport plan for the state but, at the time of writing, the preliminary documentation offers little indication that any resiliency report issues will be addressed in the plan. This reflects a pattern common to US responses to petroleum security concerns: little is imagined other than a need to increase the efficiency of existing vehicles rather than a systemic shift in the way that transport is organized.

Some US local governments have responded to energy security concerns through local oil vulnerability strategies. Portland, Oregon, was the first US municipality to undertake such policy work, via the city's Peak Oil Task Force report (PCC 2007). That report anticipated a peak in oil production in the mid-2010s, assessed the consequences for the city, and made a series of recommendations about how to respond to this possibility. The main recommendation proposed an ambitious 50 percent reduction in Portland's oil and gas consumption by 2032, accompanied by a series of recommendations for education, industry engagement and transport infrastructure investment. Little information is available to assess whether the Portland recommendations were successfully implemented.

Further municipalities have developed insights into peak oil, including the cities of San Francisco and Berkeley, California; Bloomington, Indiana; and Spokane, Washington. The San Francisco City Council (2009) report was extensive, covering almost all aspects of the city's operations and function that might be affected by constrained petroleum supplies. The recommendations offered by the report addressed an array of issues, including such measures as increasing the electrification of the municipal rail network, actively constraining car use through parking control measures, and requiring future infrastructure investment plans and models to account for petroleum costs. This plan for San Francisco can be broadly viewed as fitting within a "resilience" framework, addressing an array of sectors beyond transport and offering a holistic view of use, within the city, of petroleum and other energy sources.

Compared to that prepared for San Francisco, the Bloomington City Council (2009) report is a comparably extensive assessment of peak oil and its consequences for the city. One of the notable features of the Bloomington report was the detailed assessment of the city's land-use structure and the way that it intersected with oil vulnerability problems. Accompanying the land-use discussion was a similarly detailed assessment of local household oil vulnerability and impacts of higher oil prices on travel demand. This included discussion of the inadequacy of public transport networks in Bloomington and potential directions for improvement – reflecting certain directions identified by Stone and Mees (Chapter 10 in this book) – although not comprehensively.

Overall, the few US efforts to assess oil vulnerability at the local scale have been extensive and do offer useful directions for municipal responses to future petroleum constraint. The key problem is that such assessments are so rare. Although they might offer some local respite to oil vulnerability in the municipalities to which they apply, these plans are unlikely to have much impact beyond this very local scale.

Australian subnational policy development

If the Australian federal government has been largely ambivalent or dismissive of concerns about declining global petroleum security, state governments have been nearly equally disinterested. There are few state policies or documents that address petroleum security issues or the consequences of a decline in petroleum security for cities. The single exception to this pattern is in the state of Queensland, which has demonstrated a cautious willingness to engage with petroleum security questions and to establish a modest policy effort to address the problem. This work program has taken two forms: a statewide strategy accompanied by application within the South East Queensland Regional Plan (SEQRP).

The Queensland oil vulnerability policy work was initiated by the then state member for Harvey Bay, Andrew McNamara. An Oil Vulnerability Taskforce (Queensland Government 2007) was set up to investigate the background to energy security questions, the potential for global depletion of oil reserves, petroleum production constraints and some initial impact scoping for cities. In 2007, the task force recommended that Queensland develop an oil vulnerability strategy. Initially, subsequent work on this strategy, with Mr. McNamara as minister, was under the auspices of the Environmental Protection Agency. After McNamara's departure from government at the March 2009 state election, the Department of Main Roads and Transport furthered that work.

The initial strategy work comprised of a research and background paper prepared by external consultants (Waller 2008). The Waller paper argued that Queensland's vast mineral and gas energy resources would provide a "natural hedge" against a decline in global petroleum security through improved terms of trade and economic activity. The challenge of oil vulnerability, according to Waller, was to offset the adverse impacts on sectors unable to pass on costs to consumers and especially exposed sectors, such as low-income households or households disadvantaged through location. The background report advocated detailed attention to the effects of higher petroleum prices across various industry sectors. However, the Waller report offered few dedicated measures to reduce urban oil vulnerability. A formal Queensland oil vulnerability strategy was never released; policy work on this problem was mainstreamed within existing policy efforts in transport and land-use planning.

Queensland's approach is unique, and potentially innovative, within subnational government policy making in Australia. No other state in Australia has recognized the shifting global petroleum context as a matter deserving further consideration, nor have any states instituted any such consideration as part of their strategic policy framing. Florida appears to be the only comparable US example. Furthermore, the Queensland government is the only state in Australia to have incorporated oil vulnerability mitigation into its metropolitan planning processes, which will be discussed further shortly.

Some metropolitan plans have begun to address oil vulnerability concerns. Dodson and Sipe (2008a) reported that metropolitan plans prepared in the 2000s – for Brisbane (the capital of Queensland) and adjacent cities in South East Queensland (SEQ), Sydney (the capital of New South Wales) and Melbourne (the capital of Victoria) – all failed to adequately account for exposure to higher oil prices and, consequently, did not apply sufficient policy leverage to reduce this vulnerability. However, since 2008, metropolitan plans have been released for Brisbane/SEQ and Melbourne that offered their state governments the opportunity to redress the failures to mitigate the oil vulnerability of their metropolitan regions.

The Melbourne @ 5 Million plan (DPCD 2008), released by the Victorian state government, sought to update the Melbourne 2030 Metropolitan Strategy, which had been in place since 2002. The Melbourne @ 5 Million plan focused on three main factors: a shift to a polycentric urban structure; the expansion of development into growth corridors to the west, north and southeast of the city; and the imposition of infrastructure levies on land rezoned for development. None of these factors addressed issues of oil vulnerability, either directly or indirectly. While planned activity concentration might have contributed to reduced oil reliance for residents in the new precincts, dependence on slow and cyclically sensitive processes of land-use change implied long time frames for achieving effective change (Dodson 2008). Employment corridors have risked dispersing activities concentrated in central districts. Furthermore, the expansion of urban growth has threatened to exacerbate the problems of oil vulnerability – as identified by Dodson and Sipe (2007, 2008b) – and there has been no indication that growth corridor infrastructure charges might be usefully dispersed to fund oil resilient transport modes. The more recent Plan Melbourne (DTPLI 2014) largely reiterated the content of Melbourne @ 5 Million, although with somewhat less specific proposals around planning directions. Certainly the latter scheme offered no consideration of oil vulnerability nor of urban energy use in general.

Perhaps the clearest statement on oil vulnerability in any major metropolitan plan, both in Australia and internationally, is the 2009 SEQRP (DIP 2009). In early 2016, it remained the operative statutory plan for the SEQ region but was under review. As with Melbourne @ 5 Million, this 2009 plan was prepared after the sharp rise in global oil prices in the mid-2000s and the surge in global interest in climate change mitigation and adaptation. Urban growth management was the over-determining concern of the SEQRP, which nonetheless included considerable discussion of climate matters and a substantive policy on oil vulnerability. The climate policy in the SEQRP incorporated a number of components, including mitigation measures relating to design guidelines to make new developments more energy efficient, reducing transport fuel consumption, renewable energy generation, carbon bio-sequestration and many adaptation elements. Some of these had an energy vulnerability dimension.

The substantive oil vulnerability content of the SEQRP is found in the section "Responding to Oil Supply Vulnerability," and is clear about the possibility of petroleum depletion:

> Most of the world is now dependent on a diminishing number of oil-producing countries for their oil needs. Current rates of global oil production are predicted to decline within the next five years.
>
> *(DIP 2009: 46)*

The report recommends the use of the VAMPIRE oil vulnerability assessment methodology as developed by Dodson and Sipe (2008c, and their Chapter 11 in this book) where policies offered to address this challenge are explicit and include:

1. Manage risks and reduce impacts on people, economic sectors and areas from the effects of oil supply vulnerability;
2. Design development areas to encourage walking, cycling and public transport use to get to local shopping facilities and employment locations, and early provision of public transport services;
3. Ensure transport infrastructure and service investment actively reduces oil dependence, particularly for trips that could be undertaken by public or active transport;
4. Reduce the length of trips and dependence on oil by localizing access to goods, services and employment opportunities.

While the first two policies are expressed in general terms, the third and fourth suggest a stronger policy intent.

The third policy proposal seeks to ensure a reduction in oil dependence from transport investment that, if implemented, would have far-reaching consequences for metropolitan planning. For instance, it would be difficult to justify the expansion of road capacity because of the robust evidence that such expansion encourages and facilitates car, and thus petroleum, dependence (DIP 2009: 46). Active application of the policy would imply curtailing almost all the road schemes within the SEQ Infrastructure Plan (SEQIP), expected to cost around A$49.8 billion over the 2009–31 period. Such curtailment has not occurred; the region continues to expand road capacity, primarily via construction of major tollway tunnels, although these failed financially due to overestimation of demand and, thus, revenue.

Further adherence to the oil vulnerability policies within the SEQRP would imply investment in public transport infrastructure in oil vulnerable areas. A methodology for determining oil vulnerable zones within the region is identified. However, perversely, the majority of transport infrastructure planned and constructed since has been directed to the least oil vulnerable, such as the inner-city Northern and Eastern busway projects (estimated to cost A$2.6 billion and A$3.6 billion, respectively) and the A$8.5 billion Brisbane Inner City Rail. Active redress of the extensive oil vulnerability experienced in outer suburban zones, as identified by Dodson and Sipe (2008b), is not canvassed by either the SEQRP or the SEQIP.

The inclusion of a dedicated section on oil vulnerability within the SEQRP places it far ahead of metropolitan plans for Australia's four other major capital cities, and beyond most other urban regions in the world, in responding to this problem. While the inclusion of these policies is laudable, the real test lies with how these policies are implemented. To reduce oil vulnerability in Australian cities, especially within car dependent outer suburbia where public transport services are infrequent and poorly integrated with land uses, is a major challenge (see Stone and Mees, Chapter 10 in this book). While much effort will be required to transition to a mode of metropolitan planning in SEQ that takes full account of oil vulnerability, say spatially, for other major cities where this challenge has scarcely been recognized in their plans, the task will be much harder in the future.

Local planning

In the absence of national or state metropolitan leadership on oil vulnerability issues, some local governments in Australia have begun the task of responding to oil vulnerability problems. This response is not widespread. As listed in Table 5.1, only a few councils across Australia have addressed energy security

TABLE 5.1 Oil vulnerability responses by Australian local governments

LGA	*Date*	*Action*
Brisbane City (Qld)	2007	Climate Change and Energy Taskforce Final Report: A Call for Action
Marrickville City (NSW)	2007	Oil Depletion Protocol
Maribyrnong City (Vic.)	2008	Peak Oil Contingency Plan
Darebin City (Vic.)	2009	Climate Change & Peak Oil Adaptation Plan
Gold Coast City (Qld)	2010	Peak Oil Desk Top Study
Sunshine Coast Regional Council (Qld)	2009	Peak Oil Background Study
	2010	Sunshine Coast Climate Change and Peak Oil Strategy, 2010–20
Whitehorse City (Vic.)	2011	Peak Oil Action Plan
Manningham City (Vic.)	2012	Climate, Peak Oil and Food Security

issues. They include Marrickville City Council (Sydney), which agreed to an "oil depletion protocol" promoted by international advocates of action on peak oil. The protocol commits its adherents to reduce their oil consumption to match global depletion rates. The Marrickville City Council has committed to a 3 percent annual reduction in operational oil consumption and, in fact, achieved a 6 percent reduction over the year from mid-2008 to mid-2009.

One potential early leader, the Brisbane City Council (BCC), governs Australia's most populous municipality. In 2006, the BCC established a Climate Change and Energy Taskforce that produced a report to council in 2007 (see Table 5.1). That report included extensive recommendations on both climate and oil vulnerability challenges, and advocated a raft of actions to reduce Brisbane's exposure to climate change and oil depletion. The transport components of this mix included dramatically increasing support for public transport and active travel, accompanied by private motor vehicle and petroleum fuel demand constraint. However, subsequently, the BCC avoided many of the task force's recommendations, preferring to leave oil depletion adaptation responses to citizens. Consequently, little obvious shift has occurred in BCC policy. As shown in Table 5.1, within the adjacent cities of SEQ, the Sunshine Coast Regional Council released a Climate Change and Peak Oil Strategy, following on its 2009 Peak Oil Background Study. The Gold Coast City Council also undertook a desktop study of peak oil in 2010, which assisted in decision making and pilot studies undertaken to date, including alternative fuel and electric vehicle trials.

Melbourne seems to be the main metropolitan area where oil vulnerability has been addressed in practical terms at the municipality level. Maribyrnong, Darebin, Whitehorse and Manningham councils have all prepared peak oil strategies, often alongside consideration of climate change and, sometimes, food security. Resilience is a theme that arises within the most recent plans. The plan by Darebin City Council (Taygfeld and Burton 2009) provides a detailed assessment of peak oil and the problems it poses for the council's operations with an array of operational and strategic actions. Similar content is found in the plans of Maribyrnong City Council (Fishman et al. 2009), the City of Whitehorse (Klindworth et al. 2011) and Manningham City Council (MCC 2012). This material is too extensive and detailed in terms of operational actions to present here. In summary, however, the content of these plans is highly consistent with contemporary approaches to climate mitigation and adaptation, reduced household and government energy demand, promotion of sustainable transport and compact urban planning.

Despite the willingness of this small number of Australian local governments to recognize and address questions of oil vulnerability, overall local government action remains weak. At the time of writing, we know of only eight (1.4 percent) of Australia's 565 local governments that have prepared assessments responding to oil vulnerability concerns. This limited effort signals that local government is unlikely to be a major site of strategic policy action to respond to the urban impacts of declining global energy security, preferring to focus on adaptive measures instead.

Most local governments within Australia's major cities – except for those in SEQ – are too limited in their technical capacity and their policy and spatial scope to effect meaningful resilience beyond the local scale. Given their inner-city locations, a number of those that have responded to energy security problems (such as Darebin or Maribyrnong) are not even at great direct risk from a decline in global energy security. Beyond its own operations, Brisbane – the largest local government in Australia, with the greatest capacity to effect change – has pursued an approach relying largely on voluntary efforts toward mitigation and resilience. The Sunshine Coast Regional Council and Gold Coast City Council have the capacity to act, and their plans represent an important first step in recognizing petroleum vulnerability. While they are large councils, many key transport decisions remain to be made by state and federal governments.

Whether in the US or Australia, actions that local and regional councils might incorporate into an oil dependency reduction strategy could include:

- Shifting infrastructure spending away from road construction and toward sustainable transport options, or halting major road construction;
- Only approving new fringe developments with viable public transport alternatives to private motor vehicle travel;
- Increasing space for walking and cycling within road reservations;
- Reducing car parking requirements for new developments;
- Designing new developments to minimize car use by providing local centers/villages;
- Encouraging local food production through community gardens and farmers' markets;
- Planning for renewable energy recharge requirements for electric vehicles;
- Encouraging state governments to improve public transport provision.

Beyond immediate council actions, questions remain about integration with, and support from, metropolitan documents provided to local governments. Councils within SEQ, at least, have a policy mandate (or requirement) set out in the SEQRP to which they can link local policy. Councils in other metropolitan jurisdictions have little or no guidance, or authority, to enact oil vulnerability mitigation plans except to the limited extent that any policy or action might be supported by current policy and legislation. Local governments control few of the key metropolitan policy levers, particularly transport policy or the provision of transport services. Many local governments in the most oil vulnerable outer-suburban or new growth zones have limited rates and resource bases to deploy. It is unlikely, therefore, that local governments will play more than an advocacy or local adaptation role in responding to declining global energy security. The exceptions to this limited role are likely to be found in SEQ, which has large local government institutions with considerable planning, service provision and infrastructure planning powers to effect structural urban change. In short, generally, effort relies on action originating with state and federal governments.

Conclusions

What can planners learn from this survey of federal, state and local government responses to mounting concerns about declining global energy security in the US and Australia? A number of conclusions can be drawn about the state and context of planning to reduce the oil vulnerability of cities in both the US and Australia.

There is a major policy deficit nationally that impedes the development of state and local responses. This is evidenced, in the US, by a focus on efficiency and technology solutions rather than any effort to manage demand in a systemic sense. In Australia, major assessments from federal bodies downplay the significance of energy security and barely consider the implications of potential petroleum supply scenarios for cities. Consequently, Australia does not even approach the modest federal efforts pursued in the US.

In both countries, federal policy is captive to a view that global energy markets present few concerns and that any such concerns can be allayed by exploitation of new resources, technological innovation and small demand reductions. Urban problems do not appear to register on federal policy dials. Agencies with strong roles in long-term planning for cities and that need to take account of energy security questions for long-term viability of investment decisions – such as the US Government Accountability Office (2007) or Infrastructure Australia (2008) – have been more skeptical in their outlook on petroleum security and have been more concerned that petroleum supply constraints may strengthen during shorter time frames than those envisaged by federal energy agencies.

The weak federal attention to the oil vulnerability problem in the US and Australia can, in turn, be identified as a major reason underpinning limited metropolitan and local planning attention. Without federal guidance on what is partly a foreign policy problem, urban managers in both countries are left to undertake their own analyses of strategic petroleum challenges. Typically, they are neither trained nor experienced in undertaking such a task. Therefore, planners have had to look to the underdeveloped scholarship and wider advocacy literature on the oil vulnerability of cities for advice.

Furthermore, metropolitan and city planners seem reluctant to address petroleum security issues, and their planning schemes are generally inadequate, typically due to weak articulation between land-use plans and metropolitan transport plans, and disjuncture between the strategic planning of public transport infrastructure and the ongoing operational planning of public transport networks, especially beyond trunk services. While the efforts of a few local governments to address oil vulnerability are welcome, their impact is slight given that the strategic content and implications of most metropolitan plans support ongoing oil dependence.

In summary, current policy and planning prescriptions are inadequate to meet the challenge of reducing the vulnerability of US and Australian cities to global petroleum supply constraints. The analytical, institutional, political and motivational deficits that have so far limited government action must be overcome if cities in the US and Australia are to reduce their oil vulnerability.

References

ACIL Tasman (2009) *An Assessment of Australia's Liquid Fuel Vulnerability*, Canberra: Department of Resources, Energy and Tourism.

Australian Senate (2007) *Inquiry into Australia's Future Oil Supply and Alternative Transport Fuels: Final Report*, Canberra: Australian Senate.

Bloomington City Council (2009) *Redefining Prosperity: Energy Descent and Community Resilience*, Bloomington: Bloomington City Council.

Climate Change Authority (2014) *Light Vehicle Emissions Standards for Australia: Research Report*, Canberra: Australian Government.

DIP (2009) *South East Queensland Regional Plan 2009–2031*, Brisbane: Department of Infrastructure and Planning.

DIS (2011) *Energy White Paper*, Canberra: Department of Industry and Science (Australian Government).

Dodson, J. (2008) "The Wrong Place at the Wrong Time: Why the Structure of Housing Markets Means Urban Consolidation Can't Equitably Solve Our Urban Planning Challenges," paper presented at *Third Australasian Housing Researchers Conference*, Rydges Hotel, Carlton, 18–20 June, accessed 30 January 2016 — www98.griffith.edu.au/dspace/bitstream/handle/10072/28070/55026_1.pdf?sequence=1

Dodson, J. and N. Sipe (2008a) "Planned Household Risk: Mortgage and Oil Vulnerability in Australian Cities," *Australian Planner* 45(1): 38–47.

——— (2008b) "Shocking the Suburbs: Urban Location, Homeownership and Oil Vulnerability in the Australian City," *Housing Studies* 23(3): 377–401.

——— (2008c) *Unsettling Suburbia: The New Landscape of Oil and Mortgage Vulnerability in Australia's Cities*, Research Paper 18, Brisbane: Urban Research Program, Griffith University.

——— (2007) "Oil Vulnerability in the Australian City: Assessing Socio-Economic Risks from Higher Urban Fuel Prices," *Urban Studies* 44(March): 37–62.

DOT (2015) *Highway Trust Fund Ticker*, Washington, DC: Department of Transportation, accessed 20 January 2016 — www.transportation.gov/highway-trust-fund-ticker

DPCD (2008) *Melbourne @ 5 Million: Melbourne 2030 — A Planning Update*, Melbourne: Department of Planning and Community Development (Victorian Government).

DRET (2011) *National Energy Security Assessment*, Canberra: Department of Resources Energy and Transport.

——— (2009) *National Energy Security Assessment*, Canberra: Department of Resources, Energy and Tourism.

DTPLI (2014) *Plan Melbourne: Metropolitan Planning Strategy*, Melbourne: Department of Transport, Planning and Local Infrastructure (Victorian Government).

EIA (2014) *Five States and the Gulf of Mexico Produce More Than 80% Of U.S. Crude Oil*, Washington, DC: Energy Information Agency, accessed 16 January 2016 — www.eia.gov/todayinenergy/detail.cfm?id=15631

Fishman, E., Hart, P. and J. Hurley (2009) *Maribyrnong City Council Peak Oil Contingency Plan*, Fairfield (Victoria): Institute for Sensible Transport.

Florida Regional Councils Association (2013) *Florida Energy Resiliency* Report, Tallahassee: Florida Regional Councils Association, with the Florida Department of Agriculture and Consumer Services Office of Energy, and the US Department of Commerce Economic Development Administration.

Government Accountability Office (2007) *Crude Oil: Uncertainty about Future Oil Supply Makes It Important to Develop a Strategy for Addressing a Peak and Decline in Oil Production*, Washington, DC: US Government.

Hughes, L. (2014) "The Limits of Energy Independence: Assessing the Implications of Oil Abundance for U.S. Foreign Policy," *Energy Research & Social Science* 3: 55–64.

IA (2008) *A Report to the Council of Australian Governments*, Canberra: Infrastructure Australia (Australian Government).

Klindworth, A., Campbell, A. and V. Berzkialns (2011) *Whitehorse Peak Oil Action Plan*, Melbourne: AECOM Australia Pty Ltd.

MCC (2012) *Securing the Future: Responding to Climate Change, Peak Oil and Food Scarcity*, Doncaster, Victoria: Manningham City Council.

PCC (2007) *Descending the Oil Peak: Navigating the Transition from Oil and Natural Gas*, Portland, Oregon: Portland City Council.

Queensland Government (2007) *Queensland's Vulnerability to Rising Oil Prices: Taskforce Report*, Brisbane: Queensland Government.

San Francisco City Council (2009) *Report of the San Francisco Peak Oil Preparedness Taskforce*, San Francisco: San Francisco City Council.

Senate Hansard (2009) "*Oil*," 18 November (3.44 pm) Parliament of Australia: 8235.

Taygfeld, P. and D. Burton (2009) *Climate Change and Peak Oil Plan*, Preston, Victoria: Darebin City Council.

Waller, M. (2008) *Oil Vulnerability Strategy/Action Plan for Queensland: Research Paper*, Brisbane: Heuris Partners and Queensland Government.

White House (2011) *Blueprint for a Secure Energy Future*, Washington, DC: White House.

6

POST-PETROLEUM URBAN JUSTICE

Wendy Steele, Lisa de Kleyn and Katelyn Samson

The idea that cities might be depleted of oil is, for many, a vision of the end of the world. At its most visceral, oil is an energy source that underpins everyday urban practices in the form of transport, lighting, heating and cooling networks and is the foundation of many of the products we use. Less clear, at least for some, are the multifarious ways that oil operates as resource capital through largely invisible circuits – to drive and sustain currently unsustainable and inequitable processes of ownership, production and consumption in the globalized city. As Jameson (2003: 76) argues, "It is easier to imagine the end of the world than it is to imagine the end of capitalism." At stake is not just the need for a radical transition away from petroleum-based urban environments but a fundamental shift away from the oil industry–supported urban economy, and associated patterns of power and privilege. Arguably, good planning is implicit for an equitable transition to build a post-petroleum urban future.

But here's the rub. Planning has long been intimately linked with the dominant regime of capital accumulation and distribution (Harvey 1989). Despite the many aspirations of planners, there is a poor record generally of planning achieving just urban outcomes (Winkler 2009). For Hall (2002: 4), planning is a "response to the capitalist system and the problem of organizing production." Yet planning, as an intervention, can exacerbate the very issues it sets out to resolve. As Kenny and Meadowcroft (1999: 1) observe, "the history of planning is littered with unintended consequences and undesired outcomes." Planning, as currently practiced, serves to compound rather than improve urban conditions of poverty and access to basic services such as energy and ecological integrity (Steele et al. 2012).

The purpose of this chapter is to focus on two questions: Who is served by the transition to a post-petroleum city? What is the role of planning in this transition? The chapter begins with the key tenets underpinning urban climate justice and the relationship to energy equity and oil justice in cities. The next section focuses on the role of planning and how this role can usefully inform and guide cities. Drawing on climate justice as a lens for better understanding energy (and oil) equity, we highlight the role of local movements in re-shaping the broader project of planning the oil constrained city (politically and practically). In particular, the diverse work of "planners" is highlighted in terms of working through collective local and community action – a coalition of alternative stories and practices – that seeks to create new post-petroleum possibilities for cities.

Energy equity in the climate-just city

Energy equity, as part of a broader urban climate justice agenda, attends to the social and equity implications of energy and low carbon objectives, including energy access, oil distribution and social power. Yolande Strengers (2013) describes a powerful energy ontology that relates to the everyday practices, critical urban infrastructures and technologies that both mask and mediate ways of being in the city. Energy related poverty emerges within this context and works at multiple scales, from the level of local everyday practices to national economic policy, to multinational corporations and their global regimes. Energy injustice is perhaps best understood as "an assemblage of human and nonhuman presences and absences, practices, norms, and possibilities, with considerations of energy justice at the local, national, and global levels" (Day and Walker 2013: x).

Energy equity in cities fundamentally involves justice questions around planning for provision and accessibility, and the availability and affordability of modern energy services both in the short term and the long term. This may range from replacing inefficient and coal-fired power plants with renewable sources of energy, to expanding networks to maximize distributed generation technologies, and establishing institutions capable of preserving the longevity of energy resources, thereby providing opportunities for intergenerational and intragenerational equity (Sovacool et al. 2013). Planning is necessary to reduce the injustices that can be associated with transitions to lower carbon energy systems or post-petroleum cities. This process, however, involves enlarging the boundaries of planning dialogue and action "so that demands for equity are no longer marginalized as a first step towards reversing the current tendency that excludes eco-social justice from the aims of urban policy" (Fainstein 2006: 26).

Three key domains of energy policy action and responses related to building momentum around energy equity and climate justice, and energy efficiency and alternative energy, are outlined by Byrne and Portenger (2014) as part of a broader urban climate justice agenda. This approach recognizes that socio-political, economic and material processes drive and frame the experience of energy injustice and vulnerability (Bickerstaff et al. 2013). To this end, Byrne and Portenger (2014: 334) highlight that "there is an urgent need to rethink the capacity of spatial planning to respond to climate change and to recognize that some energy-related strategies may inadvertently harm marginalized and vulnerable populations."

The urban justice consequences span the spatial, biophysical and socio-demographic realms. The domains identified by Byrne and Portenger (2014: 333) revolve around the trifecta of climate change impacts, geographical scale and social vulnerability to include: (1) socio-economic issues, such as energy/fuel poverty; (2) governance issues, such as energy security management, demand monitoring and prediction, and policy implementation; and (3) planning issues, such as regulating land and development, building codes and direct investment in urban infrastructure. Energy and oil injustice is

> not just a matter of the juxtaposition of certain kinds of people with an energy supply and a pricing system, but rather a matter of how these all interact: how people inhabit their house, how the type of energy is supplied and can be used; and how consumers are directed to make energy choices.
>
> *(Bickerstaff et al. 2013: 11)*

These decisions and pathways are part of a broader agenda related to climate justice in cities, particularly the distributive and procedural justice implications and their urban governance and policy antecedents and planning outcomes.

Urban climate justice

As a lens for engaging with socio-political and environmental crisis, urban climate justice offers a framework for highlighting those most marginalized sectors of society, often rendered largely invisible in mainstream policy and planning discourses. We argue that this lens offers insights for better understanding the more specific issues of energy equity and justice in the transition to the post-petroleum city – related to fuel security, rising oil prices and peak oil, and deliberate action to mitigate the broader agenda of climate change – and its impacts at the nexus of nature and society. With respect to energy and oil equity issues, our approach follows in a similar vein to Jamieson (2001: 289), who observes that

> debates about climate change are as much about the distribution of wealth, power and authority as they are about whether or not scientists have accurately depicted the natural and human systems that have contributed to climate change.

Yet urban planning processes remain largely focused on technocratic responses to global environmental change while hardwired to an emphasis on economic, not equitable, models of growth within a climate of change (Steele et al. 2012).

Urban climate justice has its antecedents in the environmental justice movement. Environmental justice sees the inextricable link between people and the environment. Traditionally, it has focused on the negative environmental effects of commercial and state activities, including planning, development and consequent distributive impacts, particularly those experienced by communities marginalized due to race and class. The environmental justice movement originated in the US in the 1970s, out of the civil rights movement and in response to African American communities fighting hazardous waste landfills and polluting industries being sited near their communities. As a consequence, it has also been termed and analyzed through "environmental racism" – see Brulle and Pellow (2006) and works of Robert D. Bullard (2015). The concept developed in the US to address not only distributive injustice and pollution but also other environmental issues, including "public health, worker safety, land use, transportation, housing, resource allocation and community empowerment" (Bullard 1999: 37), and procedural and recognition justice as means of addressing injustice.

Environmental justice has been uniquely fought, defined and applied by communities, reflecting their histories, cultures, needs, vulnerabilities and strengths, social institutions and the specific cases of injustice that they face. It has been described as "political ecology from the bottom up" (Martinez-Alier et al. 2014: 21) and is necessarily contextual in nature. Currently, environmental justice approaches provide a language, principles and methods for understanding and addressing injustice. Communities and state organizations apply these approaches through policy and regulation. Environmental justice approaches examine power, politics and capitalism (Arcioni and Mitchell 2005) and, increasingly, affects not only on humans but also the environment.

Urban climate justice analysis builds on environmental justice principles to span diverse spatial and temporal scales (Walker 2012), from global to local, including understanding past conditions and actions to predictions and visions of the future. Climate justice considers emissions, sources and impacts and their relationship with economic development, exploitation of people and environments, social and environmental vulnerability, the ability to mitigate emissions and adapt to impacts of climate change, and the various responsibilities across all of these areas.

Climate justice differs from capitalist and technocratic responses that seek to re-engineer the use of oil or alternative sources of energy such as nuclear power, and continue growth with short-term thinking around profit margins over long-term environmental impacts. The latter is exemplified by governments focused on three- or four-year election cycles, centered on state and national agendas, and individuals with immediate self-interests and a concept of community that is bound to their immediate locality. This works to reinforce patterns of power and authority within urban politics, policy, institutions and systems.

Within urban climate justice literature there has been a strong focus on questions related to the "who" and the "how" of urban climate justice within the nexus of distributive justice (i.e. the allocation of benefits and burdens); procedural justice, that is, how processes and procedures recognize those interests (Steele et al. 2015); and recognition of justice as both a "subject and condition of justice" (Walker 2012: 50). This turn involves greater engagement with social and natural systems as co-contributors to both the conditions of urban climate resilience and vulnerability and how urban climate justice is contested and understood in practice.

Urban vulnerability and responsibility are key aspects to understanding conditions that demand attention to questions of energy and oil equity as part of a climate justice agenda. Barnett (2006: 115) has identified the following such conditions: the responsibility for climate change is not equally distributed; climate change will not affect all people equally, with some people and groups more vulnerable; this vulnerability is determined by politico-economic processes that benefit some more than others; climate change will compound underdevelopment because of the processes of disadvantage embedded within the neoliberal politico-economic status quo; and climate change policies may themselves create unfair outcomes by exacerbating, maintaining or ignoring existing and future inequalities. Better recognizing these conditions assists in shifting flashpoint issues, such as "peak oil," away from narratives that fetishize environmental resource constraints and reliance on technological fixes, and toward a critical engagement with the complexity of society-nature relations and the links to capitalist growth and urban development underpinning oil constrained cities.

As both a critique and call for action, urban climate justice seeks to support and promote diverse and creative climate experiments for sustainable transitions in practice (Bulkeley et al. 2014; MacCallum et al. 2014). This includes recognizing, mapping and indexing as a basis for prioritizing urban climate justice action, and involves both human and nonhuman stories, relations, politics and practices to radically destabilize and unsettle technocratic responses to anthropogenic climate change (De Broto and Bulkeley 2013; Strengers 2013). In this way, urban climate justice works as a critical conceptual tool to engage both the collective practices and imagination necessary to create a transition to a more sustainable and equitable urban future (Steele et al. 2015). In short, this is a critical practice-based approach to "seeing" the post-petroleum city.

In the following section we examine the challenges and possibilities of planning for a post-petroleum future through the urban climate justice lens. As urban policy around peak oil shifts from risk to opportunity, there is growing recognition that the plight of the most vulnerable human and non-human communities must lead the agenda of progressive planning practice. Brenner et al. (2012: 1) argue that cities "are not only sites of capitalist accumulation" but "also arenas in which the conflicts and contradictions associated with historically and geographically specific accumulation strategies are expressed and fought out." We would add to this analysis the significant tensions and contradictions inherent in planning policy and practice for an oil constrained urban future.

Planning just oil constrained cities

There is an increasing disconnect between energy security discourses and the multifarious differences in human (and non-human) circumstances, vulnerability and adaptive capability within cities. Oil depletion and scarcity is bound up in a broader narrative around energy and is a key part of this matrix, which includes deeply ingrained equity issues associated with extractive industries through to consumption, urban design and waste. The impacts of oil depletion and scarcity for the most marginalized include – but are not limited to – the capacity to access clean and affordable modern energy and infrastructure-related services within the urban realm.

The centrality of oil as energy in cities and the implications of this for urban vulnerability, transport and suburban disadvantage have been extensively highlighted from the international down to the everyday household scales (see Dalby 2009; Dodson and Sipe 2008b; Strengers 2013). Equitable responses move from providing charity to marginalized and vulnerable communities to upholding human rights, enacting polluter-pays principles and providing compensation in cases of adverse impacts (Walker 2012). For Huber (2015: 481), "there is perhaps no better example of the inescapably political nature of nature-society relations than energy." This political complexity includes "big oil" corporations and electricity companies, with monopolies over resources and control of supply and pricing; resource over-extraction; air, land and water pollution; the history of coal mining as a working class struggle over wages and conditions; and the more recent and most significant links to carbon emissions in anthropogenic climate change.

Planning for oil constrained cities involves, at heart, the redistribution of resources and creation of processes to achieve this that are democratic, fair and transparent. Fainstein (2006) highlights a central question for progressive contemporary urban planning, focused as it is on achieving improvement in the quality of human life within the context of a global capitalist political economy. In particular, in the context of oil vulnerability, Allison (2015) emphasizes the need to critically highlight the victims of energy poverty in urban settings; the ways low carbon technologies are introduced and produce energy injustices; and the importance of understanding complex techno-social systems.

As both a function and a response to multilevel governance, planning is about delivering common goods for communities. Such planning involves diverse actors and a dispersion of authority coming together to deliver collective action as part of a complex and fragmented landscape. Such action, in turn, is located within the twin meta-realms of radical democracy and political economy. Bulkeley (2009) found that spatial planning initiatives around energy supply, energy demand and adaptation in the UK remain largely on the fringe of the traditional bounds of the planning system and tend to adhere to technical rationalities or dominant lines of top-down command from the national to local levels. In Australia, urban planning and policy and services tend to emphasize growth over sustainability, shareholders/stakeholders rather than citizens, velocity over quality, and economic efficiency over equity (Gleeson and Steele 2010). Meanwhile, low carbon initiatives remain largely on the margins of mainstream planning.

The capacity for energy and oil equity in cities is constrained if planners are too closely aligned with conservative politics and private interests, serve as instruments of red tape and bureaucracy, and help to facilitate the (mis)workings of a grossly manipulated system (Goodman 1972). Brenner et al. (2012: 1) describe this widespread malaise as the

> injustice, destructiveness and unsustainability of capitalist forms of urbanization; the need to roll back contemporary profit-based forms of urbanization; and to promote alternative, radically democratic and sustainable forms of urbanism.

These circumstances call for the remit of planning to be expanded beyond just formal institutions and to be embedded within local, grassroots movements, which are rarely self-contained but work to creatively disrupt the order of things. For Palmer (in Goodman 1972: 13), the problems of inequality are deeply embedded structural problems, so at least part of "the motive force behind 'planning' must come from those people who are excluded from the machinery of government and power."

The climate-just city values powerless groups (whether human or non-human) and seeks to redress – through improvement in circumstance and capability, participation in decision making and access to resources – unfair conditions of hardship and associated misery and vulnerability. Who dominates? Who benefits? Who get left behind? Established categories of marginalization highlighted in the justice literature more broadly and applied within environmental justice research are race, ethnicity, culture, income and gender (Walker 2012). As an example, although research into gender and climate change is limited, Terry (2009) argues that women's social, political and economic experiences need to be understood within their spatial context, including their vulnerabilities and strengths, to increase their potential to adapt to climate change in a way that recognizes their empowerment. The intersection of vulnerabilities also adds complexity:

> While poor women's greater vulnerability compared with men is partly due to their relatively limited access to resources and their resulting poverty, this is not the whole story. It also arises from social and cultural norms about, for instance, gendered divisions of labour, physical mobility, and who is entitled to take part in decision-making at household and community levels.
>
> *(Terry 2009: 7)*

Planning for energy equity should not be viewed simply as a delivery mechanism (i.e. part of the problem) but, rather, the portal through which community understandings and practices around the oil constrained city can be "defined, contested and made material through processes of negotiation and conflict" (Bulkeley 2009: 294). Specifically, alternative or novel approaches to oil constrained cities are largely found from the bottom up, thereby creating alternative rationalities, discourses and practices.

A practice-based approach to energy equity

A practice-based approach to urban climate justice, as Hillier et al. (2013) outline in their work on social innovation at the local scale, focuses on critical engagement with existing urban architectures: social, political, economic and environmental imaginaries, stories, ethics and norms, and designing how effective responses to global environmental change can be collectively imagined and made manifest in cities at the local scale. Therefore, planning oil constrained cities involves:

- Identifying constraints and potentials for more contextually relevant responses to planning oil constrained cities;
- Shifting emphasis from a goal- to a process-oriented planning approach to spatial strategy making, involving the consideration of alternative planning and policy instruments appropriate to local situations;
- Broadening the scope for understanding around planning for post-petroleum urban futures and addressing change at the local scale to include ecological and social justice concerns and multi-scalar networks;
- Explicitly broadening the scope of planning responsibility to include non-institutional actors, to enable more integrated, holistic and socially innovative planning–related energy justice policies and practices;
- Focusing on spatial planning practice and the realm of actors who can make a difference, and provide an antidote to disempowering discourses that alienate the problem of oil vulnerability from the everyday at the local scale (adapted from Hillier et al. 2013).

Using the urban climate justice framing for energy and oil equity, the responsibility and remit of planning post-petroleum cities becomes expanded and shared between different stakeholders at different scales to create equitable change at the local level. Global organizations such as 350.org (2015) work to connect local climate actions with global politics through the construction of a new public commons that promotes clean energy and transport, such as cycling, in 2,000 events in 175 countries. Planning oil constrained cities becomes, then, "a process of learning involving a collective of human and more-than human actors – a process of co-transformation that re/constitutes the world" (Roelvink and Gibson-Graham 2010: 1).

In Australia, for example, local community-led initiatives in cities include: community gardening to combat growing food insecurity and preserve agro-biodiversity; blocking coal trains from delivering their payload to coal-fired power stations as a form of protest and intervention; working in solidarity with indigenous groups to effect change (i.e. building networks with traditionally marginalized groups); farmers joining with city folk to try to block coal-seam gas projects; and trade unions joining with schools, community groups, environmental organizations and tertiary institutions to map out transitions to a low carbon economy (Hillier et al. 2013).

Across Australia, local activist groups have united to develop an alternative collaborative vision to the government-led National Food Plan. The Peoples' Food Plan emphasizes that "people are hungry in the suburbs, whilst supermarkets are throwing away food." The aim of the Australian Food Sovereignty Alliance (2012: 1) is to implement "a more coherent and cohesive food justice/fair food movement around the country" rather than "a business-as-usual plan to suit the needs of big agri-business, major retailers and commodity exporters." Activities have included performance and speaking events, and the active engagement of social and other media around alternative food visions to integrate alternative ideas for the ways that food is consumed and distributed into mainstream urban policy and planning processes, and to serve as a critique and counterpoint to top-down, technocratic and managerial approaches (such as the development of the National Food Plan).

Innovative work and initiatives with communities to enact change is also occurring at the local government planning scale. In Melbourne (Australia), an inner-city Darebin Council initiative has been the *Climate Change and Peak Oil Adaptation Plan* (Taygfeld and Burton 2009), under review in 2015. Darebin has one of Australia's most diverse communities, with a large number of pensioners (both aged and socially disadvantaged), low-income households, and people who are socially isolated and disadvantaged. One of Darebin Council's key adaptation initiatives is the Solar $aver Program, which enables pensioners to install solar power to their homes with no upfront cost and, instead, pay for the system along with council rates over ten years and free of interest. Darebin Council's adaptation focus is on heat stress response, which reflects the vulnerable status of many social groups within its shire who are unable to afford and access air conditioning or heat stress management options in times of extreme heat. To this end, the council has committed several million dollars of its budget (including A$1 million in 2015) to the Solar $aver Program, which will continue into 2016 (Solar Programs 2015).

In her analysis of urban energy policy systems, Bulkeley (2009) highlights that planning needs to be enmeshed in multilevel responses at all scales and that the challenge of spatial planning is finding locally contextualized strategies best suited for particular people and places. As Sennett (2013: x) describes, this involves building an imaginative and material sense of "togetherness" that involves creatively confronting the planning maladaptation nemesis by exploring co-operation as a craft:

> It requires of people the skill of understanding and responding to one another in order to act together, but this is a thorny process, full of difficulty and ambiguity and often leading to destructive consequences.

Planning for energy and oil equity, particularly at the local scale, involves learning lessons drawn from the injustices and inconsistencies of everyday life as a way of informing and guiding future action around oil constrained cities – and these must be confronted wherever they occur.

Conclusion

Who is served by the transition to a post-petroleum city in the current urban imaginary? From a climate-just city perspective, it is not enough for those planning cities after petroleum to simply attend to the fringes of an oil-constrained urban future by tweaking at the edges of the status quo with centralized, top-down climate adaptation plans and policies. Currently, planners' power and authority is perpetuated through reverence for property rights; incentives supporting development itself, rather than developments that are designed to meet community needs; transport reliant on roads and, therefore, oil (a heightened problem for people with high social vulnerability on the urban fringes); access to jobs, services and healthy environments that favor the already privileged; extending urban growth boundaries that encroach upon agricultural land and biodiversity; and prioritizing anthropogenic uses over non-human species and environmental protection.

Rather, what is required is a quite different planning rationality that serves those most vulnerable in the transition to a post-oil society, particularly with respect to alternative energy supply, distribution, efficiency and justice. As urban policy around peak oil shifts from risk to opportunity, there is growing recognition that the plight of the most vulnerable communities (human and non-human) must lead the agenda of progressive planning practice. This is important if planning is to reverse the current centralized planning tendency that excludes or marginalizes issues of climate and energy justice in the pursuit of economic growth and market development.

Diverse community actions, both globally and locally, suggest that there is already evidence of innovative and active work underway. Central to such grassroots planning activities are processes of social learning and interconnection. We need to hear the stories and practices of world citizenry, and we must enlarge the boundaries of dialogue and action. The focus instead must be to create new planning possibilities for cities in the transition to an oil constrained future. Urban climate justice work already underway reflects stories and practices that shape the future of planning the post-petroleum cities yet to come.

References

Allison, J. (2015) "Book Review Essay – Energy Justice, Climate Change and the Challenge of Energy Governance," *Global Environmental Politics* 15(1): 123–28.

Arcioni, A. and G. Mitchell (2005) "Environmental Justice in Australia: When the RATS Became IRATE," *Environmental Politics* 14(3): 363–79.

Australian Food Sovereignty Alliance (2012) *A People's Food Plan for Australia*, accessed 1 October 2015 — www.australianfoodsovereigntyalliance.org/wp-content/uploads/2012/10/20120914-PeoplesFoodPlan-DiscussionPaper-Overview.pdf

Barnett, J. (2006) "Climate Change, Insecurity and Injustice," in W. Adger, J. Paavola, S. Huq and M. Mace (eds) *In Fairness to Adaptation to Climate Change*, Cambridge: Massachusetts Institute of Technology Press, 115–29.

Bickerstaff, K., Walker, G. and H. Bulkeley (2013) *Energy Justice in a Changing Climate: Social Equity and Low-Carbon Energy*, London: Zed Books.

Brenner, N., Marcuse, P. and M. Mayer (eds) (2012) *Cities For People, Not For Profit. Critical Urban Theory and the Right to the City*, New York: Routledge.

Brulle, R.J. and D.N. Pellow (2006) "Environmental Justice: Human Health and Environmental Inequalities," *Annual Review of Public Health* 27: 103–24.

Bulkeley, H. (2009) "Planning Governance of Climate Change," in S. Davoudi, J. Crawford and A. Mehmood (eds) *Planning for Climate Change Strategies for Mitigation and Adaptation for Spatial Planners*, London: Earthscan, 284–96.

Bulkeley, H., Edwards, G. and A. Fuller (2014) "Contesting Climate Justice in the City: Examining Politics and Practice in Urban Climate Change Experiments," *Global Environmental Change* 25: 31–40.

Bullard, R.D. (2015) Dr. Robert Bullard (site) "Publications" (pages), accessed 15 December 2015 – drrobertbullard.com

——— (1999) "Environmental Justice Challenges at Home and Abroad," in N. Low (ed.) *Global Ethics and Environment*, London: Routledge, 33–46.

Byrne, J. and C. Portenger (2014) "Climate Change, Energy Policy and Justice: A Systematic Review," *Analyse & Kritic* 36(2): 315–43.

Dalby, S. (2009) *Security and Environmental Change*, Oxford: Polity.

Day, R. and G. Walker (2013) "Energy Vulnerability as an 'Assemblage'," in K. Bickerstaff, H. Bulkeley and G. Walker (eds) *Energy and Justice in a Changing Climate*, London: Zed Books, 14–29.

De Broto, V. and H. Bulkeley (2013) "Maintaining Climate Change Experiments: Urban Political Ecology and the Everyday Reconfiguration of Urban Infrastructure," *International Journal of Urban and Regional Research* 6: 1943–48.

Dodson, J. and N. Sipe (2008) *Unsettling Suburbia: The New Landscape of Oil and Mortgage Vulnerability in Australian Cities*, Urban Research Program, Brisbane: Griffith University.

Fainstein, S. (2006) "Planning and the Just City," paper presented at Searching for the Just City conference, Columbia University, New York, 29 April.

Gleeson, B. and W. Steele (2010) *A Climate for Growth: Planning South-East Queensland*, St. Lucia: Queensland University Press.

Goodman, R. (1972) *After the Planners*, Harmondsworth: Penguin.

Hall, P. (2002) *Cities of Tomorrow: An Intellectual History of Urban Planning and Design in the Twentieth Century*, Malden: Blackwell.

Harvey, D. (1989) *The Urban Experience*, Oxford: Blackwell.

Hillier, J. MacCallum, D. Steele, W. Houston, D. and J. Byrne (2013) *Climate Justice in the Australian City*, State of Australian Cities Conference, Sydney: State of Australian Cities Research Network.

Huber, M. (2015) "Energy and Social Power: From Political Ecology to the Ecology of Politics," in T. Perreault, G. Bridge and J. McCarthy (eds) *The Routledge Handbook of Political Ecology*, London: Routledge, 481–92.

Jameson, F. (2003) "Future City," *New Left Review* 21: 65–79.

Jamieson, D. (2001) "Climate Change and Global Environmental Justice," in C. Miller and P. Edwards (eds) *Changing the Atmosphere: Expert Knowledge and Environmental Governance*, Cambridge: Massachusetts Institute of Technology Press, 287–308.

Kenny, M. and J. Meadowcroft (1999) *Planning Sustainability*, London: Routledge.

MacCallum, D. Byrne, J. and W. Steele (2014) "Whither Justice? An Analysis of Climate Change Responses from South East Queensland, Australia," *Environment and Planning C* 32: 70–92.

Martinez-Alier, J., Anguelovski, I., Bond, P., Del Bene, D., Demaria, F., Gerber, J.-F., Greyl, L., Haas, W., Healy, H., Marín-Burgos, V., Ojo, G., Porto, M., Rijnhout, L., Rodríguez-Labajos, B., Spangenberg, J., Temper, L., Warlenius, R., and I. Yánez (2014) "Between Activism and Science: Grassroots Concepts for Sustainability Coined by Environmental Justice Organizations," *Journal of Political Ecology* 21: 19–60.

Roelvink, G. and J.K. Gibson-Graham (2010) "An Economic Ethics for the Anthropocene," *Antipode* 41(1): 320–46.

Sennett, R. (2013) *Together: The Rituals Pleasures and Politics of Cooperation*, New Haven: Yale University Press.

Solar Programs (2015) "Energy Climate" (page), Darebin Council (site), accessed 15 December 2015 — www.darebin.vic.gov.au/Darebin-Living/Caring-for-the-environment/EnergyClimate

Sovacool, B. Sidortsov, R. and B. Jones (2013) *Energy Security, Equality and Justice*, London: Routledge.

Steele, W.E., MacCallum, D., Byrne, J. and D. Houston (2012) "Planning the Climate-Just City," *International Planning Studies* 17(1): 67–83.

Steele, W.E., Mata, L. and H. Funfgeld (2015) "Urban Climate Justice: Creating Sustainable Pathways for Humans and Other Species," *Current Opinion in Environmental Sustainability* 14: 121–26.

Strengers, Y. (2013) *Smart Energy Technologies in Everyday Life: Smart Utopia?* London: Palgrave Macmillan.

Taygfeld, P. and D. Burton (2009) *Climate Change and Peak Oil Adaptation Plan*, Preston, Victoria: Darebin Council.

Terry, G. (2009) "No Climate Justice Without Gender Justice: An Overview of the Issues," *Gender & Development* 17(1): 5–18.

350.org (2015) 350.org (site), "Timelines and Milestones" (page), accessed 15 December 2015 — http://350.org/timelines-milestones/

Walker, G. (2012) *Environmental Justice: Concepts, Evidence and Politics*, Abingdon: Routledge.

Winkler, T. (2009) "For the Equitable City Yet to Come," *Planning Theory and Practice* 10(1): 65–83.

PART II
Transport and land use

7

WALKING THE CITY

John Whitelegg

In a significant work on sustainable urban transport, Schiller et al. (2010: 218) stress the "long, profound and central role of walking in shaping human physical development, cognition and the formation of pre-automobile communities," advocate a central place for walking in transportation planning, and conclude that the environment created for pedestrians "defines the quality of the public realm and its capacity to support human community." In reality, walking is the Cinderella of contemporary urban and transport planning and hardly figures at all in rural transport planning. In the Global North there is a considerable mismatch between the rhetoric around walking and the reality of what happens on the ground. In the US, UK and Australia, walking is in decline and struggles to find its place in political and planning discourse that, at best, talks about the importance of public transport even if it does not deliver.

This chapter traces the decline, during the last few decades, of walking as a transport choice, then discusses an example of enlightened social urban planning and policy to enhance cycling and use of public transport as well as walking (Freiburg-im-Breisgau in southern Germany) and reviews environmental, health, social and environmental reasons for promoting walking.

Walking as a modal share of transport options

It is quite normal in the UK to see transport policy documents promoting walking and cycling within a wider sustainability agenda. On the street, however, there is a rising volume of traffic to deter the pedestrian, a general fear of road safety dangers, a year-on-year increase in parking space numbers and developments that generate even more traffic, and an inability to grasp the importance and multiple benefits associated with higher levels of walking. There is also a preference for measures and interventions that damage walking, such as altering the timings on traffic light–controlled pedestrian crossings in London to give less time to pedestrians to cross the road. In the UK the green phase for pedestrians is referred to as "Green Man time":

> Green Man time has been reduced at 568 crossings across London since 2010. Reduced crossing times encourage pedestrians to take greater risks. For other groups, particularly older and disabled people, it can affect their confidence when crossing the road. The Committee is concerned to note that there has been little analysis of the effect of reducing Green Man time on crossing behaviour.
>
> *(Transport Committee, London Assembly 2014: 24)*

There are very significant global differences, especially between the Global North and Global South, in the importance of walking as a mode of transport and the distribution of trips by mode. Globally about 37 percent of urban trips are made on foot or by bike (UN 2013). While in many African cities the proportion of walking trips is 30–35 percent of the total, in Dakar (Senegal) and Doualą (Cameroon) it is greater than 60 percent (UN 2013). In an American city the percentage of urban trips made by foot is often less than 1, when measured by distance traveled, but the percentage of trips made on foot – as opposed to distance – is much higher. In New York it is 39 percent and in Chicago it is 19 percent (Anon. 2011). In Freiburg-im-Breisgau, southern Germany, it is 24 percent (Whitelegg 2013). Kenworthy (2014) has provided a useful summary of modal split data by distance traveled (Table 7.1). A very different picture emerges if we look at modal split percentages by trip or journey rather than distance traveled (Table 7.2).

The very small modal share for walking revealed in Table 7.1 is the result of many years of neglect of this mode of transport and the associated promotion of car trips that then create an aggressive and uncomfortable environment for walking. Walking is not likely to thrive in cities where large volumes of car traffic act as barriers to pedestrian movement (Appleyard 1981) and are even less likely to thrive if the distance between destinations grows over time (Whitelegg 2013). Walking is often the victim of pervasive spatial trends that increase the distances that have to be traveled for the journey to work, school

TABLE 7.1 Daily person-kilometers by mode of transport (%) in a sample of world cities, 2005

	Car	*Transit*	*Foot*	*Bike*
Atlanta	98	1.4	0.3	0
Berlin	59.1	6.2	4.1	8.9
Brisbane	91.1	6.5	0.9	0.6
Copenhagen	72.1	13.7	2	11.5
Hong Kong	14.3	78.5	5.8	0.5
Manchester	87.5	8.1	3.2	0.8
Melbourne	89.4	8.2	1.1	0.8
Singapore	47	40	1.5	3.4
Sydney	85.8	11.7	1.8	0.5
Zurich	67	19.7	7.8	4

Note: Numbers do not add to 100 because motorcycles are not included. Motorcycles have a very small modal share, except for Singapore (8.2%).

Source: Data drawn from Kenworthy (2014: 46).

TABLE 7.2 Passenger transport modal share of trips (%)

	Walk	*Cycle*	*Private Transport*	*Public Transport*	*Other*
Berlin	29	13	32	26	–
Chicago	19	1	63	16	1
London	30	2	40	27	1
Melbourne	13	2	77	7	1
Singapore	22	1	29	44	4
Sydney	18	–	69	11	2
Vienna	28	5	31	36	–

Source: Data drawn from Anon. (2011).

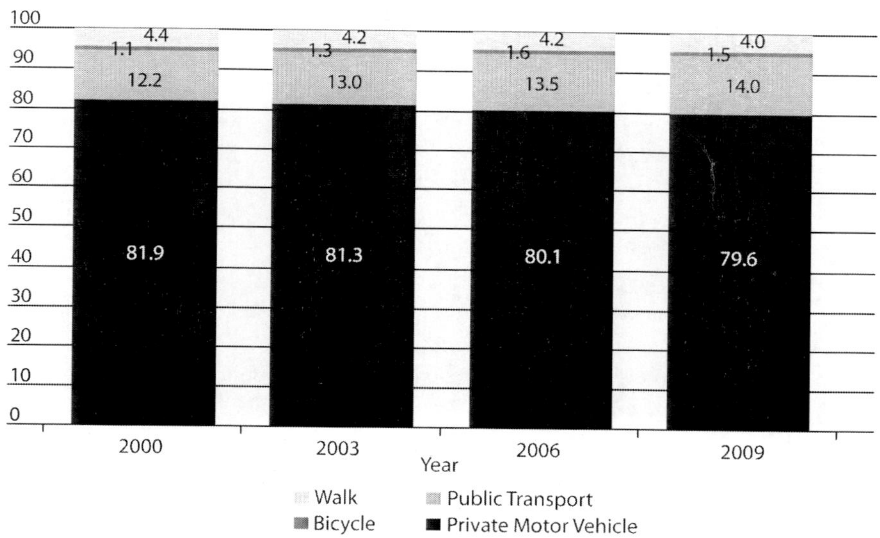

FIGURE 7.1 Modal split on the journey to work and study in Australia, 2000–2009

Source: Commonwealth of Australia (2012: 27) (using Australian Bureau of Statistics data).

or health care facility and as distances increase trips migrate to the car. Walking has declined in many countries (though not in Germany). Australian data (Figure 7.1) on the journey to work and study is fairly typical in this respect.

Germany has seen an increase in walking rates: between 2001–2002 and 2008–2009, the proportion of "any walking" in Germany increased from 36.5 to 42.3 percent (Buehler et al. 2011). It was stable at 18.5 percent in the US. In Britain walking trips continue to decline in spite of significant amounts of governmental and public health rhetoric about the importance of walking (DfT 2011). Over the long term, since the early 1970s, the average distance people travel per year has increased by 50 percent. Between 1995–97 and 2010, overall trip rates fell by 12 percent. Trips by private modes fell by 14 percent while public transport modes increased by 8 percent. Walking trips saw the largest decrease. Walking trips fell 8 percent when 2010 is compared to the previous year, making 2010 the lowest ever recorded to date at 210 trips per person each year (DfT 2011).

The decline in walking, and increases in distances traveled as accessibility declines, do not represent some kind of immutable law. These transport choices are susceptible to intelligent public policy interventions focusing on land use, mixed use and high-quality urban design. The city of Freiburg-im-Breisgau in southern Germany is an example of enlightened spatial, social and environmental planning to promote the sustainable modes, such as walking, cycling and public transport.

Freiburg-im-Breisgau

When my son first visited Freiburg-im-Breisgau (Freiburg), he asked: "Why can't everywhere in England be like this?" Indeed, Schiller et al. (2010: 280–81) point out that Freiburg

> boasts one of the most comprehensive approaches in the world to transit, walking and cycling from both a transportation and land-use perspective . . . perhaps its most striking feature is its obvious

> attention to high-quality coherent urban design throughout the city . . . Freiburg has for many years been the "pin-up" city for sustainable transportation and more generally for developing a much more sustainable city environment based on ecological building principles, prioritising public transportation, walking and cycling, and all supported by superb urban design of the public realm.

At time of writing, Freiburg is a city with a population of almost 220,000 residents. It has a strong reputation as a "green" city, with large and successful solar energy and wind components in its energy production. The city is a European center for photovoltaic research, development and manufacturing, employing 10,000 people (Stadt Freiburg 2010). Freiburg has clear strategies in place to deal with climate change, air pollution and energy efficient housing, and has a local law requiring all new residential buildings to operate to a "Passive Haus" standard of energy use. This standard results in very low energy consumption levels per square meter of living space. Climate change policy is a hugely important political priority in Freiburg. By 2009, on a 1992 base, Freiburg had reduced CO_2 emissions by 25.6 percent per capita (Stadt Freiburg 2011).

The city has an international reputation for its successful sustainable transport system (Whitelegg 2013). Walking and cycling around Freiburg makes a very strong impression on the senses. Walking accounts for 24 percent of all trips every day, cycling 28 percent, public transport 18 percent and automobiles 30 percent (Stadt Freiburg 2010). There is very little congestion. The streets are relatively calm with such large numbers of pedestrians and cyclists. The speed of the traffic, where it is allowed, is below 20 mph. There is a profusion of small retail outlets, cafes and restaurants and an absence of large supermarkets with large car parks. The streets convey a strong sense of calmness, smooth flowing traffic, high standards of design of footways, bike paths, tree planting and cleanliness, and the whole experience generates feelings of satisfaction and pleasure in sampling the delights of an urban environment.

Moving around by any mode of transport could not be easier. Quiet trams make their way through city center streets and out to the suburbs and integrate seamlessly with buses. The main railway station is an integrated transport hub with direct access to buses; a major tram stop is located at the southern end of the platforms; and a cycling center offers parking for 1,000 bikes as well as repair facilities and public transport information. Public transport fares are reasonable, and there are generous offers for commuters and families to make the cost of travel affordable as well as enjoyable.

Visitors staying in the hotels of Freiburg are given a free travel pass for the duration of their stay for all forms of public transport within a large region around the city. The trams and buses are clean and tidy and the systems are legible and easy to use; in a quiet, non-intrusive manner they send the strong message that cars are not really necessary or desirable when choices and experiences are so rich without them.

Why is walking so important?

Walking has many claims to be the most important mode of transport. It is available to a very large proportion of the population. It is inexpensive to both user and public provider. Walking is a zero carbon and zero polluting mode of movement. Cycling shares many of these advantages. Both walking and cycling can be promoted as part of an integrated approach to maximizing the use of public transport. Walking should be ranked at the top of any prioritization of transport funding and infrastructure provision.

The public policy usefulness and quality of life credentials of walking can be summarized as: intimate links between walking and livability and the promotion of sustainable cities and lifestyles; proven public health benefits of walking especially in dealing with the global obesity epidemic; and the importance of walking as a core component of policies and interventions to reduce carbon emissions, reduce energy consumption and deal with the problem of peak oil.

Walking, livability and promotion of sustainable cities

Donald Appleyard (1981) has captured the importance of the traffic environment in determining patterns of walking and social interaction. His famous diagram is reproduced in Figure 7.2. Appleyard's diagram shows that, when there is a low level of traffic, there is a great deal of street life, walking and social interactions. Residents on the low traffic street have the highest number of friends and acquaintances. As traffic levels increase from medium to heavy, so the amount of use of street space on the part of residents goes down, social interaction declines, conviviality evaporates and residents have fewer friends and acquaintances than they have on the lightly trafficked street.

Appleyard's pioneering work has been replicated in Bristol (UK) by Hart and Parkhurst (2011) with similar results. There is a clear dose-response relationship between traffic levels and people crossing the road and using street space. The higher the traffic volume, the lower is the use of that space. Hart and Parkhurst (2011: 12) reinforced the conclusions of Appleyard's 1981 book, specifically "demonstrating that casual conversations, children's play, and other street-based social life tend to be suppressed, particularly as vehicle volumes and speeds increase."

This fundamental finding has been taken up in many of the world's major cities by Danish architect Jan Gehl. His recommendations in London, Brisbane, Melbourne and many other major cities focus heavily on promoting walking and cycling. In this way, high-quality living environments are created,

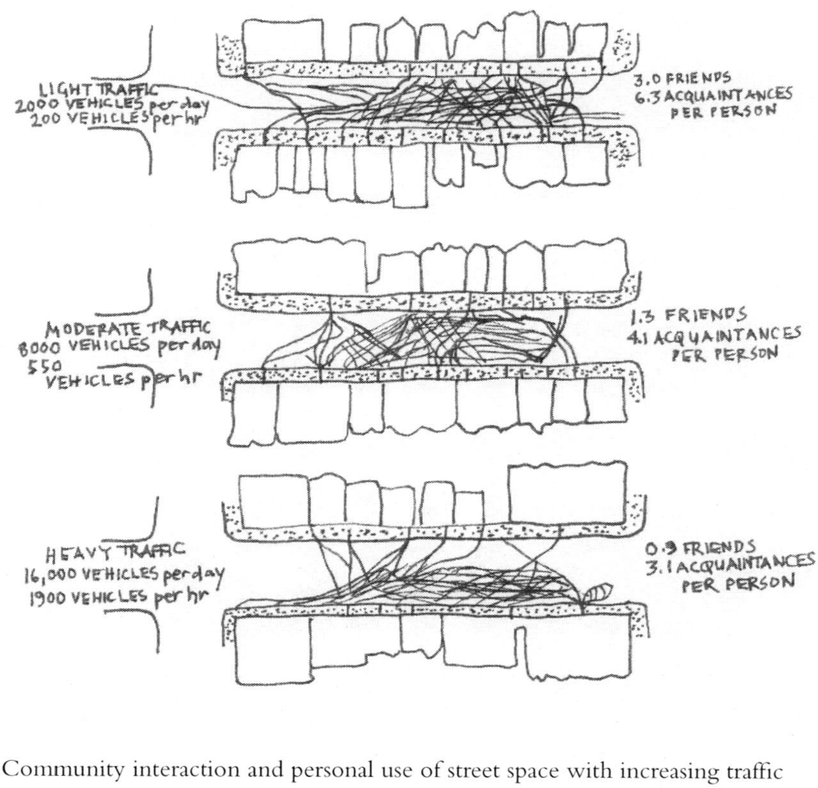

FIGURE 7.2 Community interaction and personal use of street space with increasing traffic

Source: Whitelegg in Holzapfel (2014: 20).

transforming cities so that they become more civilized, more human in scale and rich in social interaction, with children playing and all ages and social groups enjoying the street life (Gehl 2010).

Heavy traffic volumes and speeding traffic destroy community life and deter pedestrian activity. Heavy traffic triggers a perverse feedback mechanism whereby traffic levels increase to the point that walking is unpleasant and is perceived as dangerous, so car use rises yet again. The process continues to produce a city dependent on cars (even for short journeys) and heavily dependent on increasingly scarce fossil fuels. Restoring a high quality of life and a high level of pedestrian activity requires a reduction in traffic volumes and much reduced speeds. Such improvements in quality of life embrace more walking, lower energy consumption per capita and less dependence on fossil fuels.

Proven public health benefits of walking

There is a significant positive association between the density of traffic around children's homes and obesity, as measured by the BMI (Jerrett et al. 2010). A research project carried out in Atlanta (US) found that each additional hour spent in a car per day is associated with a 6 percent increase in the risk of obesity (Frank et al. 2004). The relationship between declining levels of active transport (walking and cycling) and obesity has been explored in detail by Roberts and Edwards (2010) and very clearly summarized by Pucher (2010), as reproduced in Figure 7.3.

Woodcock et al. (2009) have estimated the health effects of transport policies that would meet greenhouse gas emission reduction targets. They concluded that meeting emission targets in the transport

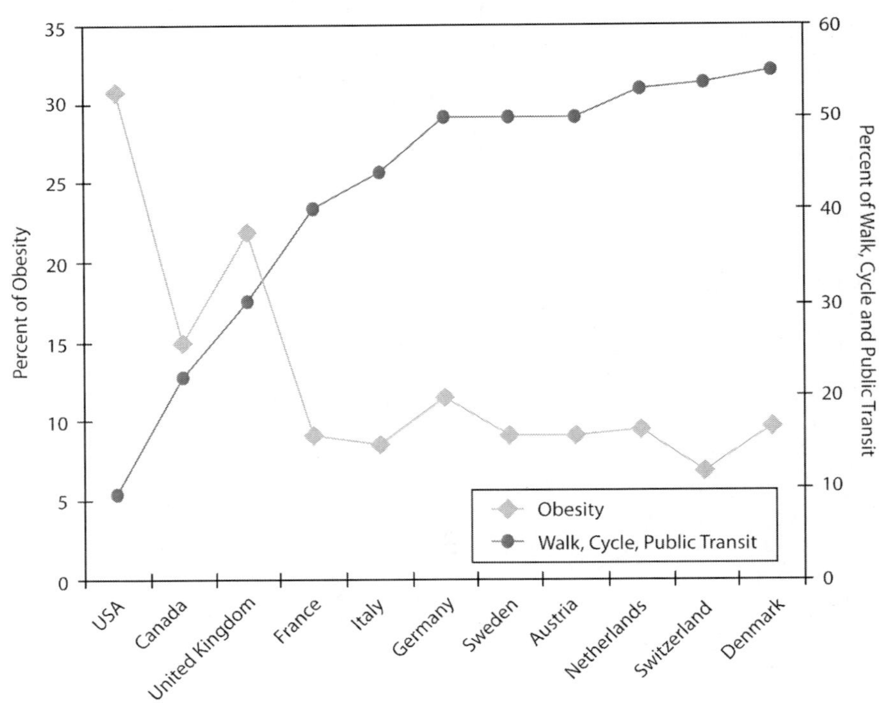

FIGURE 7.3 Relationship between active travel and obesity

Source: Pucher (2010).

sector would require substantial increases in walking and cycling, with correspondingly large reductions in car use. Based on scientific evidence linking physical activity and health, Woodcock et al. (2009) estimated that increases in walking and cycling would dramatically cut rates of chronic disease, with 10–20 percent less heart disease and stroke, 12–18 percent less breast cancer and 8 percent less dementia.

There is a very clear and useful policy opportunity in the promotion and advancement of walking and cycling. Higher levels of so-called active transport will improve population health and will also reduce greenhouse gas emissions by substituting walk and cycle trips for car trips. Roberts and Edwards (2010: 6) make explicit links between obesity and climate change:

> The human race is getting fatter and the planet is getting hotter, and fossil fuels are the cause of both . . . [O]besity is a normal human response to a sick environment, the bodily consequences of living in a world flooded with cheap energy. As a result of petroleum powered transportation and the road danger it creates, we walk and cycle less than ever before in the history of the world and our personal energy output has plummeted . . . climate change and fatness are different facets of the same basic problem.

We are much more likely to be able to solve climate change and obesity problems and deal with peak oil vulnerability if we harness the synergy that exists between public health, quality of life and climate change concerns. They are all part of the same debate and all require a serious and dramatic upgrade of our built environment to reward the cyclist and the pedestrian. The current zeitgeist is to reward the motorist, which increases climate change and obesity problems.

Carbon, energy and peak oil

The literatures on carbon dioxide emissions from the transport sector and the necessity to reduce these emissions are huge. A useful summary can be found in Davis et al. (2007). Less well known, but equally clear, is that substantial decarbonization of the transport sector is possible, achievable, desirable and presents no technical or fiscal problems (Whitelegg et al. 2010). The lack of progress with de-carbonizing transport is a function of a deeply embedded political ideology and cultural fetish that equates mobility with freedom, quality of life and endless economic growth. The vast majority of politicians, and those who make decisions about transport policy, fiscal policy and infrastructure, are captives of what Wolfgang Sachs (1992: vii–viii) elegantly captured in the title of his book, *For the Love of the Automobile*, from which the following quote is taken.

> The automobile is much more than a mere means of transportation; rather it is wholly imbued with feelings and desires that raise it to the level of a cultural symbol. Behind the gradual infiltration of the automobile into the world of pure dreams lie many stories: one of disdain for the unmendable horse . . . of the driver's megalomania, of the sense of having a miracle parked in the drive, and of the generalised desire for social betterment . . . a technological history . . . singing a devotional hymn to increasing perfection . . . where the breast first swelled with the pride of independence, where the love of speed was born, where the feeling of comfort took root, and where the automobile became allied with the clock as a "time saving machine."

If we set out on a journey to reduce carbon emissions, energy use and the problems associated with peak oil, then we should be aware of the context described by Sachs and all the cultural reactions associated with a message that says, "please drive less and walk more."

Reducing carbon emissions and energy use and dealing with peak oil concerns requires a highly integrated, synergistic approach using the full range of spatial, fiscal and behavioral policies to reduce transport's carbon emissions (Whitelegg et al. 2010). The spatial interventions include implementing best practices on pedestrian-oriented design, road-space reallocation and the promotion of compact development. These interventions are intended to maximize walking and cycling level. They are designed to reduce distances between origins and destinations so that the trip can be managed on foot or by bike. At the same time fiscal measures would need to be introduced to make car use less attractive; these include road user charges (much wider than congestion charging), workplace car parking charges, and fuel price uplift to internalize the external costs of road transport. The wide range of behavioral interventions required would center on travel plans and personalized journey planning. All these measures and interventions could collectively reduce carbon emissions from the road transport sector by 75 percent by 2050 in the UK (Whitelegg et al. 2010).

The importance of the spatial dimension, compact cities and densification has been recognized in the work of Newman and Kenworthy (1989) and in some recent work by the global public transport organization, Union Internationale Transports Publique (UITP) (Vivier 2006). This latter contribution is illustrated in Figure 7.4 and Figure 7.5. Figure 7.4 shows a very clear relationship between energy use and density (energy use tracks greenhouse gas emissions very accurately).

Increasing density in our large urban areas brings significant reductions in energy use and carbon emissions. Density and high levels of modal share for walking, cycling and public transport are closely related. Figure 7.5 shows how energy use declines as the percentage of walking, cycling and public transport grows. Figure 7.5 makes it abundantly clear that energy consumption is reduced as the proportion of trips by foot, bike and public transport increases. Reducing energy consumption is a key part of a climate change policy and the reduction of greenhouse gases. Cities with a modal share for walk, cycle and public transport above 55 percent produce on average about 2.4 tonnes less CO_2 per capita from passenger transport than cities below that level of performance (UITP 2009).

Reducing energy consumption is also a key component of a resilience strategy. Cities are much less likely to experience severe disruption from interrupted energy supplies, whatever the cause, if they are operating at lower levels of energy consumption. They will have shifted modal preferences to forms of transport that are not susceptible to shocks to the same degree as those depending on liquid fossil fuels from troubled parts of the world. This is a significant shift to a resilient future. With higher levels of

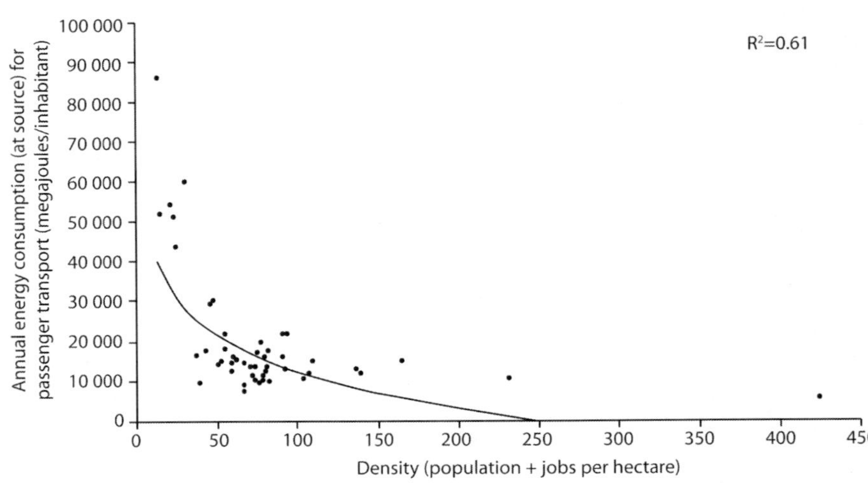

FIGURE 7.4 Energy consumption for passenger transport versus density

Source: Vivier (2006).

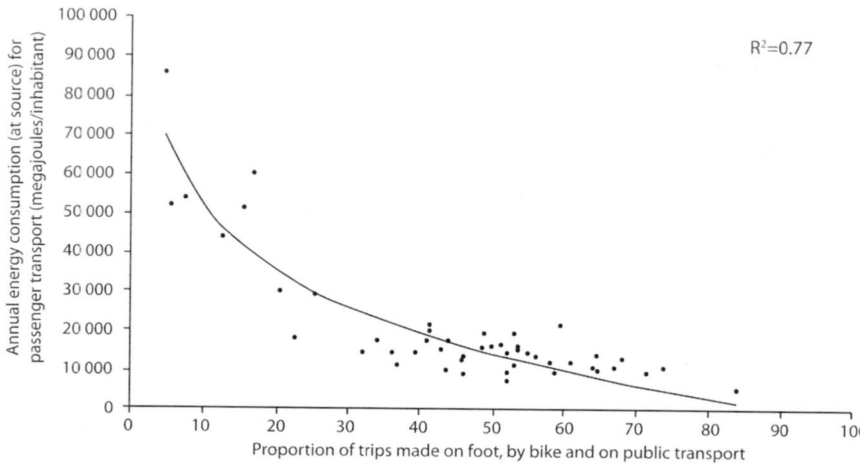

FIGURE 7.5 Energy consumption for passenger transport versus proportion of trips made by foot, bike and public transport

Source: Vivier (2006).

walking, cycling and public transport usage, residents will be able to get to work more easily than they do in a car dependent city. More staff will be available to run schools, hospitals and emergency services and more residents will be able to make visits to look after each other, neighbors and older relatives. Resilience can be built into the city as a direct result of increasing walking, cycling and public transport.

Resilience is also about affordability, budgets and fiscal prudence. Cities are prone to fiscal crisis. Cities are expensive to run and depend on uncertain income streams. It makes sense, therefore, to run a city at a lower cost, if this option is available. A city with high levels of walking, cycling and public transport can be run at lower costs than one with high rates of car use.

A report from the global organization representing public transport operators UITP has used an international database to shed light on the costs of mobility at the level of a city and the way this varies with modal split (Vivier 2006). Not surprisingly, and in line with citizen expectations and common sense, the costs of mobility are much less in a city with high levels of walking, cycling and public transport (Table 7.3).

Table 7.3 shows a number of results of importance to all those concerned with the costs of running cities and the availability of public expenditure for health, education and social welfare. It is not

TABLE 7.3 Relationship between public transport modal share and community costs

	Modal share of public transport (%)		*Cost of transport to the community (% of GDP)*	
	1995	*2001*	*1995*	*2001*
Geneva	18.8	21.7	10.2	9.4
London	23.9	26.8	8.5	7.5
Madrid	27.2	30.2	12	10.4
Paris	27.1	27.5	6.8	6.7
Vienna	43.2	46.6	6.9	6.6

Source: Data drawn from Vivier (2006).

just a matter of technical or academic interest if public budgets are consumed by growth in mobility leaving much less for other areas; it is an important matter for quality of life and democracy. Table 7.3 shows that, as the modal split for public transport increases, so the cost of transport as a percentage of GDP reduces. Cities that have managed an increase in the modal share of public transport saw a decrease in the cost of transport. Elsewhere in the report, Vivier (2006:9) estimates that transport costs in cities mainly using public transport can be "up to half the cost in cities where the private car is dominant," resulting in savings of, say €2,000 per capita so that "cities characterized by the lowest cost of transport to the community are often those where expenditure on public transport is the highest."

Vivier quantifies the GDP percentage cost of transport as it varies by the percentage of all trips by walking, cycling and public transport (Table 7.4). Figure 7.6 shows the full dataset referred to in Table 7.4, revealing a clear relationship between modes of transport and total transport costs.

The implications of the relationship between the two variables in Figure 7.6 and Table 7.4 sends a very clear signal to urban planners, economists and all those responsible for fiscal matters in any city in the world. It is essential to raise the percentage modal share of walking, cycling and public transport if city finances are to be well managed and a high quality of life achieved at a low cost. High levels of car

TABLE 7.4 Cost of transport to the community, by modal split

Modal share of walking, cycling and public transport (%)	*Community cost of city GDP (%)*
> 55	6.3
40–55	8.8
25–40	10.2
< 25	12.5

Source: Data drawn from Vivier (2006).

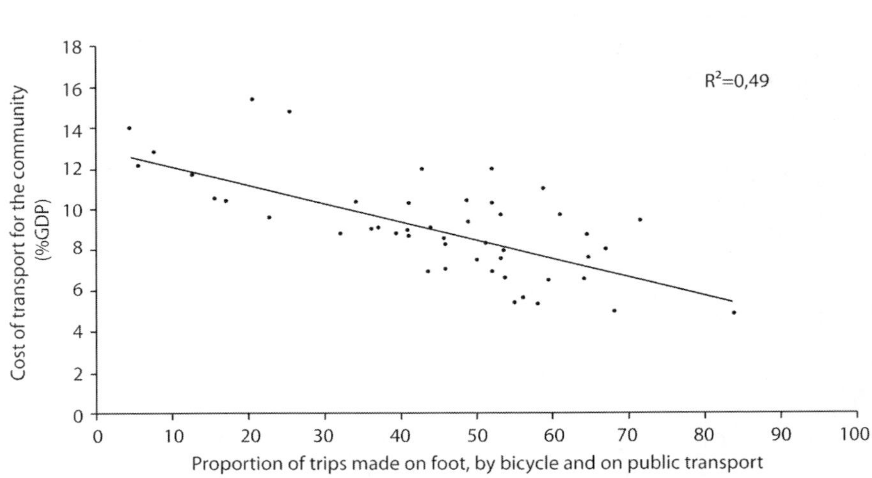

FIGURE 7.6 The relationship between percentage modal split for walking, cycling and public transport and the total cost of transport to the community as a percentage of GDP

Source: Data drawn from Vivier (2006).

use are a recipe for fiscal meltdown and the decline of cities. A future organized around much-reduced consumption of energy through diminished car use, and much-improved accessibility based on higher levels of walking, cycling and public transport, is a fiscally prudent and affordable future. It also brings a much-increased probability of enjoying the multiple benefits of living in well-run, well-funded cities, no longer living on the edge of drastic budget cuts and bankruptcy. Furthermore, it offers the potential to have the world's best education, health care, clean air and public spaces because we are no longer squandering billions on the futile quest for more mobility.

Peak oil

"Peak oil" was the subject of a special issue of the journal *Philosophical Transactions of the Royal Society*, where Miller and Sorrel (2014: 12) explained:

> The production of conventional oil must eventually decline to almost zero, because it is a finite resource. The phenomenon of "peak oil" derives from basic physical features of the oil resource that constrain the "shape" of the production cycle from an oil-producing region (i.e. the rate of production over time) and typically lead production to rise to a peak and then decline. But these physical features are mediated by multiple technical, economic and political factors that create a range of possibilities for the shape of the production cycle for a region and considerable uncertainty about the timing of any future peaks in production. The relative importance of these "below-ground" and "above-ground" factors varies between regions and over time and has become a central focus of dispute.

The peak oil discussion in Newman et al. (2009) adds a further layer of urgency and importance – should any further layer be needed – to the transformation of our fossil fuel, car-centric, space hungry, resource-greedy, polluting transport systems. There are multiple benefits for cities, demographics, fiscal prudence and climate change to improve how we move around, so that walking and cycling achieve the high level to be found in exemplars like Freiburg and car use can be proportionally reduced.

The links between greenhouse gas emissions, climate change, peak oil, the future of cities and resilient cities are thoroughly examined by Newman et al. (2009), who iterate the robust case for taking peak oil seriously, whether or not oil will "run out" by a specific date. The peak oil debate draws our attention to significant global problems, such as dependence on energy sources in politically volatile parts of the globe, as well as excessive reliance on fossil fuels for daily movement, and the need to make our cities more resilient. In this context "more resilient" means more walking, more cycling, higher density and a pattern of development that Newman et al. (2009) refer to as Pedestrian Oriented Development, Transit Oriented Development and Green Oriented Development.

In Germany a similar concern for re-engineering cities and promoting walking and cycling has emerged under the label "the city of short distances" (Holzapfel 2014). Davis et al. (2007: 3) quantify the benefits of increasing the amount of walking in the following terms:

> The decline in levels of walking over recent decades could and should be reversed. We estimated the potential of CO_2 reductions achievable by restoring walking where it has been substituted by car in the last 30 years: if today we reverted to the walking patterns of 1975, we would save 5.7 per cent of current emissions from passenger cars (4.10 Mt CO_2 out of 72 Mt CO_2); and if today all main drivers (amounting to more than 26 million people) reverted to the walking patterns they had before owning a car (i.e. miles walked by people with no car), 11.1 Mt of CO_2 could be saved,

amounting to 15.4 per cent of total emissions from passenger cars. This would require increased attention to changing personal travel behaviour. However the benefits of this are that walking for travel is readily available to most people, and changes could be achieved over a relatively short timescale.

So, what do we do next?

There are a number of clear interventions that can be organized and put in place, all in the same place and within a short, well-defined intervention period. One is the replacement of myth and inertia in speed limits by science – based on known relationships between deaths and injury and impact of a vehicle on a human being. In general, this means speed limits in urban areas and villages of less than 20 mph (30 km/h). A target of zero deaths and injuries in the road environment (the Swedish Vision Zero policy) can be set. Automatic speed control on vehicles can be introduced. A review of traffic law, to rebalance the system in favor of victims, is recommended – along with democratic renewal so that local residents can have a real say in design, speed, access and enforcement in their community.

Public open space and space allocation on the highways need to be redesigned: How much should be allocated for pedestrians, cyclists, public transport and cars? On the eradication of perverse subsidies: Why is the motorist subsidized? The re-engineering of planning – to be based on high accessibility, low carbon and the "city of short distances" – is necessary so that "facility erosion" would cease and would be replaced by "facility enhancement." These interventions can then be reviewed in more traditional benefit-cost terms and appreciated in terms of health.

In short, to be successful, planning after petroleum will reverse the decline in walking precipitated by automobiles to focus on this most environmentally sustainable, healthy and economic mode of transport.

References

Anon. (2011) "Passenger Transport Mode Shares in World Cities," *Journeys*: 60–70, accessed 20 September 2015 — www.lta.gov.sg/ltaacademy/doc/J11Nov-p60PassengerTransportModeShares.pdf

Appleyard, D. (1981) *Livable Streets*, Berkeley: University of California Press.

Buehler, R., Pucher, J., Merom, D. and A. Bauman (2011) "Active Travel in Germany and the United States," *American Journal of Preventive Medicine* 41(3): 241–50.

Commonwealth of Australia (2012) *Walking, Riding and Access to Public Transport: Draft Report for Discussion*, October 2012, Canberra: Department of Infrastructure and Transport, accessed 20 September – https://infrastructure.gov.au/infrastructure/pab/active_transport/files/active_travel_discussion.pdf

Davis, A., Valsecchi, C. and M. Fergusson (2007) *Unfit for Purpose: How Car Use Fuels Climate Change and Obesity*, London: Institute for European Environmental Policy.

DfT (2011) *Statistical Release – National Travel Survey: 2010*, July 2011, Great Britain: Department for Transport, accessed 20 September 2015 — www.gov.uk/government/uploads/system/uploads/attachment_data/file/8932/nts2010-01.pdf

Frank, L., Andresen, M. and T. Schmidt (2004) "Obesity Relationships with Community Design, Physical Activity and Time Spent in Cars," *American Journal of Preventive Medicine* 27: 87–96.

Gehl, J. (2010) *Cities for People*, Washington, DC: Island Press.

Hart, J. and J. Parkhurst (2011) "Driven to Excess: Impact of Motor Vehicles on the Quality of Life of Residents in Three Streets in Bristol," *World Transport Policy and Practice* 17(2): 12–30.

Holzapfel, H. (2014) *Transport and Urbanism*, New York: Routledge.

Jerrett, M., McConnell, R., Chang, C., Wolch, J., Reynolds, K., Lurman, F., Gilliland, F. and K. Berhane (2010) "Automobile Traffic Around the Home and Attained Body Mass Index: A Longitudinal Cohort Study of Children Aged 10–18 Years," *Preventive Medicine* 50: S50–8.

Kenworthy, J. (2014) "Total Daily Mobility Patterns and the Policy Implications for 43 Global Cities in 1995 and 2005," *World Transport Policy and Practice* 20(1): 41–55.

Miller, R.G. and S.R. Sorrell (2014) "The Future of Oil Supply," *Philosophical Transactions A* 372: 20130179.

Newman, P., Beatley, T. and B. Boyer (2009) *Resilient Cities: Responding to Peak Oil and Climate Change*, Washington, DC: Island Press.

Newman, P. and J. Kenworthy (1989) "Gasoline Consumption and Cities: A Comparison of US Cities with a Global Survey," *Journal of the American Planning Association* 5(1): 24–37.

Pucher, J. (2010) *Walking and Cycling for Healthy Cities*, PowerPoint Presentation in San Diego, Slide 9, accessed 3 September 2015 — http://bloustein.rutgers.edu/wp-content/uploads/2014/10/SANDIEGOAug2010.pdf

Roberts, I. and P. Edwards (2010) *The Energy Glut: The Politics of Fatness in an Overheating World*, London: Zed Books.

Sachs, W. (1992) *For the Love of the Automobile*, Oakland: University of California Press.

Schiller, P.L., Bruun, E.C. and J.R. Kenworthy (2010) *An Introduction to Sustainable Transportation: Policy, Planning and Implementation*, London: Earthscan.

Stadt Freiburg (2011) Personal communication from Dr. Dieter Wörner, Stadt Freiburg Umweltschutzamt (24 October).

——— (2010) *Umweltpolitik in Freiburg, Dezernat für Umwelt, Schule, Bildung und Gebaudemanagement*, Stadt Freiburg im Breisgau.

Transport Committee, London Assembly (2014) *Feet First: Improving Pedestrian Safety in London*, London: Greater London Authority.

UITP (2009) *Public Transport and CO2 Emissions*, Brussels: International Association of Public Transport.

United Nations (2013) *Planning and Design for Sustainable Urban Mobility: Global Report on Human Settlements*, Nairobi: United Nations Human Settlements Programme, UN-Habitat.

Vivier, J. (2006) *Mobility in Cities Database: Better Mobility for People Worldwide —Analysis and Recommendations*, Brussels: UITP (Union Internationale Transports Publique).

Whitelegg, J. (2013) *Quality of Life and Public Management: Redefining Development in the Local Environment*, Abington: Routledge.

Whitelegg, J., Haq, G., Cambridge, C., and H. Vallack (2010) *Towards a Zero Carbon Vision for UK Transport*, New York: Stockholm Environment Institute.

Woodcock, J. and 17 others (2009) "Public Health Benefits of Strategies to Reduce Greenhouse Gas Emissions: Urban Land Transport," *Lancet* 374: 1930–43.

8

CYCLING POTENTIAL IN DISPERSED CITIES

Jennifer Bonham and Matthew Burke

Numerous energy, sustainability and bicycle advocates have promoted cycling as a way to address oil-related transportation crises. Bicycle advocacy groups as diverse as California's East Bay Bicycle Coalition (Harvey 2008), Australia's Cycling Promotion Fund (CPF 2009) and the UK Sustrans (2011) have included oil vulnerability in their political campaigning for cycling investment. Even in rigorous academic work oil vulnerability has been built into assessments of the value of commuter cycling (Bauman et al. 2008: 25), and oil shock simulation research identifies cycling as a key adaptive response to petrol price increases (Watcharasukarn et al. 2012: 259).

Yet, if we look at government, there is little engagement with such views. Local, provincial and national cycling strategies generally identify health, environment and traffic congestion, rather than oil depletion, as main motivators for encouraging cycling. Where governments identify higher mode share targets for cycling (Austroads 2010; City of Austin 2009; Translink 2011), they are often well below levels recommended by peak oil advocates. So, what is a reasonable response?

This chapter offers an evaluation of cycling's role in low density suburbs, such as those found in many Australian, Canadian and US cities. Thus far, considered appraisals are absent from the literature. While no one is able to discern the exact role cycling is going to play in an oil constrained world, we are seeking to lay out a realistic picture as to where, and for what purposes, the bicycle may play its part. This chapter provides an overview of cycling in low density suburbs and then proceeds to a discussion of oil vulnerability, the potential role of cycling in vulnerable localities and the implications for policy and planning.

Cycling in dispersed cities

Australia and North America remain on par with the lowest proportion of cycling in the world (Pucher and Buehler 2008: 4). Census data continues to show low mode share (less than 2 percent) for cycling to work in all but a few of the largest cities (Loader 2014; McKenzie 2014; Statistics Canada 2013). Data on the decline in children's cycling is scarce, but US national figures indicate that the percentage of children living within one mile of their school who walked or cycled to school declined from 87 percent in 1969 to 63 percent in 2001 (Centers for Disease Control and Prevention n.d.). Statistics for major urban centers like Adelaide (Australia) show a similar trend, with the percentage of children cycling to school falling from 20.6 percent in 1985 to 5.4 percent in 2004 (Lewis et al. 2007: 422).

Despite the low mode share for commuting to work and school, cycling is gaining in popularity across a number of dimensions. Cycling is Australia's fifth most popular physical activity (Australian Bureau of Statistics 2015), and exercise and recreation account for 85 percent of journeys by bike (Austroads 2015). For at least fifteen years, sales of bicycles in Australia have outstripped those for cars, with the vast majority of these being adult bicycles (CPF 2015). Since the mid-1990s, environmental and health concerns have established a discursive space in which cycling is routinely identified as addressing environmental, public health and access issues (Forsyth et al. 2009; Horton et al. 2007). The findings of an increasing number of research projects on cycling have circulated through the media and, despite some continuing negative press, there is a growing positive popular message about cycling (Rissel et al. 2010).

It is now possible to suggest that the long decline in bicycle participation in North American and Australian cities has bottomed out (it couldn't have gone much lower). Cities from Perth to Chicago have been reporting steady growth in the number of people cycling (Chicago Department of Transportation 2012; Department of Transport 2012). Increases in mode share are observable within major metropolitan regions where cycle treatments are creating viable cycling networks (Buehler and Pucher 2012; N. McDonald 2012). This shift has been varied, with Sydney the laggard. Indeed, participation rates are almost twice as high in Melbourne as in Sydney and have been growing three times as fast (Loader 2014; Pucher et al. 2011). These marked increases in specific locations and across wide urban regions suggest a number of factors are at play across the built, social and policy environments, to turn around the fortunes of cycling. This includes the emergence of new thinking in government and in the community.

Australian and North American cities share dispersed urban forms and car-oriented suburban landscapes. Urban residential densities are surprisingly similar if using comparative measures (Mees 2010: 59). Design parameters in Australian suburbs have their own local peculiarities, but for the last half-century have not been too dissimilar to those employed in the US. Indeed, they were labeled "Austerica" by Robin Boyd (1963: 80) in the 1950s, when Australia was already the world's fourth highest car-owning nation (Gleeson 2006: 16). The auto-oriented Modernist design found in the middle- and outer-suburbs in Australia and the US, supported by land-use zoning and other policies, has been widely criticized for limiting sustainable transport and affecting human behavior, wellbeing and health (Frumkin et al. 2004: 2). Here "sustainable transport" simply means "low" or "no" emissions travel and generally includes walking, cycling and public transport.

The specific "failings" of these suburban forms include low densities, poor land-use mixes, homogenous zoning, poorly connected (non-grid) street and path networks, car-oriented retail (including mega-malls, strip malls and drive-thru restaurants), poor public transport, high posted travel speeds on roads, and limited on- and off-road bicycle facilities (Moriarty 2002; Saelens et al. 2003). These arrangements act to separate land uses, increase trip distances to the point where non-motorized travel is less viable, make for a hostile road environment for cyclists and are vulnerable to any changes in the availability of oil (Dodson and Sipe 2007; Handy et al. 2002).

Australian and US cities do differ in the strength of their central business districts (CBDs). Australian CBDs have retained their primacy such that there are concerns regarding over-centralization, with government attention now focusing on strategic decentralization (Burke, Dodson and Gleeson 2010). Australian CBDs never experienced the downturns and "white flight" of their US counterparts, with few "edge cities" apparent in the suburban landscape. The pulling power of the Australian CBDs has played its part in the resurgence of inner-city cycling, with commuter cyclists a key market segment, the bicycle promoted as a means to circumvent congestion on other modes (Austroads 2011) and allowing additional volumes of commuters into central places.

Cycling in low density suburbia

For many years, low residential densities have been used as an excuse as to why suburbanites cannot be provided with reasonable public transport services and cannot be expected to walk or cycle much. Yet the transport and land-use research underpinning this notion illuminates much more complex relationships (Krizek 2012). The first meta-study on the issue (Ewing and Cervero 2010) suggests that population (and employment) densities are only weakly associated with travel behavior, if one controls for such variables as access to destinations, land-use mix and street connectivity. Transport and land-use relationship studies specific to cycling are exceedingly few, as it is difficult to conduct research when current mode shares and trip rates are so low, especially if using regional or national household travel surveys. Regression studies suggest density may have an effect, but not always in expected ways and not always with cycling specifically isolated from walking (Guo et al. 2007; Pucher and Buehler 2006).

However, there are suburban landscapes, not too dissimilar to those in US and Australian cities, where residential densities are not much higher but bicycle mode shares are exponentially greater. The most striking examples are the outermost suburbs and peri-urban "exurbs" of many northern European cities. Despite densities lower than the average for locations such as inner Sydney, cycling rates are often orders of magnitudes higher in the Netherlands and Denmark (Martens 2004). In Ghent, the suburbs and "urban fringes" boast mode shares for cycling of around 13–14 percent (Witlox and Tindemans 2004). This suggests cycling can flourish in low density suburban settings.

What undermines cyclist safety in Australian and US suburbs?

The posted speed environment of Australian and US local streets has made cycling less safe. Across Australia, it is standard for local streets to have a 50 km/h posted street speed, whereas 30 km/h is common in German and Dutch cities. Street design also differs, with greater proportions of streets traffic-calmed in Europe than in Australia or the US, ensuring motor vehicles travel slowly. Further, the Netherlands does not simply rely on traffic-calming techniques and speed limits to facilitate cycling; they have well-developed design manuals that integrate walking and cycling into road design and detail appropriate cycling treatments for different road functions (CROW 2006, 2007).

There is another phenomenon likely to be at play. It is well established that the presence of greater numbers of cyclists and pedestrians, by and of itself, leads to an increase in safety for these modes (Elvik and Bjørnskau 2015). Jacobsen's (2003) comparison of cyclist deaths among world cities showed that, as levels of cycling increase, the rates of cyclist fatalities decline. Similar findings have been demonstrated at an intra-urban scale in Australia (Bonham et al. 2006). This inverse relation is largely explained by a prevalence effect whereby motorists are able to detect cyclists and pedestrians, and indeed any other road user, when they are more common (Jacobsen et al. 2015: 218).

Beyond urban form and structure: The importance of social and cultural factors

Although the built environment of Australian and US suburbia presently fails to support non-motorized travel, bicycle participation rates are not solely influenced by built form and infrastructure provision. The social environment is heavily implicated. This takes a number of forms. First, the marginalization of walking and cycling in these cities is not simply about the decline in bike riders and pedestrians but is also related to knowledge we produce about our travel (inclusions and exclusions), and how these travel practices and the people who perform them are positioned in relation to each other and addressed in our urban environments (Bonham 2006). The very tools used to create knowledge about mobility, developed largely in the post-WWII period, have operated to normalize the automobile and offer little assistance in understanding or progressing social, rather than individual, change (Schwanen et al. 2011). Knowledge about urban traffic and transport establishes a hierarchy of movement suggesting that the automobile is the only vehicle that can move people through the urban environment "efficiently" in terms of time (Bonham 2006). Cycling and walking are located at the bottom of the transport hierarchy, practices constituted as slow and often disorderly, and only seem "competitive" in congested, compacted environments.

This hierarchy has specific consequences in decision making. The way in which planners perceive the "choices" that people make to use specific modes is often operationalized according to a function of the user's preferences and the relative costs of the different available modes of travel. This approach follows modelers of micro-economic disaggregate models, focused myopically on a particular type of utility, and fails to consider the other values and benefits pertaining to cycling, including equity (Frank et al. 2003: 108–10) and health (Rissel 2015). However, these values are being embedded in the growing production of alternative "knowledges" about urban travel that challenge existing mobile identities (or subject positions) and produce "new" ones. For example, cycling and walking practices have become characteristics of the "healthy" (or "active") traveler (Mulley et al. 2013; Nazelle et al. 2011). The emphasis on the production of knowledge and the identities it makes available will be a key element in bringing about cultural change (Bonham and Bacchi 2013).

Oil depletion, vulnerability, walking and cycling

If cycling is presently marginalized, and only starting to become a significant mode in the inner areas of cities in Australia, Canada and the US, then why is it being promoted as a sure-fire "solution" to any impending oil crisis? Is the bicycle really going to "save the earth" as the aforementioned advocates suggest? These two questions are helpful in unpacking the dimensions of what is considered a difficult problem, a "problem" of the suburbs.

The research by Dodson and Sipe (2007) on the spatial dimensions of oil vulnerability in Australian cities has clearly shown that households with greater vulnerability to fuel price increases are located in the middle and outer suburbs. Yet, in the middle and outer suburbs of dispersed Australian cities, cycling has low participation rates and appears to have lower priority in policy and programs and little safe infrastructure. Local governments urging an increase in cycling as a response to peak oil tend to be in the inner suburbs, such as Moreland City Council (2011). The US appears to be further along than Australia, as many city governments acknowledge the impacts of peak oil on transport affordability and recommend cycling as a mechanism to address these impacts (City of Austin Energy Depletion Risks Task Force Report 2009). Nonetheless, in both the US and Australia a mismatch continues as localities set to bear the greatest impacts of an oil constrained future have limited scope to ameliorate those impacts through transport choice.

Cycling in prospective peak oil futures

Dennis and Urry (2009: 151–60) paint a picture of three familiar urban future scenarios. Each scenario involves distinct social and economic changes that are a response to climate change and related concerns, such as peak oil. The first, "local sustainability," envisages a kind of austere localized eco-communalism, with cycling, walking and low-technology public transport replacing the overconsumption of mass car travel. Alternatively, "barbarization" describes a dystopia, along the lines of *Mad Max* (Miller 1979), where we imagine inhabitants might use recycled, mutant bicycles to navigate the walled and gated enclaves into which they gather. Finally, in "digital networks of control," electric vehicles provide a technological fix, under some form of intelligent flexible transport system, with bicycles playing a supplementary role. Research on switching of vehicle fleets suggests this latter scenario will be impossible to achieve in a short time frame, with impacts again most likely falling heavier on the poor and marginalized of suburbia (Dodson et al. 2010).

These extreme scenarios provoke us to think harder about how suburbs will work under peak oil. We recommend a fourth, more pragmatic, scenario that realizes the opportunities of our present landscapes and technologies. It sees a transition toward suburban features and communities created in places like Ghent, through changes right across the built environment, the policy environment and the social environment that, in turn, generate strong public transport and non-motorized travel mode shares. This transition includes some limited urban restructuring (including in retail and schooling), with rapid short-term changes in transport networks, policy and priority (especially public transport networking, harnessing car-share and ride-share programs, and driverless vehicle technologies). Any demand destruction for the car is used to free up road space for walking, cycling and on-road transit. Such environmental changes are feasible in the short term and could deliver much of the necessary mobility and access for low density populations, without dependence on high volumes of oil. The pragmatism of this scenario will be influenced by just how difficult it is to retrofit particular suburban spaces – some of which will be much harder work than others.

Bicycle trip types in an oil crisis

Under the scenarios outlined, the bicycle is predicted to play specific roles, raising questions as to how transport and urban planners can better facilitate cycling. Bicycles will supply much local to mid-range trip making, and play a role as a complementary mode for access to and from mass public transport. Beyond short trips, the bicycle will also play an increased role in delivering longer trips, issues to which we now turn.

While acknowledging that distance influences willingness to cycle in complex ways (Wuerzer and Mason 2015), we suggest that local trips up to 5 km provide good prospects for mode shift. A significant percentage of all household trips in Brisbane are within a 1 km walk or a 3 km bike ride (Burke and Brown 2007). Of course, this differs by trip type, suggesting that some key trips, such as those to local activity precincts (schools, retail-café-commercial strips and community facilities) can be readily facilitated through minor changes to the regulation of streets (such as reducing speeds, limiting on-street parking and cycle parking at destinations) and by addressing barriers such as access across arterial roads (Koorey 2015). Economic forces under peak oil may well reduce the viability of mega-malls and "super schools" (extremely large schools servicing subregional catchments). A return to a more localized distribution of both grocery shopping and primary and secondary schooling would bring a greater proportion of such trips within cycling and walking distance.

There is also significant opportunity in terms of journeys-to-work. Analysis of 2011 Australian Census data shows a median commute to work distance of 2.5 km for Melbourne's inner suburbs (DEDJTR 2015). This finding suggests that if we can align environmental factors, there is significant potential to increase cycling beyond the current metropolitan averages. Wight and Newman (2010) argue that decentralized centralization of employment and other activities would also reduce transport energy use. Such a strategy has merit, but only if there is planned decentralization to activity centers, and if cycling and walking are given priority in road infrastructure design and regulation to prioritize access by these modes.

Mid-range and longer trips

The bicycle will be an option for some mid-range (5–15 km) trips, but it will play a lesser role in longer journeys. Although a high proportion of cycling journeys in Northern European cities are certainly less than 5 km, figures reported by Fietsberaad (2009: 10–11) show that, in the Netherlands, 34 percent of journeys up to 7.5 km and 15 percent of 7.6–15 km journeys are made by bicycle. Research from Sweden indicates that the bicycle is "competitive with car and public transport travel up to a distance of 10 km from the city center" (Eriksson et al. cited in Gustafson and Archer 2013: 286). The average trip distances for users of Brisbane's King George Square Cycle Center are 10 km, with some traveling more than twice as far (Burke, Sipe and Hatfield 2010: 20). Recognizing the viability of the bike for mid-range journeys, the City of Vancouver is aiming for a bicycle mode share of 15 percent of all journeys up to 8 km (Translink 2011: 27). Facilitating these mid-range trips requires high-quality, metropolitan-wide cycle network planning and will entail the systematic review and implementation of treatments – from changing posted speeds to installing segregated on-road tracks – to accommodate cyclists in safety and comfort (Koorey 2015).

In the middle and outer suburbs of Australian and US cities, where household (particularly journey-to-work) trips often exceed 15 km, we expect public transport to provide more for long-distance travel. The extremely high mode shares for cycling in some Dutch and northern German cities (such as Groningen), where bicycle journey-to-work mode shares are particularly large, might relate in part to those city's more modest public transport services and to their small city size. In more dispersed but public transport–focused cities, bicycle access trips to transit will be more important. A metropolitan average of 15 percent of all trips by bicycle alone is possible in dispersed cities with extensive public transport, albeit with lower rates in outer suburbia where commutes tend to be longer.

The access trip to and from public transport

Public transport systems will change under oil scarcity, with a likely focus on public transport networking in cities where this does not yet presently occur. Public transport networking – knitting together individual routes and modes to form a network that services diverse spatial and temporal travel patterns – as detailed by Mees (2010) – appears to produce higher public transport mode shares by unlocking public transport's potential to connect more origins and destinations in the city, especially in middle suburbs and outer suburbia. This is notably lacking in most Australian, New Zealand and US cities. In integrated public transport networks there may be less need for longer-distance bicycle trips to key destinations, particularly for cross-suburban trips, as public transport extends its services. But bicycle travel will play an important role as a feeder mode (Pucher and Buehler 2012) and well-designed end-of-trip facilities at public transport stops and stations will be crucial to its success.

Questions also arise regarding the need and supports for bicycles at the "other end" of public transport trips. In dispersed cities, and especially for cities where commerce, retail, health and other services are not clustered into designated activity centers, there may be more use for the bicycle to go that "last kilometer" or two, especially in the short term before any significant change to urban structures is possible under oil scarcity. There are three ways to achieve this. The first is storage of "second bikes" at central rail stations, as pursued in Amsterdam (Pucher and Buehler 2007: 19). The second is on-board carriage of bicycles on public transport. The third is public bicycle hire schemes, such as those operating in Paris, Barcelona and London, where bike hire stations are found throughout the city (Beroud and Anaya 2012). Alternatively, hire schemes located at railway stations, like *OV-Fiets* in the Netherlands, provide subscription or single-use hire for rail travelers. Hire schemes combined with bike storage and repair services may provide opportunities for existing bike shops to expand their business.

Though on-board carriage policies are often promoted by cycle advocates and are appropriate for specific trip types in some circumstances, this feature is not considered necessary in many cycling nations. For instance, the Netherlands does not encourage bicycles on trains because they have significant secure bike parking available at train stations. Japan has similar policies. Perhaps where public transport systems are as poor as they presently are in the outskirts of most Australian and US (yet not all Canadian) cities, there may be stronger consideration given to on-board carriage of bicycles. Regardless, if cities do opt for carriage of bicycles on public transport, it will be necessary to retrofit train carriages and install bus racks on the front of buses, to properly accommodate cyclists and minimize inconvenience to other passengers. The option for city centers is the public bicycle hire or share schemes that have proliferated in the cities of North America (Bixi in Montreal, Bikeshare in Washington) and, to a lesser extent, Australia (City Bikes in Adelaide, City Cycle in Brisbane) over the past decade. In Australia, a compulsory helmet law has threatened to undermine unplanned bicycle use so that Melbourne Bike Share introduced free helmet hire (Melbourne Bike Share 2015).

Which cyclists must be accommodated for effective mitigation of peak oil?

A simulation study of university staff and student responses to oil shocks in Christchurch (NZ) found that 33 percent of participants would shift to cycling (Watcharasukarn et al. 2012). These results bode well for the future, but a broader range of users will need to be accommodated if the mode is to provide any mitigating effect for transport in an oil crisis. The paucity of cycling-related data means the nature of cyclists and cycling in Australian and US suburbs remains largely unexplored, yet it is clear that inner urban male commuters are over-represented in the current bicycle travel market. Others need to be accommodated to make any difference under peak oil.

First, middle- and outer-suburban workers must be targeted. Mapping of intra-urban differences of commuter cycling across Adelaide, Brisbane, Melbourne and Sydney gives an overall picture of higher rates in inner (rather than middle or outer) suburban areas. However, this pattern is not straightforward. In Adelaide, the statistical local areas included in the "North-West Corridor," stretching 16 km from the city center to Port Adelaide, have higher levels of cycling for the journey-to-work than some statistical local areas much closer to the city center (Bonham and Suh 2008). Higher employment densities in both the inner and middle suburbs along this corridor are likely to explain these higher levels of cycling, at least in part, by creating better relationships between jobs and housing. Getting more middle- and outer-suburban workers onto bicycles, at least as part of a trip involving public transport, would be a priority.

Second, in Australian, Canadian and US cities, women generally have much lower levels of cycling than men, especially for the journey-to-work, where figures consistently show that only one in four commuter cyclists is a woman – slightly lower in Australia and slightly higher in Canada (Pucher et al. 2011). However, it is well known that these national averages hide differences in the ratio of female to male cyclists both across and within cities. Women's willingness to cycle has been demonstrated in metropolitan local government areas where, with high overall rates of cycling, 30–40 percent of cyclists are women (Garrard et al. 2012). These high levels of female cycling usually occur in the inner suburbs and contrast markedly with lower levels of cycling found in middle suburbs and outer suburbs, where ratios of males to females are closer to national averages. The reasons for women's lower rates of cycling are complex (Garrard et al. 2012), but the gendering of cycling as masculine (associated with sport, high-level fitness and bravado in traffic) seems to be eroded where everyday cycling is accommodated (Bonham et al. 2015). Under peak oil, we will need to learn from the low traffic speeds in Unley (Adelaide), the inverted road hierarchy in Yarra (Melbourne) and investments into appropriate on- and off-road infrastructure, policy and programs such as exist in Portland and Vancouver (A. McDonald 2012; Pucher et al. 2011).

Third, children's cycling will be particularly important. Although children's bicycle use – generally measured by cycling to school – has been declining, in Australia it seems that children in the middle and outer suburbs are still more likely to cycle than adults (Transport NSW 2003; VicRoads 1999). Two important issues emerge from this finding. First, the lower level of adult cycling is likely to be linked, at least in part, to lifecycle stage (Jones et al. 2014). The presence of children aged 14 years and under in a household has been negatively related to cycling for the journey-to-work for both men and women (Bonham and Suh 2008). Given that families with young children tend to be concentrated in the middle suburbs and outer suburbs of Australian cities, it appears that juggling familial and work responsibilities – and the complex travel patterns that result – make cycling a difficult choice for parents. Second, while parents express concern about distances and safety during travel, parental convenience is also a major factor in parents chauffeuring children (McDonald and Aalborg 2009). Parents may chauffeur children relatively short distances if it is part of a trip chain. Peak oil may force us to facilitate children's independent bicycle travel (Sharpe and Tranter 2010), "freeing up" parents to cycle as well, and providing children with bicycle awareness throughout their lives.

Finally, seniors are over-represented in Dutch bicycle markets, yet are very much the minority in Australian and US cities (Pucher and Buehler 2008: 10). Mobility for seniors may be particularly impacted in an oil crisis and increased cycling (and walking) participation may help increase not only access to goods and services but also contribute to critical social engagement and participation across age groups (Leyden 2003).

Which bikes?

An oil crisis might have implications for the bicycle fleet through the type of bikes available to consumers and deemed to meet Australian and North American demands and standards. This includes both powered and non-powered bicycles. Power-assisted bikes (specifically electric bikes) can be useful for people with some types of disabilities and those recovering from illness and injuries or trying to build fitness (Johnson and Rose 2013; Lumb 2013). Power-assisted bikes provide riders with options when cycling in difficult conditions (such as headwinds, hills and long distances) and are growing in popularity in Europe, China and Japan. Recent studies of electric bike owners in the US and Australia found a major motivation for purchasing an electric bike was to replace car trips with cycle trips (Dill and Rose 2012). Such research points to the significant role electric bikes will play in an oil constrained future.

A return to utilitarian cycling in an oil crisis will also require cyclists to carry items such as clothing, laptops, papers, books, and shopping and other goods (Daley et al. 2007). Bikes equipped with panniers can address this problem, but step-through (upright) bikes, complete with baskets and carriers, are gradually making their way back into mainstream cycle retailing in Australia and North America. For much of the last three decades, "mountain bikes" and road racing styles have dominated adult bicycle sales in Australia, with mountain bikes strong in discount and department store sales. Until recently, upright bikes of Japan and much of Europe have not been seen in great numbers while the cargo bikes of the type used in northern Europe are still quite rare.

Despite the apparent lack of research on the phenomenon, the proliferation of "Cycle Chic" websites, both in North America over the past decade and in Australia more recently, has occurred alongside the re-emergence of utility bikes in cities from Austin to Vancouver and Perth to Brisbane. These bikes provide for carrying goods and generally allow riders to sit upright rather than lean forward. The upright position seems to afford greater visibility for the cyclist and, perhaps, for other road users to see the cyclist but this observation needs to be tested through research. However, if the upright position does allow greater visibility and awareness it may overcome the fear and insecurity many people feel in relation to bike riding.

Implications for policy and planning

There are many implications that emerge from this analysis, some of which raise more questions than answers. Nonetheless, it is clear that under any oil scarcity scenario, cycling is likely to play a number of significant roles, especially for trips up to 10 km. Cycling will probably be less important than mass public transport for long-distance, cross-metropolitan journeys, but it will be an important access mode to and from transit stops and stations. With the many other benefits of cycling, in terms of public health and urban amenity, the low cost of providing for cycling makes it an obvious way forward. Planning can facilitate the mode transition, and reduce the shock of predicted oil shortages and their transport impacts.

A planning priority is to create more appropriate policy and built environments. This task involves necessary strategic planning, preparing targeted cycling strategies across infrastructure, policy and programs. Cycle network development must integrate metropolitan and local cycle networks with public transport nodes to unlock multiple destinations. Infrastructure development must be in line with network planning and investment rather than ad hoc link development (Parkin and Koorey 2012). The success of cycling strategies and network planning will also rely on embedding cycling into local level development policies and regulations (Bell and Ferretti 2015).

Broader policy environments are also important. Decision makers will need to go beyond new infrastructure to consider the litany of low cost improvements that are readily available. These improvements include making for safer cycling by reducing posted street speeds, providing improved regulation for electric bicycles and micro-electrics, reclaiming road space from vehicle lanes and car parking, and rethinking compulsory helmet laws (in Australia and New Zealand).

Knowledge and skills development are core elements of facilitating cycling. Tertiary institutions and professional associations will need to build capacity in built environment professionals to equip urban planners, designers, and traffic and transport engineers to design, construct and regulate environments that enable cycling (Rose 2015). Improving research methods and data collection will also help with the transition to new mobility futures. At present, the methods that capture travel, such as household travel surveys and questions on journey-to-work in household censuses, have many limitations that underreport cycling. Travel surveys must include a broader range of questions, such as health, as journeys do not simply concern getting from "A to B" but play multiple roles in an individual's life.

Improved data may help track the transition and allow for improved responses by planners as cycling increases. Researchers must seek to understand cultural differences in different parts of the city in relation to cycling. Socio-cultural studies are required to open up the possibility of cycling rather than reproducing non-cycling norms. One of the most obvious discourses relates to children's independent journeys, given their potential to not only replace children's motorized travel but their potential to release parents from broader motorized travel patterns.

Finally, in the long term, there is still likely to be a strong role for broader urban environmental modification to restructure our cities and discourage long commutes and other trips in a post-oil future. Yet much can be done in the short term to gain the meaningful mitigation impacts the bicycle could provide as peak oil's effects emerge.

References

Australian Bureau of Statistics (2015) *Participation in Sport and Physical Recreation, Australia, 2013–2014*, 4177.0 released 18 February, accessed 20 October 2015 — www.abs.gov.au/AUSSTATS/abs@.nsf/DetailsPage/4177.02013–14?OpenDocument

Austroads (2015) *National Cycling Participation Survey – National Results*, Sydney: Austroads.

——— (2011) *Australian Cycling Participation: Background*, Sydney: Austroads.

——— (2010) *National Cycling Strategy 2011–2016*, Sydney: Austroads.

Bauman, A., Rissel, C., Garrard, J., Ker, I., Speidel, R. and E. Fishman (2008) *Cycling: Getting Australia Moving: Barriers, Facilitators and Interventions to Get More Australians Physically Active Through Cycling*, Melbourne: Cycling Promotion Fund.

Bell, W. and D. Ferretti (2015) "What Should Planners Know About Cycling?" in J. Bonham and M. Johnson (eds) *Cycling Futures*, Adelaide: University of Adelaide Press, 321–57.

Beroud, B. and E. Anaya (2012) "Private Interventions in A Public Service: An Analysis of Public Bicycle Schemes," in J. Parkin (ed.) *Cycling and Sustainability*, Bingley: Emerald Group, 269–301.

Bonham, J. (2006) "Transport: Disciplining the Body that Travels," *Sociological Review* 54(Supplement 1): 57–74.

Bonham, J. and C. Bacchi (2013) "Cycling 'Subjectivities' in *Ongoing-Formation:* Interviews as Political Interventions," paper presented at the *Foucault and Mobilities Research Symposium*, University of Lucerne, 6–7 January, Lucerne (Switzerland) (access, authors).

Bonham, J., Bacchi, C. and T. Wanner (2015) "Gender and Cycling: Gendering Cycling Subjects and *Forming* Bikes, Practices and Spaces as Gendered Objects," in J. Bonham and M. Johnson (eds) *Cycling Futures*, Adelaide: University of Adelaide Press, 179–202.

Bonham, J., Cathcart, S., Petkov, J. and P. Lumb (2006) "Safety in Numbers: A Strategy for Cycling?" *Proceedings of the 29th Australasian Transport Research Forum*, 27–29 September, Gold Coast: Queensland Government.

Bonham, J. and J. Suh (2008) "Pedalling the City: Intra-Urban Differences in Cycling," *Road and Transport Research* 17(4): 25–40.

Boyd, R. (1963) *The Australian Ugliness*, Ringwood: Penguin Books.

Buehler, R. and J. Pucher (2012) "Cycling to Work in 90 Large American Cities: New Evidence on the Role of Bike Paths and Lanes," *Transportation* 39: 409–32.

Burke, M. and A.L. Brown (2007) "Active Transport in Brisbane: How Much is Happening and What are its Characteristics?" paper presented at *State of Australian Cities National Conference 2007*, 28–30 November, Adelaide.

Burke, M., Dodson, J. and B. Gleeson (2010) *Employment Decentralisation in SEQ: Scoping the Transport Impacts*, Brisbane: Urban Research Program, Griffith University.

Burke, M., Sipe, N. and E. Hatfield (2010) *Evaluation of King George Square Cycle Centre*, Research Paper 30, Brisbane: Urban Research Program, Griffith University, accessed 24 August 2016 — www.academia.edu/2873467/Evaluation_of_King_George_Square_Cycle_Centre

Centers for Disease Control and Prevention (n.d.) *Kids Walk-to-School*, Atlanta: Centers for Disease Control and Prevention, US Department of Health and Human Services.

Chicago Department of Transportation (2012) *Chicago Streets for Cycling Plan 2020*, Chicago: Chicago Department of Transportation.

City of Austin (2009) *Austin 2009 Bicycle Plan Update*, Austin: City of Austin.

City of Austin Energy Depletion Risks Task Force (2009) *The City of Austin Energy Depletion Risks Task Force Report*, Austin: City of Austin.

CPF (2015) *Bicycle Sales Boom but Cycling Flat-Lining*, Melbourne: Cycling Promotion Fund.

——— (2009) *Oil Depletion*, Melbourne: Cycling Promotion Fund.

CROW (2007) *Design Manual for Bicycle Traffic*, Delft: Faculty of Architecture, Delft University of Technology.

——— (2006) *Urban Design and Traffic: A Selection from Bach's Toolbox*, Delft: Faculty of Architecture, Delft University of Technology.

Daley, M., Rissel, C. and B. Lloyd (2007) "All Dressed Up and Nowhere to Go? A Qualitative Research Study of the Barriers and Enablers to Cycling in Inner Sydney," *Road and Transport Research* 16: 42–51.

DEDJTR (2015) *Census 2011 Analysis*, Department of Economic Development, Jobs, Transport and Resources, Melbourne: Victoria State Government.

Dennis, K. and J. Urry (2009) *After the Car*, Cambridge, MA: Polity.

Department of Transport (2012) *Western Australian Bicycle Network Plan*, Perth: Government of Western Australia.

Dill, J. and G. Rose (2012) "Electric Bikes and Transportation Policy: Insights from Early Adopters," *Transportation Research Record* 2314: 1–6.

Dodson, J., Li, T. and N. Sipe (2010) "Urban Structure and Socio-Economic Barriers to Consumer Adoption of Energy Efficient Automobile Technology in a Dispersed City: A Case Study of Brisbane, Australia," *Journal of Transportation Research Board* 2151: 111–18.

Dodson, J. and N. Sipe (2007) "Oil Vulnerability in the Australian City: Assessing Socioeconomic Risks from Higher Urban Fuel Prices," *Urban Studies* 44(1): 37–62.

Elvik, R. and T. Bjørnskau (2015) "Safety-in-numbers: A Systematic Review and Meta-Analysis," *Safety Science* — http://dx.doi.org/10.1016/j.ssci.2015.07.017

Ewing, R. and R. Cervero (2010) "Travel and the Built Environment: A Meta-Analysis," *Journal of the American Planning Association* 76(3): 265–94.

Fietsberaad (2009) *Cycling in the Netherlands*, Den Haag: Ministerie Verkeer en Waterstaat.

Forsyth, A., Krizek, K. and D. Rodriguez (2009) "Hot, Congested, Crowded and Diverse: Emerging Research Agendas in Planning," *Progress in Planning* 71: 153–205.

Frank, L.D., Engelke, P.O. and T.L. Schmid (2003) *Health and Community Design. The Impact of the Built Environment on Physical Activity*, Washington, DC: Island Press.

Frumkin, H., Frank, L.D. and R. Jackson (2004) *Urban Sprawl and Public Health: Designing, Planning, and Building for Healthy Communities*, Washington, DC: Island Press.

Garrard, J., Handy, S. and J. Dill (2012) "Women and Cycling," in J. Pucher and R. Buehler (eds) *City Cycling*, Cambridge: Massachusetts Institute of Technology Press, 211–34.

Gleeson, B. (2006) *Australian Heartlands: Making Space for Hope in the Suburbs*, Crows Nest, NSW: Allen & Unwin.

Guo, J., Bhat, C. and R. Copperman (2007) "Effect of the Built Environment on Motorized and Non-Motorized Trip Making: Substitutive, Complementary, or Synergistic?" *Transportation Research Record* 2010: 1–11.

Gustafsson, L. and J. Archer (2013) "A Naturalistic Study of Commuter Cyclists in the Greater Stockholm Area," *Accident Analysis and Prevention* 58: 286–98.

Handy, S.L., Boarnet, M.G., Ewing, R. and R.E. Killingsworth (2002) "How the Built Environment Affects Physical Activity: Views from Urban Planning," *American Journal of Preventive Medicine* 23(2 Supplement 1): 64–73.

Harvey, B. (2008) "Peak Oil and Bicycling," *East Bay Bicycle Coalition* (site), accessed 20 October 2015 — www.ebbc.org/node/2042

Horton, D., Cox, P. and P. Rosen (2007) "Cycling and Society," in D. Horton, P. Rosen and P. Cox (eds) *Cycling and Society*, Aldershot: Ashgate, 1–23.

Jacobsen, P. (2003) "Safety in Numbers: More Walkers and Bicyclists, Safer Walking and Bicycling," *Injury Prevention* 9: 205–9.

Jacobsen, P., Ragland, D. and C. Komanoff (2015) "Safety in Numbers for Walkers and Bicyclists: Exploring the Mechanisms," *Injury Prevention* 21(4): 217–20.

Johnson, M. and G. Rose (2013) "Electric Bikes – Cycling in the New World City: An Investigation of Australian Electric Bike Owners and the Decision Making Process for Purchase," *Australasian Transportation Research Forum Proceedings*, 2–4 October, Brisbane.

Jones, H., Chatterjee, K. and S. Gray (2014) "A Biographical Approach to Studying Individual Change and Continuity in Walking and Cycling Over the Life Course," *Journal of Transport and Health* 1(3): 182–9.

Koorey, G. (2015) "Spaces for Cycling," in J. Bonham and M. Johnson (eds) *Cycling Futures*, Adelaide: University of Adelaide Press, 251–87.

Krizek, K. (2012) "Cycling, Urban Form and Cities: What do We Know and How Should We Respond?" in J. Parkin (ed.) *Cycling and Sustainability*, Bingley: Emerald, 111–30.

Lewis, N., Dollman, J. and M. Dale (2007) "Trends in Physical Activity Behaviours and Attitudes Among South Australian Youth Between 1985–2004," *Journal of Science and Medicine in Sport* 10: 418–27.

Leyden, K. (2003) "Social Capital and the Built Environment: The Importance of Walkable Neighbourhoods," *American Journal of Public Health* 93(9): 1546–51.

Loader, C. (2014) *What Does the Census Tell Us About Cycling?* (post), *Charting Transport* (site), 27 January, accessed 20 October 2015 — http://chartingtransport.com/category/mode-share/

Lumb, P. (2013) "Everybody's Cycling. But What about Australians with Disabilities? What Are the Prospects for a More Inclusive and Diversely Mobile Society?" in *Australian Cycling Conference 'Everybody's Cycling?': Proceedings of the Fifth Australian Cycling Conference*, Adelaide: Australian Cycling Conference, 61–70.

Martens, K. (2004) "The Bicycle as a Feedering Mode: Experiences from Three European Countries," *Transportation Research Part D: Transport and Environment* 9(4): 281–94.

McDonald, A. (2012) "A Car Is 1.9m Wide: How Much Extra Space Does It Really Need?" in *Australian Cycling Conference Proceedings of the Fourth Australian Cycling Conference*, Adelaide: Australian Cycling Conference.

McDonald, N. (2012) "Children and Cycling," in J. Pucher and R. Buehler (eds) *City Cycling*, Cambridge: Massachusetts Institute of Technology Press, 235–56.

McDonald, N. and A. Aalborg (2009) "Why Parents Drive Children to School," *Journal of the American Planning Association* 75(3): 331–42.

McKenzie, D. (2014) *Modes Less Traveled – Bicycling and Walking to Work in the United States: 2008–2012*, Washington, DC: US Census Bureau.

Mees, P. (2010) *Transport for Suburbia: Beyond the Automobile Age*, London: Earthscan.

Melbourne Bike Share (2015) Melbourne Bike Share (site), accessed 20 October — www.melbournebikeshare.com.au

Miller, G. (dir.) (1979) *Mad Max* (film), Footscray, Victoria: Mad Max Pty. Ltd.

Moreland City Council (2011) *Moreland Bicycle Strategy*, Moreland: Moreland City Council.

Moriarty, P. (2002) "Environmental Sustainability of Large Australian Cities," *Urban Policy and Research* 20(3): 233–44.

Mulley, C.R., Tyson, P., McCue, C., Rissel, C. and C. Munro (2013) "Valuing Active Travel: Including the Health Benefits of Sustainable Transport in Transportation Appraisal Frameworks," *Research in Transportation Business & Management* 7: 27–34.

Nazelle, A., Nieuwenhuijsen, M.J., Antó, J.M., Brauer, M., Briggs, D., Braun-Fahrlander, C. and E. Lebret (2011) "Improving Health Through Policies that Promote Active Travel: A Review of Evidence to Support Integrated Health Impact Assessment," *Environment International* 37(4): 766–77.

Parkin, J. and G. Koorey (2012) "Network Planning and Infrastructure Design," in J. Parkin (ed.) *Cycling and Sustainability*, Bingley: Emerald, 131–60.

Pucher, J. and R. Buehler (2012) "Integration of Cycling with Public Transportation," in J. Pucher and R. Buehler (eds) *City Cycling*, Cambridge: Massachusetts Institute of Technology Press.

——— (2008) "Making Cycling Irresistible: Lessons from the Netherlands, Denmark and Germany," *Transport Reviews* 28(4): 495–528.

——— (2007) "At the Frontiers of Cycling: Policy Innovations in the Netherlands, Denmark, and Germany," *World Transport Policy and Practice*: 13(3): 8–56.

——— (2006) "Why Canadians Cycle More than Americans: A Comparative Analysis of Bicycling Trends and Policies," *Transport Policy* 13(3): 265–79.

Pucher, J., Buehler, R. and M. Seinen (2011) "Bicycling Renaissance in North America? An Update and Re-Appraisal of Cycling Trends and Policies," *Transportation Research Part A* 45: 451–75.

Pucher, J., Garrard, J. and S. Greaves (2011) "Cycling Down Under: A Comparative Analysis of Bicycling Trends and Policies in Sydney and Melbourne," *Journal of Transport Geography* 19(2): 332–45.

Rissel, C. (2015) "Health Benefits of Cycling," in J. Bonham and M. Johnson (eds) *Cycling Futures*, Adelaide: University of Adelaide Press, 43–62.

Rissel, C., Bonfiglioli, C., Emilsen, A. and B. Smith (2010) "Representations of Cycling in Metropolitan Newspapers – Changes Over Time and Differences Between Sydney and Melbourne, Australia," paper presented at *Australian Cycling Conference* 18–19 January, Adelaide: University of Adelaide.

Rose, G. (2015) "Teaching Australian Civil Engineers about Cycling," in J. Bonham and M. Johnson (eds) *Cycling Futures*, Adelaide: University of Adelaide Press, 303–20.

Saelens, B., Sallis, J. and L. Frank (2003) "Environmental Correlates of Walking and Cycling: Findings from the Transportation, Urban Design, and Planning Literatures," *Annals of Behavioural Medicine* 25(2): 80–91.

Schwanen, T., Banister, D. and J. Anable (2011) "Scientific Research About Climate Change Mitigation and Transport: A Critical Review," *Transportation Research Part A* 45A(10): 993–1006.

Sharpe, S. and P. Tranter (2010) "The Hope for Oil Crisis: Children, Oil Vulnerability and (In)Dependent Mobility," *Australian Planner* 47(4): 284–92.

Statistics Canada (2013) "Proportion of Workers Commuting to Work by Car, Truck or Van, by Public Transit, on Foot, or by Bicycle, Census Metropolitan Areas, 2011," accessed 20 October 2015 — www12.statcan.gc.ca/nhs-enm/2011/as-sa/99–012-x/2011003/tbl/tbl1a-eng.cfm

Sustrans (2011) "Written Evidence Submitted by Sustrans," submission made to the UK Parliament *Environmental Audit Committee*, 20 April, accessed 20 October 2015 — www.publications.parliament.uk/pa/cm201012/cmselect/cmenvaud/878/878vw21.htm

Translink (2011) *Cycling for Everyone: A Regional Cycling Strategy for Metro Vancouver*, Vancouver: Translink.

Transport NSW (2003) *Cycling in Sydney: Bicycle Ownership and Use*, Sydney: Transport NSW.

VicRoads (1999) *Cycling in Melbourne: Bicycle Ownership, Use and Demographics 1997–1999*, Melbourne: VicRoads.

Watcharasukarn, M., Page, S. and S. Krumdieck (2012) "Virtual Reality Simulation Game Approach to Investigate Transport Adaptive Capacity for Peak Oil Planning," *Transportation Research Part A* 46: 348–67.

Wight, W. and P. Newman (2010) "Petroleum Depletion Scenarios for Australian Cities," *Australian Planner* 47(4): 232–42.

Witlox, F. and H. Tindemans (2004) "Evaluating Bicycle-Car Transport Mode Competitiveness in an Urban Environment," *World Transport Policy and Practice* 10(4): 32–42.

Wuerzer, T. and S. Mason (2015) "Cycling Willingness: Investigating Distance as a Dependent Variable Amongst College Students," *Applied Geography* 60: 95–106.

9

CHILDREN'S ACTIVE TRANSPORT

An upside of oil vulnerability?

Scott Sharpe and Paul Tranter

Before planning for life after petroleum, it is useful to examine some more nuanced but far-reaching effects of a fossil fuel society and the structures and habits to which it gives rise. This chapter approaches this problem from the perspective of the lives of children, specifically their mobility in post-industrial, highly urbanized societies. Much has been made of the costs borne by children in a world fueled by petroleum: international studies into patterns of decline in children's active transport and independent movement indicate the negative influence of this decline on cognitive development (Joshi et al. 1999; Risotto and Tonucci 2002), emotional wellbeing (Hillman 1993), social participation (Groves 1997; Prezza et al. 2009) and physical health (Department of Health 2004).

While these problems form the context out of which the analyses of this chapter arise, our concern is with the nexus between the problem of oil vulnerability and the idea of the child-friendly city. Following even the most conservative estimates, today's children are likely to experience the peak in global oil production in their lifetimes, feeling the effects of a life without cheap oil for a longer duration than other citizens. In the context of highly suburban cities, this presents complex and urgent challenges to our habits of apprehending the place of children within society at large.

In this chapter we argue that, despite some of the doom-saying that attends the discourse of peak oil, the very fact of the end of cheap oil provides an opportunity for us to explore other modes of living. We do this with particular reference to children in the urban setting. We begin our analysis by examining some of the deleterious effects of an oil dependent society on the lives of children, namely a reduction of their opportunities for active transport and the associated restrictions of movement as independent citizens. Several decades of cheap available oil have had a series of implications for children living in cities. There has been a marked decrease in children's independent mobility (CIM), brought about by urban sprawl that cities built for the automobile, rather than the pedestrian, necessitate. This, in turn, has made the streets more dangerous for children, as cars rather than cyclists and pedestrians top the priority list for urban planners.

We show that this skewed priority results in legislating for unsafe traffic speeds and some surprising paradoxes, including a misplaced belief in the time efficiency of car transport and the self-fulfilling prophesy that we must protect our children by driving them to organized activities. Our argument does not call for a return to CIM, since this necessitates a conceptual separation of children from adults, which results in car ownership being seen as a right of passage to adulthood, and does nothing to allay parent's justifiable fears of traffic danger. Rather, we suggest that a more general push for active transport for both adults and children may give rise to child- and adult-friendly cities.

Children's independent mobility

Historically, the over-utilization of private motorized transport associated with relatively low oil prices has enabled a series of changes in Western urban lives. The compartmentalization of land uses, the centralization of services and infrastructure, the heightened perception of risks and the increased conception of certain citizens as vulnerable can all be directly, or indirectly, related to the availability of cheap oil. The sprawl of our suburbs and the consequent centralization and homogenization of service spaces – the replacement of the high street with the infamous strip mall is a case in point – increases the catchment that these spaces support, simultaneously increasing trip distance. The reduction of the streetscape to a thoroughfare for motor vehicles has heightened the perception of the streets as unwelcoming for pedestrians. Certainly, attempts have been made to critically address this narrative of Western urbanization. Planning professionals have long been familiar with a diversity of discourses seeking to address these problems, including New Urbanism, Traditional Neighborhood Design, Walkable Cities, Transit Oriented Development and, especially germane to this chapter, Child Friendly Cities.

The issue of oil vulnerability raises important questions about how livable modern cities currently are for children, and how this livability would be affected by a radical decrease in the availability of cheap oil. Although the topic of child-friendly cities has received increasing attention in urban research (Gleeson and Sipe 2006; Riggio 2002), there is a dearth of consideration of children and child-friendly cities in research on oil vulnerability. This is both surprising and a lost opportunity for planners, as child-friendly perspectives are closely aligned with the attributes of cities resilient to oil vulnerability (Freeman and Tranter 2011). This chapter examines the commonalities between a child-friendly city, on the one hand, and a city that is relatively resilient to oil vulnerability, on the other.

If somewhat populist in its phrasing, we argue that the following statement from Enrique Peñalosa, the former mayor of Bogotá, Colombia, should be seriously considered: "A city should be so constructed so that it is safely navigable by any seven-year-old on a bicycle." If this statement appears idealistic, then we should ask ourselves: Why?

The mobility of children throughout cities is more than a transport issue, and the degree of child-friendliness of the city says much about the way that children are conceptualized in contemporary society. Does our conceptualization of children position them as vulnerable, incompetent, dependent and passive? Or, alternatively, are they valued as participating members of society capable of ever-increasing degrees of autonomy and activity? In considering these questions more fully, this chapter builds on earlier research on children and peak oil in the context of (sub)urbanized populations facing the decreasing availability of cheap oil (Sharpe and Tranter 2010; Tranter and Sharpe 2007). As children are shunted back and forth by car across cities of the Global North and, increasingly, of the Global South, the prospect of oil vulnerability provides an opportunity to reflect on the place of children in the urban environment and in society more widely.

What is notable about the growing body of international literature on the altered levels of children's activity in their daily lives is the tendency to frame this problem via the discourse of CIM. While we are sympathetic to the goals of increasing CIM as they mature, part of the process of recognizing children as the bearers of human rights – and especially participation rights – involves a recognition of a child's difference from an adult. This might require a rethink of the much-vaunted concept of independence, since, as we argue, it might not be as high as other values on the list of children's priorities. In directly pragmatic terms, it may also be that encouraging CIM is more difficult than encouraging active travel.

FIGURE 9.1 Active travel can provide playful experiences for children

Photographer: Maya Spitz.

Our concern is that the conflation that often occurs between CIM and children's active travel can have a paralyzing effect on addressing the activity and transport needs of children. We argue that this inaction is unwarranted, since fostering children's active transport is a more immediately achievable and perhaps no less laudable aim. Children walking or cycling to school with an adult does not constitute "independent mobility." However, it does provide physical activity through regular daily mobility, as well as providing contact with nature and people (including children), helping to build social capital. Active travel can also provide children with a playful experience, even under the watchful eye of their parents (see Figure 9.1).

Much discourse on active transportation for children has tended to assume that the most pragmatic and cost-effective strategies will encourage mixed land use, to maximize families' access to essential services (shops, schools) and the kinds of neighborly communities that go with them. Yet it may be that we significantly underestimate children's capacity to travel more than the most minimal distance. There is much to be said for changing our perceptions of what children can do and for taking measures to allow them to exercise their capacities for active travel. Given the multiple other factors decreasing the chance of full CIM – parents' anxieties about children's safety not being the least among them – focusing efforts first on children's active transport may be a more achievable strategy for building resilience for both cities and children in an era of oil vulnerability.

The downside of cheap oil for children

The availability of cheap oil and the subsequent extravagance in the use of private motor vehicles has increased adult dependent mobility (ADM) and reduced children's active transport. The latter has been sharply curtailed in less than a generation in several nations, including Australia, Canada, New Zealand and the US (Alparone and Pacilli 2012; Buliung et al. 2009; McDonald 2007; Van der Ploeg et al. 2008; Witten et al. 2013). The percentage of children transported by car has steadily increased while those walking or using public transport has decreased.

Alarmingly, in Australia this has occurred not just for younger primary school students (5–9 years) but later primary and junior high school students (aged 11–14 years) as well. In 1971, 57 percent of Australian children aged 5–9 years walked to school, but by 1999–2003, only 26 percent did so. During the same period, the proportion being driven had risen from 22 percent to 67 percent. For 11- to 14-year-old children, in 1971, 44 percent walked to school, but by 1999–2003, only 21 percent walked – while those driven jumped from 12 percent to 48 percent (Van der Ploeg et al. 2008: 62). Recent international comparisons indicate that CIM in Australia continues to decline, with levels lower than other developed nations of the Global North, such as England and Germany (Carver et al. 2013), even though their levels of CIM are also declining (Barker 2006, 50; Hillman et al. 1990; O'Brien et al. 2000). The switch from active modes to car dependent travel is unlikely to be related to changing preferences among children, who still prefer walking and cycling to school (O'Brien and Tranter 2006).

Through their role as car passengers, children are effectively major consumers of petroleum. During school holidays, in cities throughout the world, traffic congestion is noticeably alleviated (Mackett 2002). In Melbourne, trips to take children to school make up 21 percent of all trips in the morning peak (Morris et al. 2001).

With increasing use of cars to access the spaces that children use, the average distance to school has increased over recent decades. In Britain, the average distance from home to primary school increased by 19 percent from the early 1990s to 2005. Children also travel much longer distances to visit friends. This is partly a response to the increasing size of school catchments, as local schools are amalgamated, or as parents choose the "best" school. The creation of "super schools" in many Australian cities has been designed with "economies of scale" and "maximum consumer choice" arguments in mind. In capital cities such as Adelaide, Brisbane, Canberra and Sydney, super schools have served to centralize services. The argument here is that economies of scale will facilitate schools in offering a wider choice of services, including a greater range of subjects, more specialized teachers and more sophisticated facilities to more students. Yet, this development has given scant attention to the burden created by increased travel distances and the demise of local communities. Super schools draw from a more widespread catchment, reducing – but not eliminating – the opportunity for children to use active means of transport such as walking and cycling.

"Serve passenger trips" to take children to a range of destinations, such as friends' houses, playgrounds and sports, have been one of the fastest growing types of urban car travel over the last ten years. In Sydney, the number of serve passenger trips grew at the rate of 2.8 percent each year between 2001 and 2006 – much faster than the growth in total driver trips (1.6 percent), which itself was much higher than the population growth (0.9 percent) (Shaz and Corpuz 2008). Gilbert and O'Brien (2005: 10) note that in Toronto, Canada, the increase in car trips for children aged 11–15 years was 83 percent between 1986 and 2001, in comparison to an increase of only 11 percent in car use overall by adults.

One can understand, to some extent, the conflation of CIM and children's active transport in discourse on the potential shape of post–peak oil urban life; many of the deleterious effects of a lack of active transport are identical to many of the deleterious effects of a lack of CIM. Certainly, the effects of reduced active transport and reduced CIM are well documented. Anxiety around children's health associated with car travel focuses on a range of acute and chronic, physical and mental conditions. The leading cause of death in Australian children aged 1–14 years is external causes (36 percent) and the major component in this category is traffic accidents, which accounts for around 15 percent of the total mortality for that age group (ABS 2008).

Data on the increased risk for pedestrians with higher vehicle speeds understates the risk for children. Children are particularly vulnerable as pedestrians – they are more likely to be dragged under a vehicle than adults, who are more likely to bounce off or over a vehicle when hit. Nor do child-cyclists fare particularly well on Australian roads. In the mid-2000s males aged 10–19 years made up the highest proportion of road cyclist fatalities, the majority of which occurred during 3 pm and 6 pm on weekdays, with trips from school or to after-school activities featuring highly in this time period (ATSB 2006). (See Freeman and Quigg (2009: 395) for similar time-of-day statistics for pedestrians in New Zealand.) The child-passenger inside a vehicle fares even less well; child-passengers are twice as likely as child-pedestrians, and four times as likely as child-cyclists, to be killed as a result of car accidents (Garrard 2009: 4).

Increased car-dependency for children has also led to a series of chronic conditions. Mounting evidence links childhood overweight and obesity with a lack of physical activity (Shannon 2014). In Australia 25 percent of children are overweight or obese (Lowe 2014). A series of local, state and federal government–funded initiatives and campaigns have been launched to counter the "obesity epidemic." Importantly, lack of active transport has been identified as an important factor in obesity: As Whitzman and Pike (2007: 12) argue, "walking to and from school expended more calories than school organized physical education per week." Child passengers are also exposed to higher levels of air pollution than those outside vehicles because the lack of dispersal concentrates pollutants (European Commission 2002).

Legal, but unsafe, speeds

Both speed and state-sanctioned driver attitudes are major components in pedestrian mortality figures. The sharp rise in fatal pedestrian outcomes, relative to the increasing speed of vehicles involved in car-pedestrian accidents, is an important consideration in child-unfriendly urban landscapes. Speed increases beyond 30 km/h significantly increase the risk of death or serious injury. The pedestrian fatality risk is a function of the impact speed, "with the fatality risk at 50 km/h being more than twice as high as the risk at 40 km/h and more than five times higher than the risk at 30 km/h" (Rosén and Sander 2009: 536). As well, the likelihood of avoiding any collision is much greater at lower speeds due to the much shorter stopping distances at 30 km/h compared with 50 km/h.

Most motorists have little appreciation of the huge increase in risk associated with even slight increases in driving speed. Svenson et al. (2012: 488) provide the following example, illustrating the difference between a child not being struck by a car (30 km/h) and being killed or seriously injured (50 km/h):

> We assume a reaction time of 1 second and at a speed of 30 kilometers per hour a car will travel 8.33 meters (30 000/3600) during that time before the brakes start to apply. If the speed is 50 km/h the corresponding distance is 13.89m. This is a little longer than the total stopping distance from 30 km/h (12.75m). This means that a driver who could stop from 30 km/h in front of an obstacle would hit that obstacle at a speed of 50 km/h if she drove at 50 km/h under the same conditions.

This study also identified that drivers were "overly optimistic" about their ability to stop quickly, and showed little understanding of the impact of higher speeds on their stopping ability. The authors suggested that this was an important consideration in attitudes to speed limits (Svenson et al. 2012).

Speed limits of 30 km/h or lower have been introduced in increasing numbers of cities throughout Europe, starting in Graz, Austria, in 1992. Graz provides a valuable case study of a city where the majority of residents were not in favor of the lower speed limits, but the politicians and planners took a risk and implemented lower speed limits, arguing that people could not validly judge the value of the 30 km/h limit if they had not experienced it. On 1 September 1992, Graz became the first city in Europe to implement a citywide 30 km/h limit on all roads, except some major ones with a 50 km/h limit. By 1994, the lower speed limits were seen as a positive change, and more than 80 percent of the population supported them, seeing benefits in noise pollution, safety, and livability. By 1994 "even 2/3 of car drivers were in favor of this measure – as compared to 1/3 in June 1992" (Hoenig 2000). The "20's Plenty Where People Live" movement has been successful in establishing 20 mph zones in Britain. More than 12 million people in the UK now live in cities where 20 mph limits cover the majority of roads.

When streets are seen as being safer for children, parents are more likely to allow them to walk and cycle to school and to other places. The introduction of 30 km/h zones in residential areas in Britain "cut vehicle crashes with child pedestrians and cyclists by 67%" (WHO 2013, 14).

The speed paradox

Reducing speed limits, even to 30 km/h, on residential streets may not lead to a loss of time. Indeed, it may even lead to a reduction in time pressure. Garrard (2008: 9) explains:

> Evidence from studies in several countries indicates that the main (publicly articulated) reasons for opposing reduced speed limits in urban areas; namely, increased travel time and costs, are substantially overstated. Small travel time benefits associated with higher speed limits (an average of 9 seconds/km in one study) come at substantial cost in terms of the health and wellbeing of individuals and communities.

In Bristol, sign-only 20-mph pilots resulted in increased walking and cycling, reduced road speeds, and no impact on journey times or bus reliability (Ingamells and Raffle 2012).

In addition to the trivial loss of time in actual trips made by car drivers in areas with low speed limits, there is also evidence that attempts to save time through increasing trip speeds is a futile exercise (Tranter 2010). For the majority of motorists, the main time demand of driving is not the time spent in cars; it is the time spent earning the money to pay for the multitude of costs associated with motor vehicle use. When these costs are considered, the "effective speed" of any mode of transport can be calculated. This shows that cycling is effectively faster than cars in most urban areas (Tranter 2012).

Not only do cars not provide the time savings many people believe they do, but when cars become the dominant mode of transport, local shops, schools and services are more likely to be closed, necessitating longer distances to be driven. Evidence of this can be found in Melbourne, where the number of land uses within 800m of people's homes has fallen dramatically in the last fifty years, as local shops, schools and services such as post offices have closed. In 1951, more than 70 percent of the Melbourne urban area had five or more land uses within 800 m. By 2005, this had fallen to just over 40 percent (Kelly et al. 2012). Along with other factors, the longer distances to schools have contributed to a decline in the proportion of children allowed to walk or cycle to school (Van Der Ploeg et al. 2008).

Contradictions and traps: A prisoner's dilemma

The prospect of peak oil brings to the fore some of the contradictions and traps that we seem to have laid for ourselves. Children's mobility is now not determined by their age or physical ability as much as it is by the ability of their parents to provide car transport for them. The notion of a responsible and caring parent, the idea of what constitutes immediate risk, and the role that the media plays in creating an atmosphere of fear concerning children, all have a role in shifting the expectations of what children and adults expect in contemporary society. However, we should be careful not to use the physical vulnerability of children as pedestrians and within car accidents as a justification of the conception of children as more vulnerable citizens per se.

Put differently, we are cautious not to reproduce – as many researchers in the field do – the perception that children are by nature dependent members of society, whose growing independence requires a particular type of active mobility, namely independent mobility. Rather, the neoliberal ideal of the independent individual notwithstanding, we should acknowledge the interdependence of all members of society. While the common assumption in arguments for CIM is that "hyper-protective parental attitudes" (Prezza et al. 2009: 21) position children as vulnerable, this may also be a presumption of the discourse of CIM itself. As Romero Mikkelsen and Christensen (2009: 40) argue, literature on CIM invariably positions children as "essentially dependent" and also "views mobility from the perspective of adults and not from children's own meanings of it." The problem here is not simply that the discourse of CIM is poorly defined (Frauendienst 2011), but also that the valorization of independence as a good taken for granted (cf. Loo and Lam 2015) reflects a highly cultural-specific and perhaps even gender-specific value.

In considering, then, whether the issue of oil vulnerability will force a change in the way we view ourselves and the children in our societies, we are not convinced that a "return" to the days of untrammeled CIM is either possible or desirable. Nor are we certain that independent mobility is necessarily the best means by which to achieve many of the advantages that come with active travel. Current social expectations, norms and perceptions of risk have very real implications insofar as the increasing dependence of children on adults and their motor vehicles are concerned. Advocates of CIM regularly cite the risks of traffic danger and stranger danger as the reasons for limiting their children's license to roam the streets independently. It is possible that at least some of these risks are more perceived than actual. Yet, that does not stop them having real effects in inhibiting the use of active transport in urban spaces or in inhibiting the use of certain modes of public transport. Baslington (2009), for example, found that there was a heightened perception of risks in train travel, due to highly publicized and sporadic media reports of train wrecks. She contrasted this to widespread desensitization to the risks of car travel, created by the media depicting frequent stories and images of car accidents (Baslington 2009).

We argue that these perceptions of risk cannot be dismissed simply because they are inconsistent with the actuality of risk. Some advocates of CIM de-emphasize parental concerns about child pedestrian fatalities, citing for example the reduction of child pedestrian fatalities since early the 1970s (Whitzman and Pike 2007: 11). While there is acknowledgement that this reduction is probably due to the lack of child pedestrians on the road, it is also important to recognize that the perception that roads are more risky is to some degree a self-fulfilling prophesy. Garrard (2009) points out that in the UK, child pedestrian fatalities per 100,000 are the second highest in Europe, almost double those countries with higher levels of CIM, such as Germany, Italy, Sweden and the Netherlands. A combination of an intensely pedestrian-unfriendly environment combined with children's inexperience of dealing with risks, heightens dangers when children do take to the streets. In a theme to which we shall briefly return at the end of the chapter, children have been denied the experience to learn and derive meaning from the environment, and are thus inadequately "street-schooled" in how to recognize, avoid or mitigate risk.

At least one reason for this may be the emphasis in planning on creating "children's spaces" in cities. This is a difficult area that goes to the heart of children's place in society and in urban environments. Do the urban spaces dedicated to children's recreational activities serve to separate children from wider society? While dedicated play areas are frequently recognized as evidence of a child-friendly city, it may in fact be the case that parks and children's playgrounds are indicators of child-unfriendly cities, segregating children as they do from the spaces and activities of the citizenry as a whole (Ward 1977). Among other things, the integration of children into society through active transport might allow for the mixing of children with strangers. Koskela (1997) concludes that children less used to dealing with strangers and members of broader society are more likely to grow up fearful of public spaces. If children rarely have contact with other citizens, they are less likely to be active participants in planning.

There is a real anxiety about children's participation in society. While rights discourses figure children as full participants, anxieties about their safety and their futures end in the denial of their basic rights to sociality as they are shuttled around in cars. Making urban areas child-friendly will involve more than the end product of zoning: Children's participation in the planning process must also be taken into account. The obvious argument against the granting of children's participation rights is that they simply lack the cognitive, emotional and social development to understand what is best for them. Childhood, this argument runs, is simply the means to arrive at adulthood in the best possible state to compete in contemporary society.

However, as Freeman (2007) argues, even researchers underestimate the capacity and maturity of children to represent themselves and other children. As we have suggested, it may be that the discourse of CIM, valorizing as it does the growing independence of children, reproduces the very perception of children as dependent that justifies their exclusion from planning processes and rights discourse more broadly. Indeed, it could even be argued that the push for independent mobility may itself constitute a form of social control, which negates the difference of children and ultimately restricts their autonomy. To the extent that independence connotes a separation from others (cf. Romero Mikkelsen and Christensen 2009), the discourse of CIM may translate "child-friendly planning" into "segregated urban spacing," thus perpetuating the denial of children's participation rights into modern society.

The changes toward an increase in ADM and a decrease in CIM, although relatively rapid and dramatic, may also prove to be more intractable than we might imagine, since they have brought about a degree of cultural and political inertia. ADM is a social trap: parents drive their children as a response to the dangers of traffic, thereby contributing to the problem of traffic danger (Tranter and Pawson 2001: 42–43). Whitzman and Pike (2007: 13) also refer to the problem as "a classic 'prisoner's dilemma,' wherein no parent wants to be the first to let their child roam freely," thus suggesting that it will only be solved by collective action. We argue that it may be more helpful to redefine the problem, de-centering the issue of CIM and focusing instead on children's active transport.

Children's active travel: A feasible response to oil vulnerability?

We have suggested that it may be strategically wise to question the assumption that CIM is both a "crucial dimension of the growing up process" (Brown et al. 2008: 399) and "for the betterment of the future population" (Loo and Lam 2015: 90). In pursuing the argument that increasing children's active transport is an equally laudable aim, a mixture of measures would need to be taken to this end. As Mammen et al. (2012: 3) point out,

> research has shown that being escorted while walking may also produce the associated benefits of independent mobility in children. For example, Sissons-Joshi and colleagues discovered that children who walked escorted by an adult possessed greater environmental knowledge in relation to the children who travelled to school unescorted.

Numerous formal programs have been established to increase these modes of active transport. Walking school buses (WSBs) have been so popular an initiative that it has surprised their inventor, David Engwicht (2003). The WSB is an arrangement whereby parents (the "drivers") take it in turns to walk to school, "picking up" children along the way.

There is a tendency to elide the difference between independent mobility and active transport and to assume that dependent mobility means being driven in cars with parents. WSBs go some way to challenge this assumption. However, WSBs suffer due to their formalization and associated bureaucratization, especially in increasingly litigious and risk-averse societies and in light of the time pressures and complex schedules of parents (Engwicht 2003). For us, the salient lesson from these problems is not simply that excessive regulation is to be avoided, but rather that measures need to take into account the needs and desires of adults as well as children. Whitzman and Pike (2007) suggest that a collective response is called for to address children's car dependent mobility. Whatever the approach, the concerns of parents — such as fears and time pressures – must be taken into account, even if they are ironically and collectively self-inflicted.

In Engwicht's model, WSBs were always supposed to be an intermediate step. In his eyes the WSB has not evolved because of lack of focus on the goal of CIM. But are these goals, and indeed the notions of community on which they are founded, those of adults or children? The literature is ambiguous on this issue. Malone's (2007) survey of 40,000 Australian schoolchildren found that typical children's activities, such as organized sport, although well patronized by children, ranked lower in their list of prioritized activities: Ranking first on the list of favored activities was "spending time with family." So, how do we apply these findings in planning for a change in urban lifestyle appropriate to a post–peak oil world?

Stanley et al. (2009) are enthusiastic that the trip to school is amenable to active modes of travel. We agree with this sentiment, but would like to rethink some of their assumptions on how to make this trip more welcoming. Stanley et al. (2009) suggest that separating cyclists and pedestrians from traffic is a way of reducing the risks of active travel, making it more attractive. While we are sympathetic to the safety issues of child-cyclists, we are also concerned about the narrative this constructs about the place of children (and of active transport) in our cities. Is this a similar strategy to isolating children in parks and other "designated play areas," carrying the implicit message that the streets are unwelcoming to any other than motorized transport? Bicycles – and other wheeled forms of transport, such as scooters, skateboards and in-line skates – are associated with bodily pleasure. Harnessing this pleasure may be an important strategy to both encourage cycling and deal with the time pressures of adults during morning rush hour. As Baslington (2009) suggests, those promoting active transport need to harness the pleasure that is currently used to market cars.

There is an assumption that we cannot work with more active modes of transport because our residential densities are not high enough and trip distances are too great. Yet, residential density may be an overstated factor in determining an over-reliance on car transportation. Indeed, there is evidence that children are far more adept at negotiating greater distances by active means of transport than we might otherwise credit them. Garrard (2009) argues, from a study on Belgium youth, that distances of up to 5 km are not too great for active transport modes such as cycling. Garrard (2009) notes that this distance puts 77 percent of Australian families within the range of a school. As Stanley et al. (2009: 5) note:

> Overall, about 40 percent of trips in Melbourne are less than two kilometers in length, suited to walking or cycling, while another 22 percent are between two and five kilometers, well suited for cycling, provided suitable infrastructure is in place.

How could we encourage active transport as a feasible response to oil vulnerability? The simplest, and perhaps the most effective measure on a number of fronts, is to reduce the speed limit to 30 km/h in residential areas. Jacobsen (2003) describes the positive association between increased speed and the disincentive to pursue active modes of transport, such as walking and riding. Garrard (2008, 2009) links the lowering of speed with conveying an attitude that active modes of transport are welcome on streets.

Lowering speeds would also reverse induced traffic. While building "faster" roads leads to increased traffic levels through induced traffic, the lowering of speed limits increases the use of active modes of transport, as the relative perceived trip speed advantages of cars are reduced.

Another advantage of reducing speeds for cars is that it will lead to reductions in fuel use. Lower car speeds reduce fuel use for two reasons: less fuel used per car and fewer cars. On highways, cars traveling at 60 mph will use far less fuel than if traveling at 70 mph (10–20 percent less, depending on car type). In urban contexts, although modeling of individual car performance might suggest that slower traffic has higher fuel use, empirical research demonstrates the opposite: The higher the speeds of cars in built-up areas, the greater the acceleration and braking, which increases air pollution and fuel use (Tranter 2010). The fuel savings are also boosted by land-use changes that arise from lower vehicle speeds. As local services undergo resurgence in response to greater use of active modes of travel, catchment areas shrink. Hence the pressure to use cars is further reduced.

Another strategy in encouraging active transportation modes – particularly the trip to school – is to have the school better integrated into a networked transport system. As McDonald (2012) argues, secure bicycle parking at schools can lead to significant increases in the number of children riding to school (see Figure 9.2). Schools could even serve as bicycle "parking stations" and bus nodes for the wider community, creating a more welcoming image of public service space than schools currently convey. Thus

FIGURE 9.2 Secure cycle parking in schools can significantly increase the number of student cyclists

Photographer: Paul Tranter.

a proper reconnection between adults and children could be possible, which is not just on adult terms. Having adults go into school spaces to park their bicycles in the morning, or collect them in the evenings as they alight from the bus, might force a re-evaluation of the separation of adults and children.

Opportunities from peak oil

To conclude, oil vulnerability and the end of cheap oil need not be greeted with unbridled angst or dread. It may also provide us with the opportunity to take stock of the way we configure our urban areas as well as how we conceptualize children within the urban environment. Anxieties about children's health are usually dealt with by trying to involve children in more extracurricular exercise or sport, so exercise is yet another activity that has to be factored into overcommitted car-ferrying schedules. But more incidental activity such as cycling to school (even up to 5 km distances) would help to respond to a number of issues – including health and oil vulnerability, and the problem of the increasingly disembodied nature of children's sociality and education.

In planning for a post–peak oil world, the model of the school needs to be challenged. According to Robinson (2006), our current modern schooling system arose to meet the needs of fossil fueled industrialism. Schools still operate within a mechanistic model on the assumption that the industrial and consumerist society has a future. This mechanistic model includes such features as the built form of the school, the regulation of bodies within the school environment, and segregation at various levels: children from other children, based on age and gender; children from adults; the school from the community; the classroom learning from nature. Yet, not unreasonably, Robinson asks: Why do we separate children out on the basis of age groups? And, how can we justify the growing preoccupation with the standardization of children, as reflected in standardized testing and the devaluation of some forms of learning (such as the arts) with respect to others (such as the sciences)? While these questions clearly raise issues that are irreducible to the history of industrialism to which Robinson links education, we still have to acknowledge that the much-vaunted quality of "independence" with regard to children might have very dubious political foundations.

Planning for a post–peak oil society should promote an ecological model of schools, where learning occurs throughout children's lives and across children's environments, not just at school (Steen 2003). We need research that examines, and policy that responds to, some of the connections between children's environments and their capacity to learn. The way children travel to school also has implications for children's physical, social, emotional and cognitive development. Children traveling in the back seat of a car learn little about their own environment, and miss out on the opportunities for exercise that come with active transport to school. They also miss out on valuable opportunities for contact with other people that provide important social learning experiences (Hillman 1993).

Research needs to be conducted into the time gained in parental schedules if sociable forms of exercise were integrated into routines of getting to and from school. The separation and compartmentalization of land uses is mirrored in the political process by the separation of ministerial portfolios and departments. What is eschewed in the current policy process is holism. A decision to centralize schools, for example, was made in relative isolation to transport decisions to determine how children travel to and from school. And, as we have argued in this chapter, the trip to and from school might be as rewarding and as educational as what occurs at the destination. By taking a more holistic perspective on both transport and children's wellbeing, peak oil may well provide opportunities for creating cities that are more livable for all city residents. Irrespective of which scenario for peak oil might play out, developing more child-friendly cities will be an attractive option for policy makers in dealing with the challenges.

The implication for policy and planning is that coping with oil vulnerability is literally a matter of "child's play." If planners create more child-friendly cities – cities where children are freer to safely and playfully explore their neighborhoods and cities in ever-increasing circles as they mature – they will also create cities that are more resilient in terms of the looming scarcity of oil. Child-friendly cities will be less reliant on private motor vehicles, will have stronger local communities, and will help develop resilient and adaptable citizens.

Our practical recommendations for the first planning steps to make our cities more child-friendly are: to reduce motor vehicle speeds in urban areas to 30 km/h, as has been successfully employed in European cities; to develop policies to increase children's active transport, such as safe routes to school, and parents (or adults) walking and cycling to school and other places with children; and to re-evaluate the closure of local schools and services (and centralization of facilities) in the name of economic rationality and efficiency.

Once these changes are in place and children's active transport is facilitated, it is possible that children's independent mobility will also increase – or better still, the interdependence of all members of society will be acknowledged and facilitated. This will be as a consequence of a move to an urban landscape where children's safety from traffic is assured by lower speeds and fewer cars, and children's personal safety is supported by the presence of larger numbers of people on the streets as pedestrians and cyclists. The recommendations we outline here are likely to provide significant co-benefits making cities more child-friendly, more livable and healthier for all city residents, and more resilient in the face of oil vulnerability. From a child-friendly perspective, then, oil vulnerability may not need to be regarded with angst. In peak oil we can espy opportunities for urban-dwelling children and adults.

References

Alparone, F.R. and M.G. Pacilli (2012) "On Children's Independent Mobility: The Interplay of Demographic, Environmental, and Psychosocial Factors," *Children's Geographies* 10(1): 109–22.

ATSB (2006) *Deaths of Cyclists Due to Road Crashes*, ATSB Road Safety Report, July, Canberra: Australian Transport Safety Bureau, accessed 28 September 2015 – https://infrastructure.gov.au/roads/safety/publications/2006/pdf/death_cyclists_road.pdf

Australian Bureau of Statistics (2008) *Health of Children in Australia: A Snapshot, 2004–2005*, accessed 2 September 2015 — http://abs.gov.au/AUSSTATS/abs@.nsf/mf/4829.0.55.001/

Barker, J. (2006) *"Are We There Yet?": Exploring Aspects of Automobility in Children's Lives*. Department of Geography and Earth Sciences PhD Thesis, London: Brunel University.

Baslington, H. (2009) "Children's Perceptions of, and Attitudes Towards, Transport Modes: Why a Vehicle for Change is Long Overdue," *Children's Geographies* 7(3): 305–22.

Brown, B., Mackett, R., Gong, Y., Kitazawa, K. and J. Paskins (2008) "Gender Differences in Children's Pathways to Independent Mobility," *Children's Geographies* 6(4): 385–401.

Buliung, R.N., Mitra, R. and G. Faulkner (2009) "Active School Transportation in the Greater Toronto Area, Canada: An Exploration of Trends in Space and Time (1986–2006)," *Preventative Medicine* 48: 507–12.

Carver, A., Watson, B., Shaw, B. and M. Hillman (2013) "A Comparison Study of Children's Independent Mobility in England and Australia," *Children's Geographies* 11(4): 461–75.

Department of Health (2004) *At Least Five a Week: Evidence on the Impact of Physical Activity and its Relationship to Health. A Report from the Chief Medical Officer*, London: Department of Health.

Engwicht, D. (2003) *Is the Walking School Bus Stalled in an Evolutionary Cul-de-Sac?* Brisbane: Creative Communities.

European Commission (2002) *Kids on the Move*, Luxembourg: European Commission.

Frauendienst, B. (2011) "Changes in Children's Independent Mobility: Implications For Interpreting German Road Safety," paper presented at the Association For European Transport Conference, 11 October, Glasgow, accessed 2 September 2015 — http://abstracts.aetransport.org/paper/index/id/3832/confid/17

Freeman, C. and R. Quigg (2009) "Commuting Lives: Children's Mobility and Energy Use," *Journal of Environmental Planning and Management* 52(3): 393–412.

Freeman, C. and P. Tranter (2011) *Children and Their Urban Environment: Changing Worlds*, London: Earthscan.

Freeman, M. (2007) "Why It Remains Important to Take Children's Rights Seriously," *International Journal of Children's Rights* 15: 5–22.

Garrard, J. (2009) *Active Transport: Children and Young People (An Overview of Recent Evidence)*, Victorian Health Promotion Foundation, accessed 13 May 2010 — http://apo.org.au/sites/all/modules/pubdlcnt/pubdlcnt.php?nid=20310&file=http://apo.org.au/files/Resource/active_transport.pdf

——— (2008) *Safe Speed: Promoting Safe Walking and Cycling by Reducing Traffic Speed*, Report Commissioned by the Safe Speed Interest Group, Melbourne.

Gilbert, R. and C. O'Brien (2005) *Child- and Youth-friendly Land-Use and Transport Planning Guidelines*, Toronto: Centre for Sustainable Transportation.

Gleeson, B.J. and N. Sipe (eds) (2006) *Creating Child-Friendly Cities*, London: Routledge.

Groves, B.M. (1997) "Growing Up in a Violent World: The Impact of Family and Community Violence on Young Children and their Families," *Topics in Early Childhood Special Education* 17(1): 74–102.

Hillman, M. (1993) *Children, Transport and the Quality of Life*, London: Policy Studies Institute.

Hillman, M., Adams, J. and J. Whitelegg (1990) *One False Move: A Study of Children's Independent Mobility*, London: Policy Studies Institute.

Hoenig, M. (2000) *The Graz Traffic Calming Model and its Consequences for Cyclists*, accessed 15 September 2015 — www.velomondial.net/velomondiall2000/PDF/HONIG.PDF

Ingamells, S. and A. Raffle (2012) "20 mph Speed Limits Are the Right Policy for Sociability, Safety, and a Healthy Population" [letter], *British Medical Journal*: 345 (18 September). www.bmj.com/content/345/bmj.e6165/article-info.

Jacobsen, P.L. (2003) "Safety in Numbers: More Walkers and Bicyclists, Safer Walking and Bicycling," *Injury Prevention* 9(3): 205–9.

Joshi, M.S., Maclean, M. and W. Carter (1999) "Children's Journey to School: Spatial Skills, Knowledge and Perception of the Environment," *British Journal of Developmental Psychology* 17(1): 125–39.

Kelly, J.F., Breadon, P., Mares, P., Ginnivan, L., Jackson, P., Gregson, J. and B. Viney (2012) *Tomorrow's Suburbs*, Melbourne: Grattan Institute.

Koskela, H. (1997) "'Bold Walk and Breakings': Women's Spatial Confidence Versus Fear of Violence," *Gender, Place and Culture* 4(3): 301–19.

Loo, B.P. and W.W. Lam (2015) "Does Neighbourhood Count in Affecting Children's Journeys to Schools?" *Children's Geographies* 13(1): 89–113.

Lowe, M. (2014) "Obesity and Climate Change Mitigation in Australia: Overview and Analysis of Policies with Co-Benefits," *Australian and New Zealand Journal of Public Health* 38(1): 19–24.

Mackett, R.L. (2002) "Increasing Car Dependency of Children: Should We Be Worried?" *Municipal Engineer* 151(1): 29–38.

Malone, K. (2007) "The Bubble-Wrap Generation: Children Growing Up in Walled Gardens," *Environmental Education Research* 13(4): 513–27.

Mammen, G., Faulkner, G., Buliung, R. and J. Lay (2012) "Understanding the Drive to Escort: A Cross-Sectional Analysis Examining Parental Attitudes Towards Children's School Travel and Independent Mobility," *BMC Public Health* 12: 862, accessed 2 September 2015 — http://link.springer.com/article/10.1186%2F1471-2458-12-862

McDonald, N.C. (2012) "Children and Cycling," in J. Pucher and R. Buehler (eds) *City Cycling*, Cambridge: Massachusetts Institute of Technology Press, 235–55.

——— (2007) "Active Transportation to School: Trends Among US Schoolchildren, 1969–2001," *American Journal of Preventive Medicine* 32(6): 509–16.

Morris, J., Wang, F. and L. Lilja (2001) *School Children's Travel Patterns – A Look Back and a Way Forward. Papers of 24th Australasian Transport Research Forum: Zero Road Toll – A Dream or a Realistic Vision*, Hobart: Tasmanian Department of Infrastructure Energy and Resources.

O'Brien, C. and P.J. Tranter (2006) Planning For and With Children and Youth: Insights from Children About Happiness, Well-being and Walking, Walk 21: Seventh International Conference on Walking and Liveable Communities, Melbourne, October 23–25, accessed 2 September — www.walk21.com/paper_search/results_detail.asp?Paper=89

O'Brien, M., Jones, D., Sloan, D. and M. Rustin (2000) "Children's Independent Spatial Mobility in the Urban Public Realm," *Childhood* 7(3): 257–77.

Prezza, M., Alparone, F.R., Renzi, D. and A. Pietrobono (2009) "Social Participation and Independent Mobility in Children: The Effects of Two Implementations of 'We Go to School Alone'," *Journal of Prevention and Intervention in the Community* 38(1): 8–25.

Riggio, E. (2002) "Child Friendly Cities: Good Governance in the Best Interests of the Child," *Environment and Urbanization* 14(2): 45–48.

Risotto, A. and F. Tonucci (2002) "Freedom of Movement and Environmental Knowledge in Elementary School Children," *Journal of Environmental Psychology* 22: 65–77.

Robinson, K. (2006) Do Schools Kill Creativity? TED Talks, accessed 19 May 2010 — www.ted.com/talks/lang/eng/ken_robinson_says_schools_kill_creativity.html

Romero Mikkelsen, M. and P. Christensen (2009) "Is Children's Independent Mobility Really Independent? A Study of Children's Mobility Combining Ethnography and GPS/Mobile Phone Technologies," *Mobilities* 4(1): 37–58.

Rosén, E. and U. Sander (2009) "Pedestrian Fatality Risk as a Function of Car Impact Speed," *Accident Analysis & Prevention* 41(3): 536–42.

Shannon, C.S. (2014) "Facilitating Physically Active Leisure for Children Who Are Overweight: Mothers' Experiences," *Journal of Leisure Research* 46(4): 395–418.

Sharpe, S. and P. Tranter (2010) "The Hope for Oil Crisis: Children, Oil Vulnerability and (In)dependent Mobility," *Australian Planner* 47(4): 284–92.

Shaz, K. and G. Corpuz (2008) *Serving Passengers – Are You Being Served?* 4th Annual PATREC Research Forum, Edith Cowan University, Perth.

Stanley, J., Hensher, D. and C. Loader (2009) "Road Transport and Climate Change: Stepping Off the Greenhouse Gas," *Transportation Research Part A*. Institute of Transport and Logistic Studies, Working Paper, ITLS-WP-09–21.

Steen, S. (2003) "Bastions of Mechanism, Castles Built on Sand: A Critique of Schooling from an Ecological Perspective," *Canadian Journal of Environmental Education* 8(1): 191–203.

Svenson, O., Eriksson, G. and Gonzalez, N. (2012) "Braking from Different Speeds: Judgments of Collision Speed If a Car Does Not Stop in Time," *Accident Analysis & Prevention* 45: 487–92.

Tranter, P.J. (2012) "Effective Speed: Cycling Because it's Faster," in J. Pucher and R. Buehler (eds) *City Cycling*, Cambridge: Massachusetts Institute of Technology Press: 57–74.

——— (2010) "Speed Kills: The Complex Links Between Transport, Lack of Time and Urban Health," *Journal of Urban Health* 87(2): 155–66.

Tranter, P.J. and E. Pawson (2001) "Children's Access to Local Environments: A Case-study of Christchurch, New Zealand," *Local Environment* 6(1): 27–48.

Tranter, P.J. and S. Sharpe (2007) "Children and Peak Oil: An Opportunity in Crisis," *International Journal of Children's Rights* 15(1): 181–97.

Van der Ploeg, H.P., Merom, D., Corpuz, G. and A.E. Bauman (2008) "Trends in Australian Children Travelling to School 1971–2003: Burning Petrol or Carbohydrates?" *Preventive Medicine* 46: 60–62.

Ward, C. (1977) *The Child in the City*. London: Architectural Press.

Whitzman, C. and L. Pike (2007) *From Battery-Reared to Free Range Children: Institutional Barriers and Enablers to Children's Independent Mobility in Victoria, Australia*, Initial Report to Australasian Centre for Governance and Management of Urban Transport, Melbourne: University of Melbourne.

WHO (2013) *WHO Global Status Report on Road Safety 2013: Supporting a Decade of Action*, World Health Organization.

Witten, K., Kearns, R., Carroll, P., Asiasiga, L. and N. Tava'e (2013) "New Zealand Parents' Understandings of the Intergenerational Decline in Children's Independent Outdoor Play and Active Travel," *Children's Geographies* 11(2): 215–29.

10

PUBLIC TRANSPORT NETWORKS IN THE POST-PETROLEUM ERA

John Stone and Paul Mees

Oil depletion scenarios explored in earlier chapters of this book imply substantial declines in mobility, drastic declines in auto-mobility, increased localization and correspondingly greater roles for walking and cycling, as scoped in other chapters in this Part II. However, to encourage more people to travel using their own effort, there is much that planners, legislators and communities will need to do to repair the damage done to the public realm by automobile dependent policies. For longer journeys, urban mobility will require some form of external power. Electric cars will play a part, but for reasons of efficiency in the use of both energy and space, many more trips than today will need to be made in shared vehicles. In short, we will need good public transport.

Some of the changes required for public transport to play a positive role in an energy-constrained future are the province of engineers and energy-policy analysts. However, there is great scope – in a time-scale of years rather than decades – for transport planners to increase capacity to enable greater numbers and types of trips for which public transport might be a convenient option. There is no doubt that a compact and connected urban form enhances the potential for oil-free mobility through walking, cycling, and greater public transport use. Therefore, objectives for urban planners responding to oil vulnerability will focus on localized intensification of residential development achieved through inclusive democratic processes, with appropriate controls on the quality of design and construction and, perhaps more important, concentration of employment and other trip destinations.

However, we argue that it is not necessary to intensify land use across an entire urban region before significant improvement in both patronage and economic efficiency of public transport becomes possible. This is fortunate for a number of reasons. First, cities are likely to feel major effects of peak oil within the next decade. Within this timescale, whatever urban development model is pursued, most residents of cities in the Global North will continue to live in houses and suburban subdivisions that are already built. For the majority of the urban population, alternatives to the car will need to be effective at existing urban residential densities. Second, modern approaches to public transport service design in "low density" suburbs that have proven effective offer ways to break the politicized standoff between supporters of urban consolidation and residents who choose to live in a detached house on a suburban block.

Our argument for the existence of short-term opportunities to improve public transport in cities of North America and Australasia is built on a number of observations of the relationship between public transport performance and urban form. First, one can stand on major arterial roads, such as

Yonge Street at the northern boundary of the City of Toronto, and see a similar form of suburban development extending in both directions. However, mode share for public transport falls dramatically in the suburban municipalities outside the city compared with central areas that have long been under the jurisdiction of the Toronto Transit Commission. Here, public transport use is determined to a much greater degree by the quality of the service offered rather than by residential densities (Mees 2000).

Second, analysis of changes in public transport use in Australian cities since 1950 shows that decline (and growth) in public transport use has occurred at faster rates than changes in density. This point will be illustrated both through a snapshot of Melbourne during the last period of intense and sustained constraints on oil supply, and through an overview of the performance of various transport modes in seven Australian cities in the thirty-five years from 1976 to 2011.

Third, our analysis of census data shows that, when like is compared with like, residential density at the scale found in Australian and North American cities is a poor predictor of the use of public transport and private cars. If residential density is a less certain predictor of public transport performance than many believe, what provides a better explanation for variation in public transport use? We suggest that a specific approach to public transport planning, and transport planning more generally, offers hope for greater public transport use in many cities where most planners and city officials believe this is impossible. This approach is service-based "network planning." A description of network planning forms the central part of this chapter. We conclude with some comments on the forms of transport governance required to deliver "networked" public transport services.

Changing public transport patronage in Australian cities in the era of the car

The challenge of peak oil seems daunting, but it is also an opportunity. For the first time since the end of post-WWII petrol shortages, there is a serious prospect that public transport may become the dominant motorized travel mode in cities of the Global North. As an example of the role played by public transport during the last period of severely constrained oil supplies, we discuss the experience of Melbourne in the period during and shortly after WWII. It follows a pattern that will be familiar to readers in North America and elsewhere.

Petrol rationing was introduced in Melbourne during the war and remained in force until February 1950. The following year, the Melbourne and Metropolitan Board of Works (MMBW) commissioned Australia's first comprehensive travel survey. This survey found that car ownership rates in Melbourne were still low: the city average was 121 cars per 1,000 residents, with rates ranging from 61.5 in inner Melbourne to 183.7 in the wealthy Malvern-Caulfield subregion (Opinion Research Centre 1951: 32). Rates of car use were also low, as shown in Table 10.1. Although walking and cycling rates were much higher than at present, public transport dominated travel in Melbourne. The data for non-work travel, although less detailed, shows a similar pattern. The car was a relatively minor mode, confined mainly to the wealthy, a pattern seen across Australian cities and towns at this time.

The petrol-rationing era was a boom time for Australian public transport. Patronage increased rapidly, enabling a recovery in fortunes from the decline experienced during the 1930s thanks to the Depression and rising car ownership. However, the need to conserve fuel and labor saw service levels constrained, leading to overcrowding. While this ensured healthy surpluses for public transport operators, it also created public dissatisfaction. In the minds of many members of the public, trains, trams and buses became synonymous with discomfort and crowding (Davison 2004: 1–27).

TABLE 10.1 Travel to work in Melbourne (1951)

Transport mode	*Share of workers (%)*
Train	26.0
Tram	22.1
Bus	8.8
Total public transport	*56.9*
Bicycle	9.5
Walk (or work from home)	14.1
Total non-motorized modes	*23.6*
Car	16.2
Van/truck	2.0
Motorcycle	1.3
Total private motorized transport	*19.5*

Source: ORC (1951: 35).

Public transport in Australian cities declined dramatically once petrol rationing was lifted and rising incomes made cars more affordable. Within just one decade of the MMBW's 1951 survey, the car had become the majority travel mode. Despite rapid population growth, public transport patronage began to decline in absolute numbers, not just as a percentage of the travel market. The decline was most rapid for non-work and non-central travel, which worsened the economic problems of public transport operators, as each train, tram or bus might only carry one full load of passengers per day (Mees 2000: 11–45). Within two decades of the MMBW survey, most Australian public transport operators were trapped in a vicious cycle of declining patronage, rising deficits, service cuts and fare rises.

Figure 10.1 presents the share of work trips carried by public transport in seven Australian capital cities since 1976, when the question about mode of travel to work was introduced into the Australian Bureau of Statistics census. All saw steep declines over the two decades to 1996, with modest reversals in most places since either 1996 or 2001. The largest decline occurred in Hobart, but this is due to special factors. Ferry usage was unusually high in 1976, due to the closure of the Tasman Bridge following a collision with a ship in 1975. Apart from Hobart, the largest decline occurred in Melbourne. In the late 1970s, all cities suffered falls except for Adelaide and Canberra. Both of these cities saw rises in mode share between 1976 and 1981, due to pro-public transport polices adopted by reformist state and national governments, followed by long declines after the reversal of these policies.

Figure 10.2 and Figure 10.3 compare trends on trains with those for buses, ferries and trams in Sydney, Brisbane and Hobart to show the much stronger performance of fixed rail as a transport mode. The substantial increase in Perth began well before the completion of the large new southern rail line in late 2007. Melbourne has the worst performance for trams and buses. This may come as a surprise, given that the city retains Australia's only extensive tram system, but is explained by the fact that Melbourne's buses, which provide crosstown and suburban services beyond reach of trams, are a largely operated by small family businesses and, typically, offer very poor service.

The most encouraging sign is the improvement in mode share and patronage since 1996 in all cities except for Hobart and Canberra. This trend has been the subject of much excited commentary that has tended to ignore long-term trends, which suggest that the recent recovery, while welcome, is modest by historical standards. There has been relatively little analysis of possible causes. Government analysts

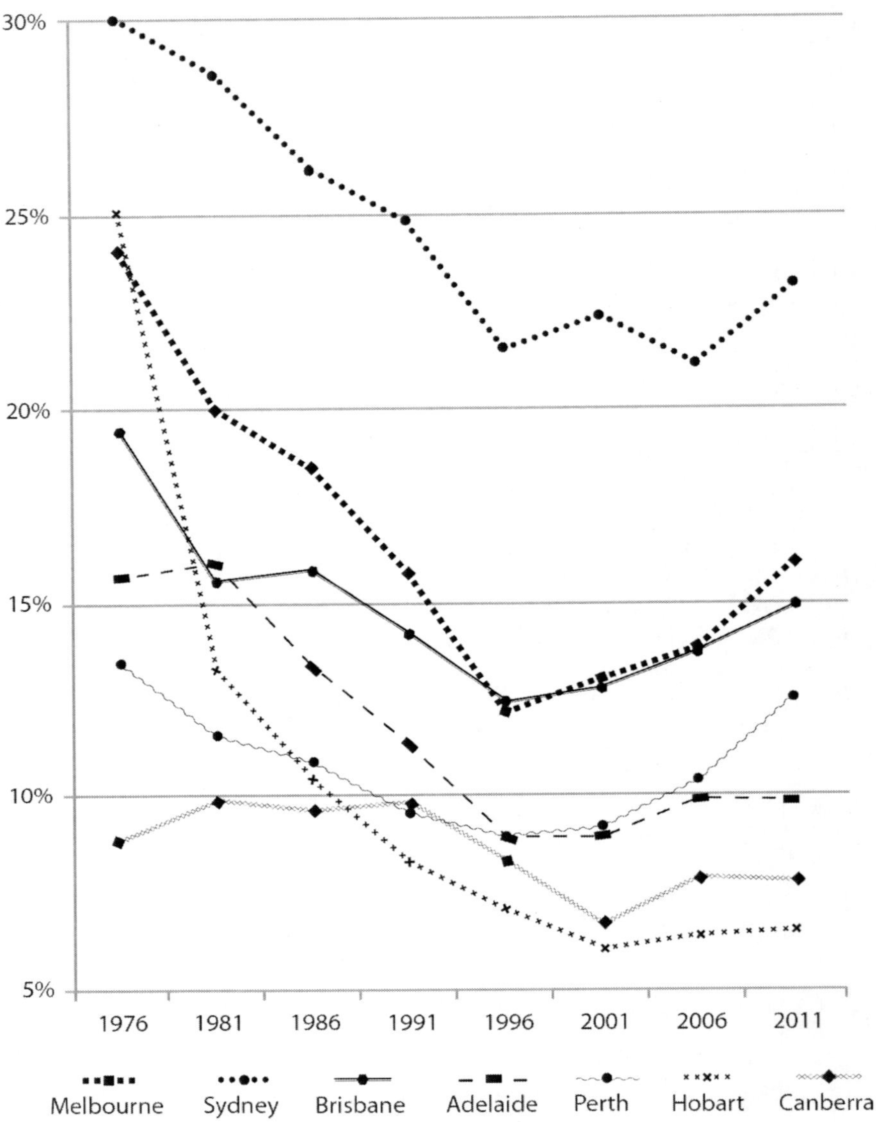

FIGURE 10.1 Share of work trips by public transport, 1976–2011

Source: Mees and Groenhart (2012: 14).

Gaymer and Kinnear (2009) suggest that the substantial rise in the share of workers employed in the central business district (CBD) – a reversal of long-term decline – and rising petrol prices and increased environmental awareness are the major causes of the growth in patronage that has continued since 2005.

Speculation is attractive that greater numbers of CBD jobs and rising inner-city residential populations and modest changes in densities are significant factors behind the observed trends. However, it is not supported by analysis of the changes in the spatial distribution of the journey to work by public transport, at least for the period between 2001 and 2006 (Stone and Mees 2011). Instead, there was some dispersal

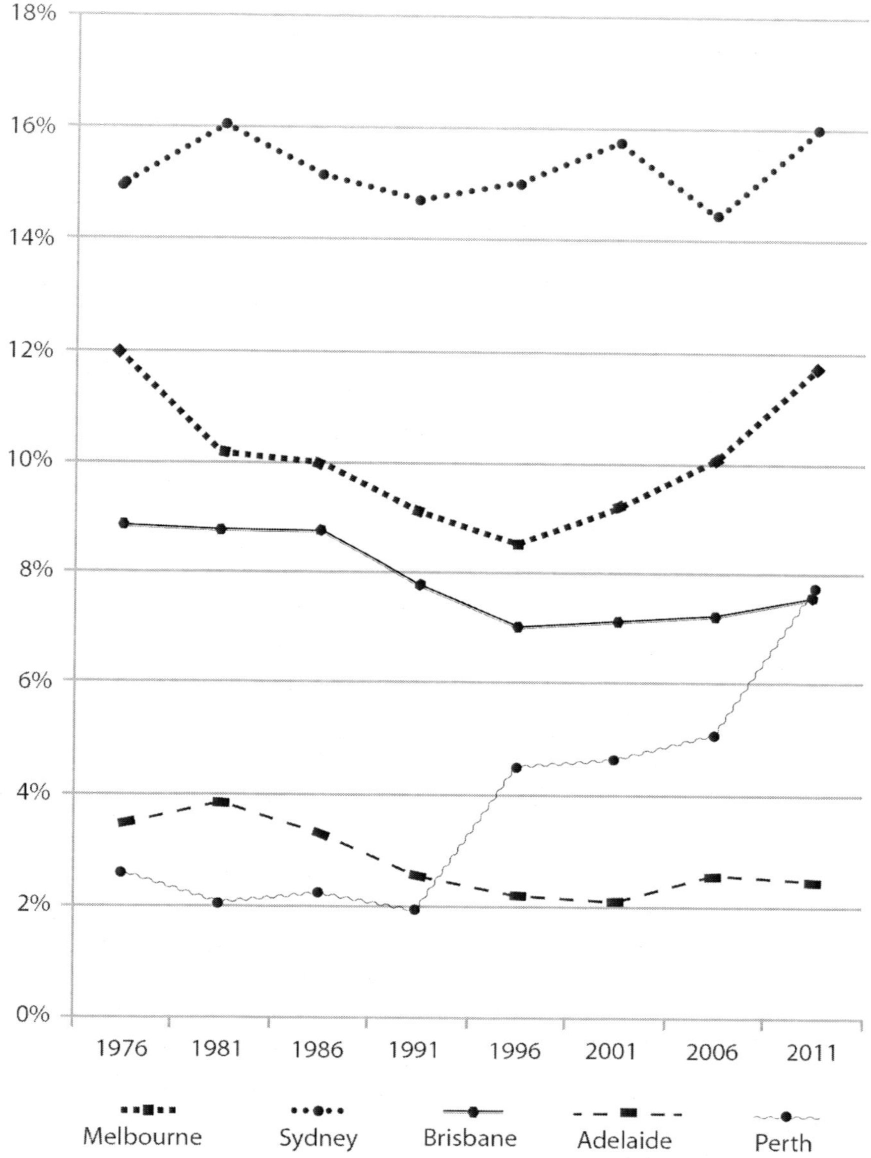

FIGURE 10.2 Share of work trips by train, 1976–2011

Source: Mees and Groenhart (2012: 15).

of workplaces away from the inner city but a greater proportion of work trips to these destinations were served by public transport. In the period 2006 to 2011, there was a slight increase in the concentration of employment in the CBD and inner city, but public transport use in larger cities continued to grow at a slower rate in the CBD than in either the inner-city fringe or in the suburban hinterland (Karanfilovski and Stone 2015). This trend is good news for public transport planners because trips to suburban destinations make less demand on the central hubs, where it is proving difficult to provide additional services.

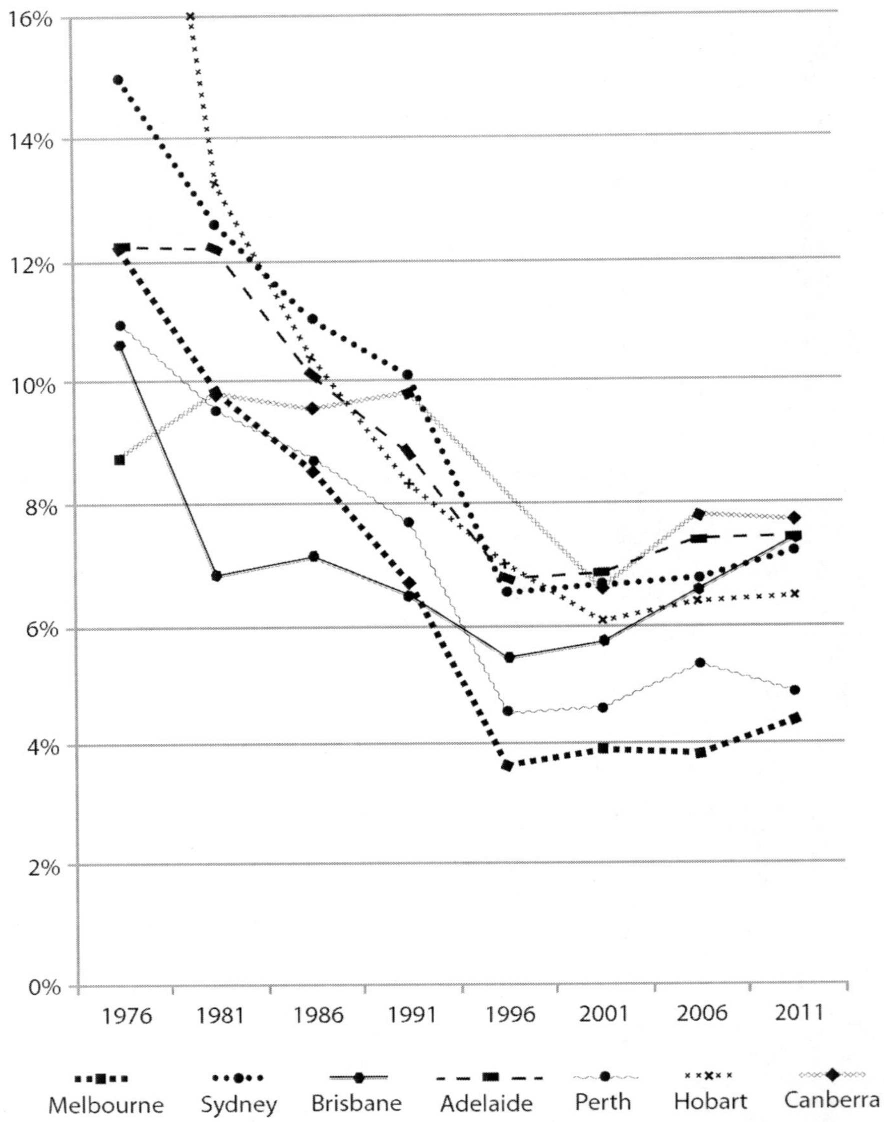

FIGURE 10.3 Share of work trips by bus, ferry and tram, 1976–2011

Source: Mees and Groenhart (2012: 15).

In addition to data on public transport, it is worth noting recent changes in, first, walking to work – where encouraging positive trends had occurred since 1996, and especially 2001–2006, reversing a downward trend since 1976 in most Australian cities – and the smaller, but more extensively reported, rise in cycling rates (see Figure 10.4 and Figure 10.5). The spatial analysis shows that this growth is concentrated in the inner zones.

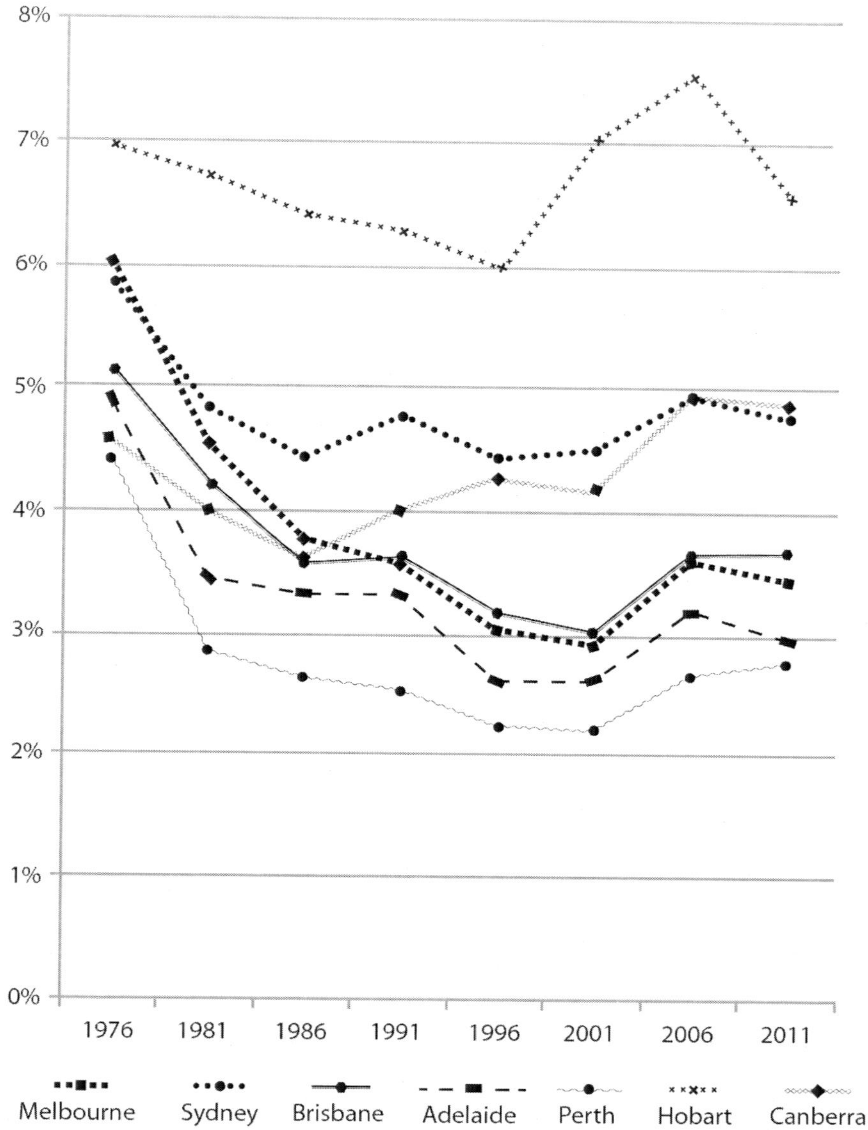

FIGURE 10.4 Share of work trips by walking, 1976–2011

Source: Mees and Groenhart (2012: 17).

Can public transport cope?

The resurgence in public transport patronage, particularly on rail systems, led to increasing public complaints about overcrowding and reliability in Australian cities, complaints that are playing an increasingly important role in electoral politics. Transport planning agencies in Melbourne, Sydney and Brisbane argue that their systems face "capacity crises" requiring multi-billion-dollar infrastructure investments

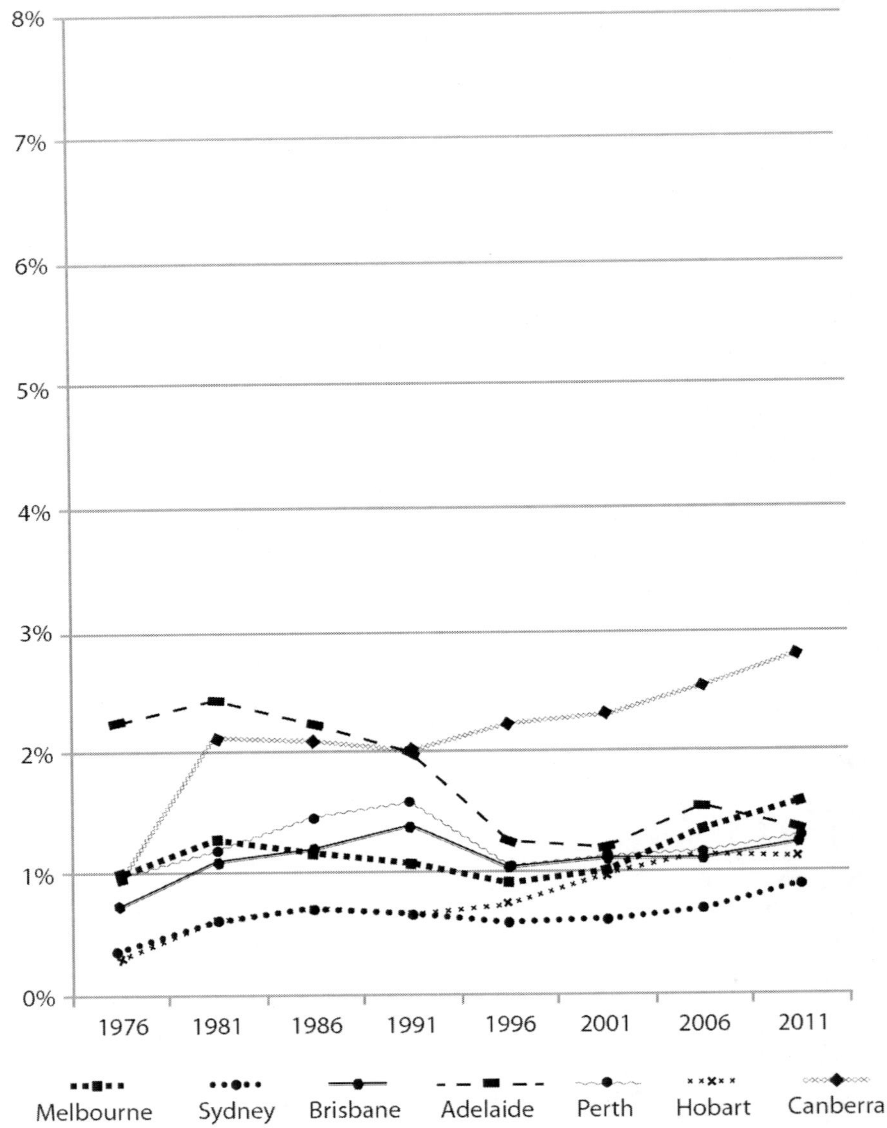

FIGURE 10.5 Share of work trips by bicycle, 1976–2011

Source: Mees and Groenhart (2012: 17).

just to keep up with growth. This is consistent with recent trends in Australian urban planning, which Dodson (2009: 109) notes show strategic urban spatial planning "replaced by a vigorous new emphasis … on large-scale urban infrastructure as a solution to urban problems." Dodson points out that this "infrastructure turn" is problematic for a number of reasons, not least the way that it narrows the focus of urban planning. Furthermore, it raises questions about the ability of public transport planning to cope with an oil constrained future.

The current "capacity crises" in Australian cities represent, by international and local historical standards, relatively modest changes in public transport mode share. For example, the Munich suburban rail system has experienced a 320 percent increase in weekday patronage since 1972, with almost no expansion of its core carrying capacity (DB Netz AG 2015). In the larger Australian cities, mode shares for public transport for work trips to central locations are many times greater than for work trips to suburban destinations: census data shows that public transport carried 74.3 percent of work trips to Sydney's CBD in 2011, falling to 38 percent in the adjacent inner suburbs and as low as 13.3 percent in remaining suburbs. Mode shares for non-work journeys also remain very low. Constrained oil supplies will require public transport to accommodate patronage volumes that are an order of magnitude higher than current levels and to serve travel to and between low density suburbs – two goals that many planners and politicians regard as virtually impossible in the dispersed cities of North America and Australasia.

The best-performing European public transport systems carry much higher patronage loads than their Australian counterparts, usually with less extensive infrastructure. For example, in the mid-2000s, Line A of the Paris RER regularly carried more than one million passengers per day over its two-track central section (RATP 2008: 25) – more passengers than the entire Sydney rail network, with its eight tracks entering the CBD. Parisian rail planners have progressively lifted hourly throughput on RER Line A, from twenty-five to thirty long double-deck trains per hour per direction in peak period.

Even international light rail lines regularly carry much higher patronage than heavy rail corridors in Australian cities. The Expo Line, the original part of Vancouver's Skytrain light rail system, carried around 200,000 passengers per weekday in 2008, achieving a similar service as the entire Brisbane heavy rail system. The Canada Line, which opened in 2009, carried 287,000 passengers on the busiest day of the 2010 Winter Olympics (Translink 2010).

Assessments of the capacity of the existing rail infrastructure in Australian cities and prospects of improvements from recent large investments are beyond the scope of this discussion. But, as we discuss elsewhere (Mees 2010a; Stone 2010), issues of skills and governance must be addressed if we are to get the best value from existing infrastructure and from future investment. Future investment is likely to be much less than the total sought by public transport infrastructure planners across the country, even though Australia, with its revenues from mineral exports, is in a better position than many to escape the worst effects of global economic conditions.

Beyond questions of the availability of infrastructure funds and the purpose to which any investment should be put, the wide acceptance of the idea that the settlement patterns of our cities and towns make greater public transport use difficult, if not impossible, is a further issue for planners seeking to improve public transport in cities in North America and Australasia.

Is density the issue?

The decline in public transport use after the end of petrol rationing happened much more quickly than changes in suburban population densities. However, it has become almost an article of faith that the reverse is impossible: many planners, and other commentators on urban issues, appear to believe that getting significantly more people to use public transport will rely on massive changes in suburban densities. Our international analysis challenges this view.

Across a range of locations, from outer-suburban "farmlets" to Hong Kong high-rises, there is a positive correlation between population density and the ability to operate efficient, attractive public transport services. Yet, within the range of population densities found in urban regions in Australia and

North America, density is not the clear predictor of public transport performance that many might expect. Following recent changes in reporting methods by US statistical agencies, it is now possible to make meaningful comparisons of densities across urban regions in Australia and North America. Data on mode share for the journey to work for the same cities is available – a full discussion of data collection methods and results can be found in Mees (2010b: 51–67). In Figure 10.6 and Figure 10.7, the overall urban density of each region is plotted against use of public transport and the private car. In a regression analysis, the R^2 values for the choice of public transport and the private car for work trips with regard to density were found to be 0.299 and 0.240, respectively. Across this sample of 41 "dispersed" cities, higher density across the whole urban region is not the explanatory variable that many might expect.

This is not to say that urban intensification plays no role. However, the "urban consolidation" that has, as described earlier, contributed to recent public transport patronage growth is modest and makes little impact on the density of the whole urban region. Localized, well-designed and democratically sanctioned urban consolidation is valuable, but it is not the whole story.

We have already mentioned that close attention to maximizing the carrying capacity of available infrastructure is a feature of many successful public transport systems. However, this is not the only lesson transport planners can learn from international best practice. Some regions are now succeeding in a task long thought to be impossible, that is to make public transport work in areas of low residential density. Without huge public subsidies, they are providing high service levels and attracting high shares of travel in and between low density suburbs and even in semi-rural locations.

The Canadian cities achieve the higher mode shares for public transport use seen in Figure 10.6 by establishing competitive bus systems in their middle suburbs, but the international exemplar for high

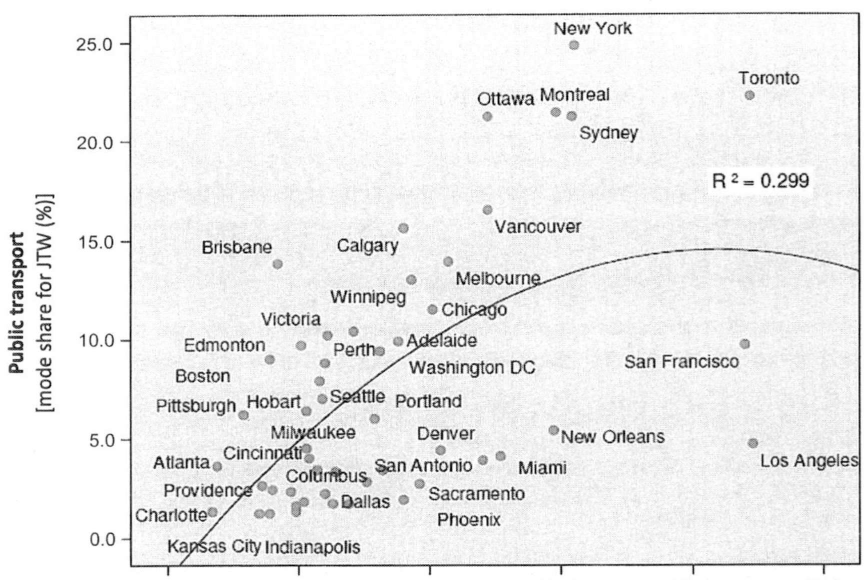

FIGURE 10.6 Population density versus public transport use in North American and Australian cities

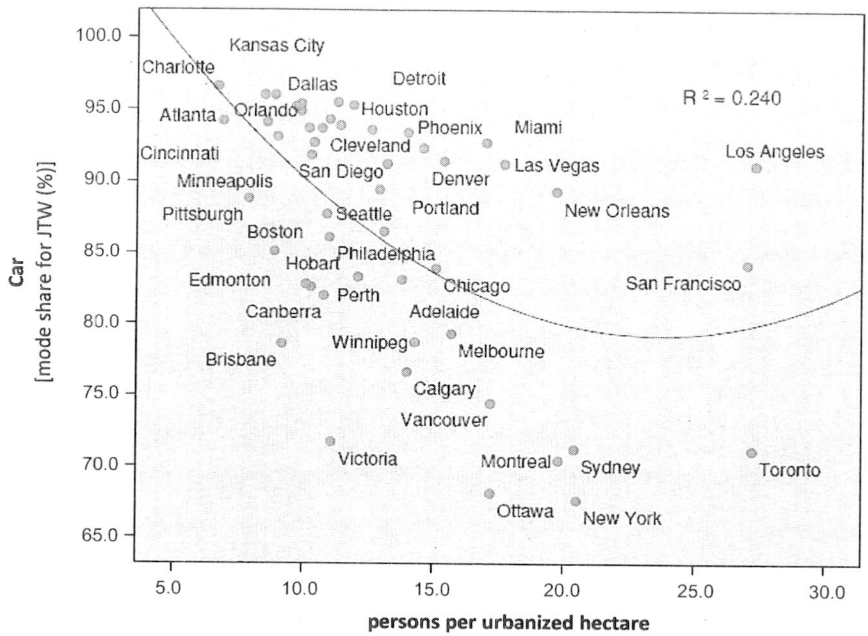

FIGURE 10.7 Population density versus private car use in North American and Australian cities

public transport performance in suburban and semi-rural locations is Switzerland. In the commuter belt around Zurich, the largest Swiss city, the 600 residents of the village of Trüllikon live more than two miles from the nearest rail station but use public transport for the journey to work at a rate higher than all but a handful of cities in North America (Petersen 2014).

To complete the range of contrasts, the largely rural Swiss canton (state) of Graubünden provides some remarkable statistics. Graubünden can be compared with the rural North Coast (Richmond-Tweed district) of New South Wales, which covers a similar area but houses 213,000 residents compared with Graubünden's 187,000. Fewer than 2 percent of North Coast workers used public transport at the 2006 census, while 8 percent walked or cycled and 87 percent traveled by car. Figures for trips to school are unavailable, but are likely to be comparable to work trips.

In contrast, across the whole of the canton, the share of workers traveling by car was just below 50 percent at the 2000 census, while the share of students was only 4 percent. Around 19 percent of workers and 31 percent of students used public transport, while 30 percent of workers and 64 percent of students walked or cycled (Mees 2010b: 179). The share of Graubünden workers using cars is much lower than in any Australian city, while the share using public transport is higher than in any Australian city apart from Sydney. The pattern for students is dramatically more environmentally friendly than can be found anywhere in Australia, and the already low car mode share is actually declining.

So, what makes public transport work so well in Graubünden and in a growing number of urban, sub-urban and semi-rural regions in Canada and in Europe? The key is in their approach to public transport planning, which offers a new perspective on balancing competing demands for attractive service levels and economic efficiency.

The network-planning alternative

Public transport is increasingly called on to serve diverse objectives, ranging from providing mobility to the disadvantaged through to alleviating traffic congestion, while making efficient use of financial resources. The challenge for public transport planners seems daunting. Public transport must cater to travelers with very different needs, ranging from peak-period access to the CBD to all-day access to local shops and community centers. Attractive service frequencies and operating hours for multiple destinations, while maintaining high occupancy rates, are required. Many observers, such as Roth and Wynne (1982), have argued that these tradeoffs present an insoluble problem, but there is evidence to counter this assertion.

The essence of public transport, reflected in its name, is carrying people with different trip origins and destinations in the same vehicle. These travelers can then be transported at lower economic and environmental costs than if they traveled separately. This is public transport's strength, but also its weakness, because people do not all have the same trip origins or destinations.

One approach to diverse travel patterns is to provide separate services for different markets: express buses and trains for peak commuters; regular buses for local trips along busy corridors; and car-like "dial-a-bus" for low-demand corridors and times. This tailor-made approach surrenders its environmental and economic advantages. A public transport system offering a direct service between every origin and destination would have low frequencies, low occupancies, high costs and high energy use per passenger.

The alternative is networks. This approach enables "anywhere-to-anywhere" travel while keeping occupancy rates high by carrying different kinds of travelers on the same services. Transfers are integral to a public transport system that offers access to a large number of potential destinations at an affordable cost (Mees 2000; Nielsen and Lange 2005). Traditionally, public transport planners have attempted to avoid transfers by designing routes that cater to the most popular travel paths and by creating circuitous bus routes that link many destinations, but the network approach embraces transfers making them the building blocks of a multi-destination system.

Two US researchers (Thompson and Matoff 2003: 298) have commented on the importance of transfers:

> Surveys asking what passengers . . . dislike about transit find that transferring is at or near the top of the list . . . [So, traditionally] transfers are avoided . . . In contrast, the multidestinational approach uses transfers to open travel paths to . . . destinations that are reachable in radial systems only by lengthy and circuitous travel.

While transfers create many new travel opportunities, they also impose inconvenience. Effective transfer-based public transport requires careful planning to ensure that the inconvenience is reduced to the minimum possible.

Four key elements underpin the creation of high-quality, transfer-based networks. The first element is a simple line structure. Simplicity makes the network easier for passengers to understand, and minimizes the resources that an operator must provide. The second element is stable line and operating patterns. As well as being simple, a network must be stable, providing a consistent, high-quality service across the network all day (rather than operating different service types in peak and off-peak periods as well as nights and weekends).

The third key element underpinning the creation of high-quality, transfer-based networks is convenient transfers. Easy transferring requires attention to timetables and physical facilities: "random" transfers are possible when all lines serving an interchange point operate frequently, generally up to or better than a standard of every ten minutes (six departures per hour); "timed" transfers are needed when services are less frequent; and the timetables for connecting lines must be coordinated (Mees 2010b: 165–81; Nielsen and Lange 2005).

The fourth key element is appropriate institutions and fare systems. Fare systems must allow free transfers, so a mechanism for the pooling of fare revenues is essential. Pooling of resources is also required to allow cross-subsidies for routes that have lower demand but provide essential components of the "go anywhere" network. On this final point, all cities that have created a successful networked public transport system have managed this through a single responsible public agency with the power to plan and share resources across the urban region (Mees 2010b). This approach has achieved positive results in London and Copenhagen and in Swiss, German and Swedish cities and towns. It is being introduced to improve integration of buses and trains in Singapore (LTAS 2008: 38–39). In Australia, these institutional arrangements are in place in Perth, where public transport patronage has grown steadily since the early 1990s (Stone 2009). Once a public agency has created a sound network plan, successful delivery of individual services can be achieved either through public bodies or via tendering to private companies.

There is clear evidence that network planning delivers improved patronage and economic efficiency. The first comprehensive comparison made between the network-planning approach and more traditional approaches was by Mees (2000) in an analysis of Melbourne and Toronto – two cities with similar populations, incomes and urban forms but very different public transport systems. Per capita public transport use in Toronto was at least twice as high as in Melbourne, despite a much smaller rail system and significantly lower public subsidies. Toronto's performance was the result of network planning by a single public agency, offering travelers frequent and direct bus services and easy transfers. Melbourne suffered from indirect, infrequent and poorly connected services that were the consequence of unproductive competition between multiple operators. This theme was further developed, with many international examples, in *Transport for Suburbia* by Mees (2010b).

Analyses of US cities have confirmed the benefits of network planning. Thompson and Matoff (2003) investigated changes in public transport service levels and patronage between 1983 and 1998 in nine urban regions. They found that cities that had adopted a network-planning approach significantly outperformed those using the traditional approach, recording higher growth in patronage and lower rises in subsidies. Similar results were found in a more recent and more extensive survey of North American cities (Brown and Thompson 2008).

The US studies show that improvements to public transport use can be achieved without changes to the urban form, such as increased density. There is clear evidence that old-fashioned planning paradigms and perverse competition between modes have stifled opportunities to make similar gains in Australasian cities (Mees et al. 2010). Achieving effective public transport networks is a political and institutional task for which planners will play a vital role.

Achieving "networked" public transport

We have seen that better public transport can be achieved with population densities similar to those found today in cities of North America and Australasia. This can be done, in part, by paying careful attention to gaining the maximum capacity from existing infrastructure. Planners can learn from cities with the most successful public transport systems and common institutional structures for the delivery of public transport services. A public planning agency is required to design the network, provide cross-subsidies and operate a multimodal fare system.

Some European urban regions have achieved enviable records in both patronage growth and economic efficiency in public transport service delivery (Stone 2011). The first moves in this direction required processes that typically began in the late 1970s: active citizen engagement in political contention over directions for urban development and the establishment of new policy networks within the relevant institutions (Bratzel 1999). Coordinated local policies of auto-restraint and support for alternatives were essential. These were later reinforced at a national level in Germany through the electoral success of the Greens. As part of a federal coalition, the Greens were able to implement policies, such as an annual incremental increase in fuel taxes (1998–2003), that helped to maintain a price advantage for public transport (Buehler and Pucher 2011).

Alongside support at a political level, new institutions for the governance and management of public transport have been central to the implementation of these coordinated policy packages. Creation of the "network effect" is central to European public transport success. The focus on the passenger is apparent in the common usage by planners of the term "service offer" (*Verkehrsangebot*, in German). In the 1980s, researchers noted the importance of service coordination delivered by organizations such as the *Hamburg Verkehrsverbund* (Dunn 1980; Topp 1989). In the mid-1990s, Baron (1995) and Köhler (1995) recognized the continued success of this coordination, and Pucher and Kurth (1996: 290) identified the vital institutional role of *Verkehrsverbünde* ("transport alliances") in establishing cooperation between large numbers of operators to achieve "more extensive, higher-quality, and better integrated services (that) have significantly increased ridership."

The transport alliances in Germany, Switzerland and Austria share a common motto: One Ticket – One Network – One Timetable. Their task is design and coordination of the overall "service offer." This does not require a big workforce: in Zurich, for example, the website of ZVV (2015) names all its thirty-three staff. Nevertheless, their form varies according to the nature of regional political institutions and past operating arrangements. The structures through which different regional alliances operate vary greatly, although almost all are set up as independent legal entities. The principal shareholders are, in various permutations, the transport operating companies and/or the responsible regional or local authorities (Stone 2011).

In Vancouver, the benefits of the network effect are actively pursued through the work of Translink, the transport-planning agency in the Vancouver urban region. This agency has emerged from decades of intense political conflict over the directions for transport and urban development policy with a clear direction and skilled staff (Stone 2014). In Australia, institutional arrangements that support networked public transport are already well established in Perth and are emerging in Brisbane, and the political momentum is building elsewhere.

Conclusion

The rapidly approaching challenges of peak oil make the kinds of changes discussed in this chapter all the more urgent. It is essential that planners understand the issues and lead the processes of political and institutional reform so that first-class public transport can take its place, alongside walking and cycling, as part of the alternative to automobile dependence.

This chapter has shown that, by changing the way public transport services are designed and delivered, reductions in car dependence can be achieved in a matter of years, even if our patterns of residential settlement change only incrementally.

By looking at the political and social histories of Canadian and European cities that have already made significant advances since the 1970s (Stone and Legacy 2013), we can see that change has come through the work of effective political entrepreneurs, who have worked with the leaders of effective social movements to open the way for the establishment of new institutional approaches to urban transport policy. The growing realization among suburban populations that the "freedom" of the automobile is increasingly illusory and expensive, is fertile ground in which to breed a new generation of urban political leaders who can translate the hopeful message of this chapter into specific reconfigurations of local public transport services.

References

Baron, P. (1995) "Transportation in Germany: A Historical Overview," *Transportation Research Part A: Policy and Practice* 29A: 9–20.

Bratzel, S. (1999) "Conditions of Success in Sustainable Urban Transport Policy: Policy Change in 'Relatively Successful' European Cities," *Transport Reviews* 19: 177–90.

Brown, J. and G. Thompson (2008) "Service Orientation, Bus-Rail Service Integration, and Transit Performance," *Transportation Research Record: Journal of the Transportation Research Board* 2042: 82–89.

Buehler, R. and J. Pucher (2011) "Making Public Transport Financially Sustainable," *Transport Policy* 18: 126–38.

Davison, G. (2004) *Car Wars: How the Car Won Our Hearts and Conquered Our Cities*, Sydney: Allen & Unwin.

DB Netz AG (2015) "2. Stammstrecke Muenchen," *Die Situation Heute (Stand2015)*, accessed 10 September 2015 — www.2.stammstrecke-muenchen.de/nutzen/situationheute

Dodson, J. (2009) "The 'Infrastructure Turn' in Australian Metropolitan Spatial Planning," *International Planning Studies* 14: 109–23.

Dunn, J. (1980) "Coordination of Urban Transit Services: The German Model," *Transportation* 9: 33–43.

Gaymer, S. and R. Kinnear (2009) "Understanding Recent Changes in Public Transport Patronage," *Victorian Department of Transport: Transport Research and Policy Analysis Bulletin* 7 (Winter), accessed 1 October 2015 — http://economicdevelopment.vic.gov.au/transport/research-and-data/transport-research-and-policy-analysis-bulletin

Karanfilovski, G. and J. Stone (2015) "The Spatial Distribution of Travel to Work by Sustainable Transport Modes in Australian Cities from 2001 to 2011," paper presented at the *37th Australasian Transport Research Forum*, Sydney, 30 September–2 October.

Köhler, U. (1995) "Traffic and Transport Planning in German Cities," *Transportation Research Part A: Policy and Practice* 29A: 253–61.

LTAS (2008) *Land Transport Master Plan: A People-Centred Land Transport System*, Singapore: Land Transport Authority of Singapore.

Mees, P. (2010a) "Planning for Major Rail Projects: The Melbourne Metro and Regional Rail," paper presented at the *33rd Australian Transport Research Forum*, Canberra, 29 September–1 October.

——— (2010b) *Transport for Suburbia: Beyond the Automobile Age*, London: Earthscan.

——— (2000) *A Very Public Solution: Transport in the Dispersed City*, Melbourne: Melbourne University Press.

Mees, P. and L. Groenhart (2012) *Transport Policy at the Crossroads: Travel to Work in Australian Capital Cities 1976–2011*, Melbourne: RMIT University.

Mees, P., Stone, J., Imran, M. and G. Neilson (2010) *Public Transport Network Planning: A Guide to Best Practice in New Zealand Cities*, Wellington: New Zealand Transport Agency.

Nielsen, G. and T. Lange (2005) *HiTrans Best Practice Guide No. 2, Public Transport: Planning the Networks* Stavanger, Norway: HiTrans (Rogaland County Council).

ORC (1951) Survey of Movement of People within Greater Melbourne, unpublished report, Melbourne: Opinion Research Centre.

Petersen, T. (2014) "Public Transport Beyond the Fringe," in B. Gleeson and B. Beza (eds) *The Public City: Essays in Honour of Paul Mees*, Melbourne: Melbourne University Press, 149–65.

Pucher, J. and S. Kurth (1996) "Verkehrsverbund: The Success of Regional Public Transport in Germany, Austria and Switzerland," *Transport Policy* 2: 279–91.

RATP (2008) *Activity and Sustainability Report 2008*, Paris: RATP, accessed 2 October 2015 — www.ratp.fr

Roth, J. and G. Wynne (1982) *Free Enterprise Urban Transportation*, New Brunswick, NJ: Transaction.

Stone, J. (2014) "Continuity and Change in Urban Transport Policy: Politics, Institutions and Actors in Melbourne and Vancouver since 1970," *Planning Practice & Research* 29: 388–404.

——— (2011) "Can Successful European Models of Public Transport Governance Help to Save Australian Cities?" *Proceedings of 5th State of Australian Cities Conference*, November, Melbourne.

——— (2010) "Turning over a New Franchise: Assessing the Health of Public Transport Management in Melbourne," paper presented at the *33rd Australian Transport Research Forum*, Canberra, 29 September–1 October.

——— (2009) "Contrasts in Reform: How the Cain and Burke Years Shaped Public Transport in Melbourne and Perth," *Urban Policy and Research* 27: 419–34.

Stone, J. and C. Legacy (2013) "Action Strategies for Paradigm Change," in N. Low (ed.) *Transforming Urban Transport: The Ethics, Politics and Practices of Sustainable Mobility*, London: Routledge, 154–69.

Stone, J. and P. Mees (2011) "Spatial Distribution of the Journey to Work by Sustainable Modes in Australian Cities," paper presented at the *34th Australasian Transport Research Forum*, Adelaide, 28–30 September.

Thompson, G. and T. Matoff (2003) "Keeping Up with the Joneses: Radial vs Multidestinational Transit in Decentralizing Regions," *Journal of the American Planning Association* 69: 296–312.

Topp, H. (1989) "Cooperation in Transit Delivery in West German Metropolitan Areas," *Transportation* 15: 279–95.

Translink (2010) "Some Canadian Line Passengers Arriving Three Years Early for Their Trips," 6 August, accessed 10 September 2015 — www.translink.ca/en/About-Us/Media/2010/August/Some-Canada-Line-passengers-arriving-three-years-early-for-their-trips.aspx

ZVV (2015) ZVV (site), "Organization" (page), accessed 20 October 2015 — www.zvv.ch/zvv/en/about-us/zuercher-verkehrsverbund/short-portrait/organisation.html

11

OIL AND MORTGAGE VULNERABILITY IN AUSTRALIAN CITIES

Jago Dodson and Neil Sipe

From 2005 onward, we undertook a series of studies of oil vulnerability in Australian cities that were reported on in working papers and journal articles during 2005–2008. This chapter draws from that work to provide the conceptual and methodological development, along with an example, of a specific approach to examine the distribution of household exposure to higher petrol prices, mortgage interest rate rises and general price inflation due to increases in global oil prices. This approach was originally deployed in Dodson and Sipe (2006), and its foundational work was presented in a subsequent journal paper (Dodson and Sipe 2007). In the 2006 paper, we presented a numerical index that we called the "vulnerability assessment for mortgage, petroleum, and inflation risks and expenditure" (VAMPIRE). It measured the extent of household exposure to the impacts of higher fuel prices and mortgage interest rates, and associated wider inflationary effects on consumer spending power. A version of this work appeared later in the journal *Housing Studies* (Dodson and Sipe 2008a).

Our initial VAMPIRE research (Dodson and Sipe 2006, 2008a) used data from the 2001 Australian Bureau of Statistics (ABS) Census, which was the most recent available at the time. The equivalent data from the subsequent 2006 Census became available in early 2008, so we have been able to update the original VAMPIRE analysis to identify changes in the distribution of oil and mortgage vulnerability over time. This chapter evaluates the early impacts of higher fuel prices, given that the 2006 Australian Census was undertaken after the global price of oil began its upward march in 2004.

This chapter has four aims. First, we review the basis for the increases in global oil prices seen since 2004. Second, we consider some of the emerging evidence of socio-economic impacts arising from higher fuel prices and mortgage interest rates. Third, we present the results of the 2006 VAMPIRE and compare them to those of the 2001 VAMPIRE. Finally, we make some observations about the policy implications of the changes in oil and mortgage vulnerability within Australian cities, including advocating for government action to address the oil vulnerability of Australian cities and suburbs, and reiterating our earlier call for more in depth research on this increasingly unsettling issue. This chapter is based on research that appeared as a working paper in 2008 — Dodson and Sipe (2008b) – that allowed longitudinal comparison of oil vulnerability. We have edited that original text for length and added some comments to update the discussion around shifts in global oil prices.

On the up

In 2004, global petroleum prices began to increase sharply from levels over the previous two decades. The price of oil was barely more than US$30 per barrel at the start of 2004 but grew rapidly over the next few years (see Figure 11.1). The "psychological barrier" of US$50 per barrel was broken in early 2005 (APF 2004). Oil prices continued to grow apace with increasingly marked volatility, reaching US$95 per barrel by the end of 2007 and hitting US$100 per barrel in February 2008. By May 2008 the price of oil had passed US$130 per barrel, and in June 2008 topped US$140 for the first time, before again dropping to US$125 per barrel in late 2008. The subsequent period has seen oil prices plateau around US$100 per barrel before declining to $30 per barrel by mid-January 2016.

There are multiple reasons behind the growth in the price of oil in the late 2000s. The global economy continued to expand prior to the global financial crisis (GFC), creating increased demand for petroleum products. Moderate economic growth within the developed nations of the Global North was accompanied by strong growth in rapidly industrializing economies, such as China and India, to produce a global growth rate of around 5 percent by 2006. China's economy had grown by more than 7 percent every year since 1991 and did not slow below 9 percent between 2002 and 2007 (International Monetary Fund 2007). India's gross domestic product growth had not been quite as great, yet remained above 4 percent from 1991 and above 7 percent after 2002 (International Monetary Fund 2007). Economic growth in China and India had been producing an expanding middle class that was demanding petroleum consumption profiles similar, although not yet as intensive, to their counterparts in developed nations. The result was large additional demand on global petroleum supplies.

In addition to strong global economic growth, concerns about the security and sustainability of global petroleum supplies heightened during the late 2000s. Capacity constraints in the oil production stream – such as old oil wells, rusting pipelines and aging refineries – had limited oil producers' abilities to keep production rates ahead of demand. In this context, it was not surprising that oil prices had increased.

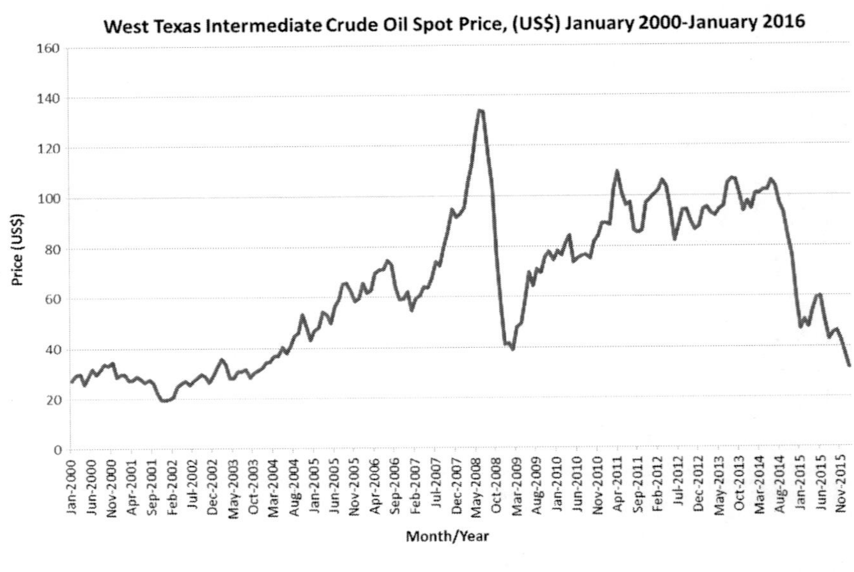

FIGURE 11.1 Spot price for West Texas Intermediate Crude Oil, 2000–2016

Source: US Energy Information Administration (2016).

It seemed, however, that this production capacity deficit would persist for some time. In the mid-2000s the cost of increasing global petroleum production capacity – including refining and distribution infrastructure – to meet the level of demand anticipated by 2030 was estimated by the International Energy Agency at US$4.3 trillion (IEA 2006: 40).

However, the source of this investment was highly uncertain given the investment risks associated with many oil-producing regions. The availability of large sums of credit was also in doubt given recent problems in global debt markets. To compound these production problems, numerous major oil-producing regions were facing declining production due to the exhaustion of oil fields. North Sea oil production had been in decline since 1999 while Mexican oil production was also declining sharply. Production losses in one oil-producing region meant increasing expectations on other zones to maintain production levels.

Another factor stimulating higher oil prices in that period was geopolitical instability in some of the major oil producing regions. The Middle East is the most significant of such zones – global oil reserves have increasingly concentrated in this region – yet this area was (and remains) wracked by conflict and political turmoil. The US had waged war in Iraq since 2003, and tensions between the US and Iran became magnified in subsequent years. For instance, oil prices jumped by US$8 overnight in April 2008 after Israel threatened to attack Iran over the latter's nuclear ambitions. In 2008, Iranian missile tests had added to the volatility of global oil markets.

On top of these pressures there was growing uncertainty over the long-term sustainability of global petroleum supplies. An expanding number of reports and official documents recognized the possibility of global petroleum production reaching a peak followed by a decline. The theory of "peak oil" gained increasing prominence in public and policy debates during the late 2000s. Institutions such as the US Government Accountability Office (2007), the US Army Corps of Engineers (Fournier and Westervelt 2005), the Australian Senate (2007) and the Queensland government (2007), among others, investigated the question of peak oil and the depletion of global petroleum supplies. All recognized petroleum depletion as a major policy issue that demands urgent attention. The bipartisan Australian Senate (2007: 55) report accepted that a global petroleum peak was likely by 2030 and argued that

> the possibility of a peak of conventional oil production before 2030 should be a matter of concern. Exactly when it occurs (which is very uncertain) is not the important point. In view of the enormous changes that will be needed to move to a less oil dependent future, Australia should be planning for it now.

Even if the world does not experience a peak in production before 2020 – and at the time of writing (January 2016) oil prices are comparatively low – the many other petroleum supply factors already discussed suggest that residents of Australian cities are likely to face higher fuel prices.

Experience has shown that global oil prices can accelerate rapidly. Higher fuel prices have a number of significant flow-on impacts on other sectors of the economy and Australia's urban systems. As argued previously (Dodson and Sipe 2008a), there is now a reasonably well-established link between the inflationary effects of higher fuel prices and Australian mortgage interest rates, mediated through the interest rate policy of the Reserve Bank of Australia (RBA). Such linkages inevitably amplify the impacts of higher fuel prices on our urban economies. As a simple matter of national importance, we need to improve our understanding of the socio-economic impacts of higher fuel prices on our urban systems. Our interest here, then, is the relative socio-spatial distribution of consequences from higher fuel prices, including their intersection with other household socio-economic pressures, such as housing costs.

The cost of petrol prices

The main economic effect of costlier oil has been higher petrol prices in Australian cities. In January 2004, the price of petrol fluctuated around A$0.85 per liter across Australia. By June 2008, the average national petrol price had increased to around A$1.60 per liter, representing an almost 100 percent gain in four years. The effect of global petroleum prices on Australian petrol prices might have been even more pronounced had the Australian dollar not appreciated so strongly against the US dollar during this period. (At the time of writing, in early 2016, the price per liter for petrol in Australia remained around A$1.14.)

Higher fuel prices have unsettled urban motorists, generating a clamor of complaint. Many households inevitably found their fuel expenses draining increasingly large proportions from their budgets. The burden of rising transport and fuel costs is shared unevenly between household income segments. Higher fuel prices tend to affect those on modest or below-average incomes the hardest. In proportional terms, the middle quintile of Australian households spends the greatest share of their income on fuel, with the next income bracket down the second most burdened (ABS 2006).

Depending on cars

Dependence on private motor vehicles for travel exacerbates the potential impact of higher fuel prices on urban households. Australian cities are highly car dependent – approximately 80 percent of trips are taken by private automobile – and, therefore, highly petroleum dependent. But, there is a wide spatial variation in automobile dependence for daily travel within our cities. In general, households in the inner zones of Australia's major cities use their cars less frequently than middle suburban households who, in turn, are not as car dependent as those in outer suburban households. For those in the inner suburbs, public transport, walking and cycling tend to substitute for car use. When the car is used in the inner suburbs, the trips are shorter. Sydney provides a good example of these patterns. One study (DIPNR 2003) showed that residents of inner eastern Sydney only used a car for 48.7 percent of their journeys, and traveled on average 10.1 km per day, so the average trip distance for households in this zone was only 5.7 km. By comparison, residents of Sydney's outer west used their car for 79.1 percent of their trips while traveling an average of 33.1 km daily, with an average trip distance of 13.7 km.

Outer suburban households tend to have higher rates of car ownership, which also increases their transport expenses. In the latter half of the 2000s, the insurance company NRMA (2007) estimated the average weekly running cost for a small car at A$144 per week; for a medium car, $237; and for a large sport utility vehicle, $323. Petrol costs made up around 18 percent of the weekly vehicle running expenses, behind depreciation, fees and the opportunity cost of vehicle capital value. Households in the outer suburbs face heavy additional costs from rising fuel prices. Based on NRMA figures, an outer suburban family with one small car and one medium-sized car, plus a four-wheel drive, could spend more than $600 per week in running costs for their three-car fleet.

Housing oil vulnerability

Australia's urban housing is another factor in the process of distributing household socio-economic vulnerability to higher fuel prices. Settled home ownership remains a major desire for many Australian households. Around one-third of households currently purchase their dwelling via a mortgage. The structure of Australia's urban housing markets strongly influences household location choices. In general, house prices decline as the distance from the central business district increases. Because household income largely determines borrowing capacity, low-income households often find their housing opportunities constrained to outer or fringe suburbs, where housing prices are lower. Around one-half of first-time homeowners in Australian cities locate in outer suburban zones (Productivity Commission 2004).

The distribution of new housing stock influences this pattern; popularly favored detached dwellings tend to be found in fringe suburban greenfield sites. Urban geographers and economists have long recognized the effects of this urban socio-spatial arrangement. Maher et al. (1992) warned of the "locational disadvantage" experienced by residents of outer suburban households who are forced to make tradeoffs between affordability and access to infrastructure and services. Dispersed urban form and infrastructure deficits, including poor access to quality public transport, compound these problems. A study by Burnley et al. (1997), for example, has shown that households who move to outer suburban areas to attain home ownership become more dependent on a car as a result of their shift.

The problems of suburban infrastructure deficits, especially in public transport, reflect the consistent failure of state governments to expand infrastructure to keep pace with the rate and scale of land development, especially during the period of rapid suburbanization in the decades after WWII. Such spatial supply deficits, in part, lie behind the large variation in travel patterns between inner, middle and outer suburban areas described earlier. These problems have been exacerbated by the planning of suburban areas around automobile travel, including the dispersion of land uses. The lack of travel alternatives and the creation of car dependence in large middle and outer suburban areas have been described as leading to "forced car ownership" (Currie and Senbergs 2007). Inevitably, such pressures impose costs on households through petrol expenses and other car ownership costs – such as depreciation, insurance, maintenance and registration.

Compounding pressures

The high concentration of households with mortgages in outer suburban zones complicates the socio-economic impact of higher petrol prices because of the association between higher fuel prices and inflation. As fuel prices increased, during 2004–2008, they contributed to general price inflation in Australia.

In Australia, higher inflation has had broader effects due to government policy responses. Under the Australian government's interest rate policy, the RBA is expected to keep inflation within 2–3 percent, by setting interest rates to control the supply of credit. Official interest rates dropped to as low as 4.25 percent in 2001, but with inflation increasing over the past four years, the RBA boosted interest rates twelve times from 2002, to reach 7.25 percent by early 2008. Oil was a factor in these interest rate increases; eight of the increases had occurred since oil prices had started accelerating in 2004. These gains resulted in higher mortgage interest rates. Standard variable home mortgage rates climbed steadily from a thirty-year low of 6.05 percent as late as February 2005 to 9.35 percent by March 2008. While these rate increases amounted to just over 3 percentage points on a standard mortgage, this equated to a 50 percent increase in interest rates in just three years.

The effect of mortgage interest rate rises was magnified by a housing boom, which saw marked house price inflation across Australian cities during the late 1990s and early 2000s. Part of this housing boom was enabled by the availability of cheap credit signaled by low interest rates. Australia's housing debt was A$139 billion in 1998, but by 2008 had more than tripled, to $448 billion. This growth was made possible by an unusual combination of historically low and stable interest rates assisted by low unemployment.

Many households took advantage of lower interest rates during the late 1990s and early 2000s to seek home ownership. In March 1998, the average Australian household was spending 4.7 percent of disposable income on mortgage interest payments. A decade later this figure had more than doubled to 9.5 percent (RBA 2008a). Over the last decade Australia's households have been spending a greater proportion of their income on housing than at any time since measurement began in 1977.

The pressure of debt has weighed heaviest on those with lower incomes. Mortgagee households in the lowest income quintile are typically more highly geared and more likely to suffer financial stress from housing debt than those in higher income brackets (La Cava and Simon 2005). Refinancing as a proportion of new loans increased from 18.7 percent in early 2000, to 30.2 percent by 2008 (RBA 2008b), suggesting that the impacts of higher fuel prices, inflation and interest costs were motivating households to rationalize their financial commitments.

Effects of recent fuel price increases

There is still little research assessing the impact of higher fuel prices on Australia's urban households – despite our previous calls for such effort. There is no publicly available dataset that records household transport activity and housing costs within Australian cities. This ongoing deficit makes comprehensive assessment of the impacts of higher fuel prices difficult. However, some insight into the broader socio-economic effects of costlier petrol and higher mortgage interest rates can be discerned from a range of sources. We review some of that evidence in the remainder of this section.

One indicator of the impact of higher fuel prices was the dramatic growth in public transport use. Brisbane was a good example of this massive patronage boost. In each of the three years when fuel prices grew strongly, from 2004 to 2006, public transport patronage increased by an average of 9.7 percent, adding 12.3 million new passengers to the region's system in 2005, followed by another 15.3 million in 2006. This sudden shift took Brisbane's transport planners by surprise, given just 3.8 percent growth in the three years leading up to 2004. Bus services struggled to cope under the strain of such massive growth. Overcrowding became widespread and chronic. In March 2007, a total of 1,749 buses were forced to abandon passengers at stops because they were simply too full. By April 2008, this kind of problem affected 1,800 services.

Meanwhile, fuel price pressures put Melbourne's tram and train network under strain. The city's rail patronage had grown by more than 10 percent annually since 2004, when it was at a level of 137 million boardings, to an estimated 200 million boardings in 2008 – although some of this apparent growth was simply a result of changes in the way rail patronage is recorded (Mees 2008). The government and private operators struggled to organize sufficient services to meet demand, but in March 2008 added another 200 weekly services.

Rail patronage in Sydney is another indicator of fuel price pressures driving modal shifts. In Sydney, rail patronage had steadily declined during the first part of the 2000s to a low of 270.3 million passengers during 2005 (RailCorp 2006). After strong fuel price growth, Sydney's rail patronage grew sharply. Eleven million extra journeys boosted annual patronage to 281.3 million by 2007 (RailCorp 2007).

For a growing number of households, the cost of mortgage repayments exceeded their capacity to pay. Mortgage repossessions in Australian cities began tracking upward. This effect was clearest in Sydney, where the housing boom was most pronounced. At the peak of the house price boom in 2004, the New South Wales Supreme Court ordered just 1,750 writs of possession. By 2007 this figure had doubled to 3,935 (Saulwick 2008). The next year, a study of mortgage delinquency by Fitch Ratings (2008) found repossessions concentrated in a set of suburbs across western Sydney.

While the aforementioned patterns indicate early responses to higher urban fuel prices, we anticipated that more profound effects were likely as higher fuel costs percolated through the urban economy, but such effects would take time to develop momentum. At the time higher fuel costs came into effect, Australia had benefited from seventeen years of continuous economic expansion, which was unlikely to slow immediately. The gradual increase in fuel prices was unlikely to instantly flow through to the urban economy; effects of higher fuel prices were more likely to be gradually realized over numbers of years. As a result – in the absence of higher-quality socio-spatial data – we had decided that our VAMPIRE Index could be a good means of assessing the spatial distribution of the socio-economic pressures from higher fuel prices.

The 2006 VAMPIRE Index: Mapping oil, mortgage and inflation vulnerability

The VAMPIRE Index was developed to calculate and understand the level of household vulnerability to socio-economic stressors at the local level, based on Australian Census data (Dodson and Sipe 2008a). It combined information on car dependence, mortgages and incomes at the ABS Collection District (CD) level of about 200 households. The index was constructed from four indicator variables obtained from the 2001 and 2006 ABS Censuses, which were combined to provide a composite vulnerability index that could be mapped at the CD level. The VAMPIRE Index assessed the average vulnerability of households within a CD rather than the specific vulnerability of individual households.

The variables the index uses are: car dependence based on the proportion of those who undertake a journey to work (JTW) by car (either as a driver or passenger) and the proportion of households with two or more cars; income level (median weekly household income); and repayment status of mortgages (the proportion of dwelling units that are being purchased either through a mortgage or a rent/buy scheme). The first two variables indicate the extent of car dependence for urban travel. The JTW variable provides a basic measure of demand for automobile travel, while the proportion of motor vehicles per household indicates the extent of household investment, and thus dependence, on motor vehicle travel. Together these two variables provide an indicator of the extent to which households are exposed to rising costs of motor vehicle travel. The mortgage variable represents the prevalence of mortgage tenure and, accordingly, household exposure to interest rate rises (i.e. mortgage stress). The income variable is used to measure the financial capacity of the locality to absorb fuel and general price increases. Together, these four variables provide a basic, yet comprehensive, spatial representation of household mortgage and oil vulnerability.

The VAMPIRE Index was constructed by combining these variables, as shown in Table 11.1. High levels of car ownership, JTW by car and mortgage tenure received high index values (points), while low levels of household incomes received lower scores. Thus, a CD with high levels of car ownership, JTW by car, income and mortgages would receive a score of 15 (5 + 5 + 0 + 5) points, as shown in Table 11.1. The four variables we have selected are not equal in their contribution to the VAMPIRE Index. The variables have been weighted according to their proportional contribution to the overall VAMPIRE

TABLE 11.1 Assignment of values (points) for the VAMPIRE Index

	Value assigned			
Percentile	*2 or more cars*	*JTW by car*	*Income*	*Mortgage*
100	5	5	0	5
90	4	4	1	4
75	3	3	2	3
50	2	2	3	2
25	1	1	4	1
10	0	0	5	0

Index. Thus, of a total possible VAMPIRE Index of 30, 5 points are provided by each of the car ownership and JTW variables, while 10 points each are provided by the income and mortgage scores.

The VAMPIRE Index permits two forms of analysis to be undertaken. A simple update of the VAMPIRE Index based on the 2006 Census data is possible. First, we present the results of the 2001 and 2006 VAMPIRE assessments. A comparison of patterns between 2001 and 2006 can be undertaken. Second, the relative increase or decline in oil vulnerability for given areas identified with results of this analysis presented.

The 2006 VAMPIRE: Distributing urban oil and mortgage vulnerability

Thematic maps were created for Brisbane, Sydney and Melbourne – based on the ABS definition of an urban area and shaded from minimal to very high vulnerability. The results of this mapping are presented in Figures 11.2–11.7. (Any slight differences between these maps and the 2001 maps in our earlier paper are the result of changes in the compilation of census data between 2001 and 2006, including changes to a small number of CD boundaries, which necessitated minor adjustments to the method to ensure intercensal comparability.) Rather than describe the distribution of oil and mortgage vulnerability suburb by suburb for each city, we consider that the patterns are sufficiently generalized that they can be grouped by VAMPIRE category. We discuss the results of the 2006 VAMPIRE assessment according to four categories, ranging from low vulnerability to high vulnerability (combining minimal vulnerability categories with low vulnerability categories for this discussion). Rather than an exhaustive recounting of the specifics of each category by city, we encourage the reader to use our discussion to guide their own perusal of the maps.

Inner-city areas of these three Australian state capital cities – Melbourne (Victoria), Sydney (NSW) and Brisbane (Queensland) – almost universally fall into categories of low or moderate vulnerability. Inner-city residents are typically wealthier than average and far more likely to use public transport, walk or cycle than those distant from the city center, in part because inner areas have some of the best public transport services. The suburbs beyond the immediate core of Australian cities are more oil vulnerable than those at the center, but only moderately so. In general, the VAMPIRE Index shows that the further from the center of the city a suburb is situated, the more likely it is to fall into the higher vulnerability categories. In Australian cities, large areas of moderate to high oil and mortgage vulnerability are found in the middle and outer suburbs. Very high vulnerability is distributed across large tracts of the outer areas of Australia's cities.

Changes in urban oil and mortgage vulnerability, 2001–2006

The mapping of 2001 and 2006 VAMPIRE levels permits assessment of changes in the spatial oil and mortgage vulnerability of households over time. This section examines how the VAMPIRE patterns changed in the period between 2001 and 2006, and identifies the areas of increased or reduced exposure to higher fuel costs and rising mortgage interest rates at the census CD level. The degree of change in VAMPIRE category has been calculated for each Australian city CD. Then we charted the number of CDs according to their degree of change in order to estimate the aggregate balance of change in spatial oil vulnerability within each city. A discussion of trends for each city follows.

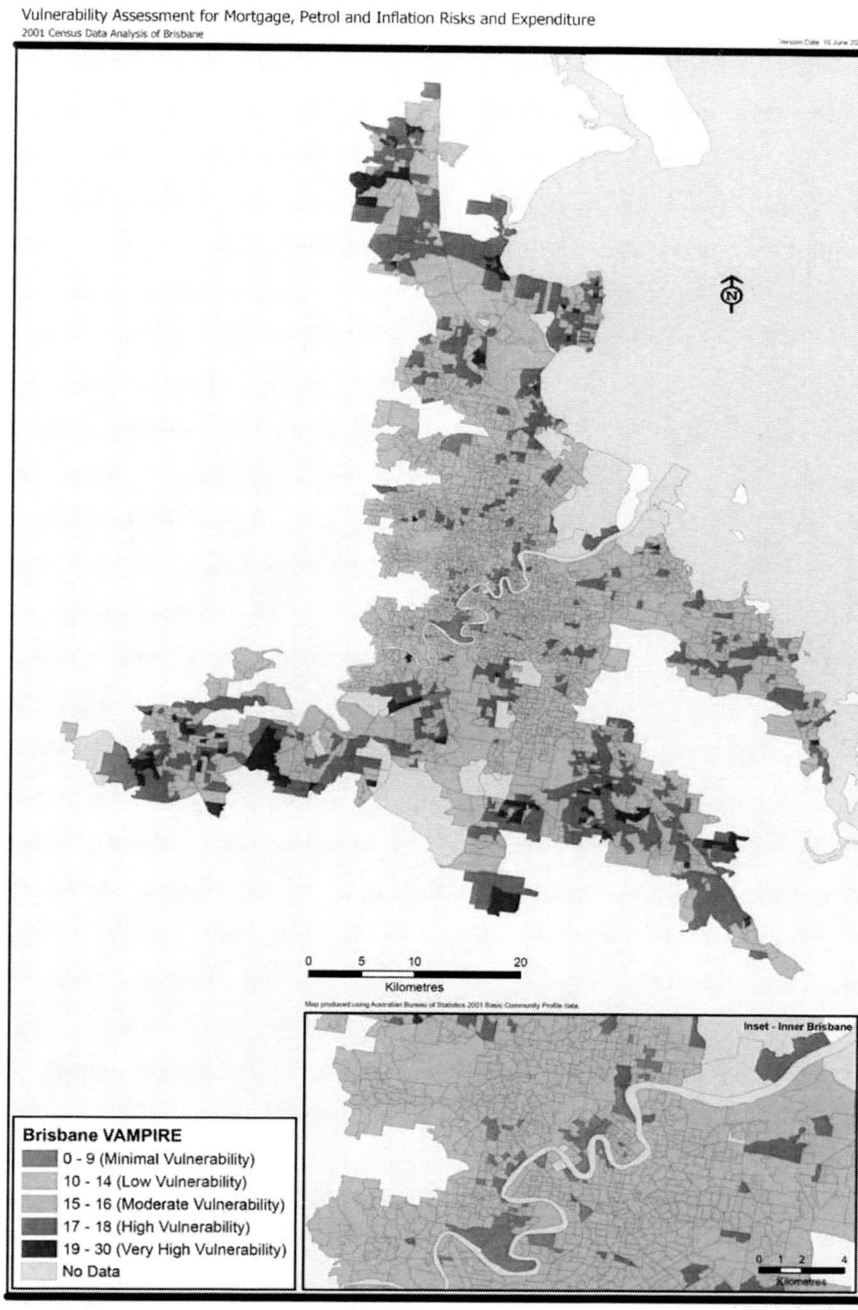

FIGURE 11.2 Oil and mortgage vulnerability in Brisbane, 2001

Source: Dodson and Sipe (2008b: 14).

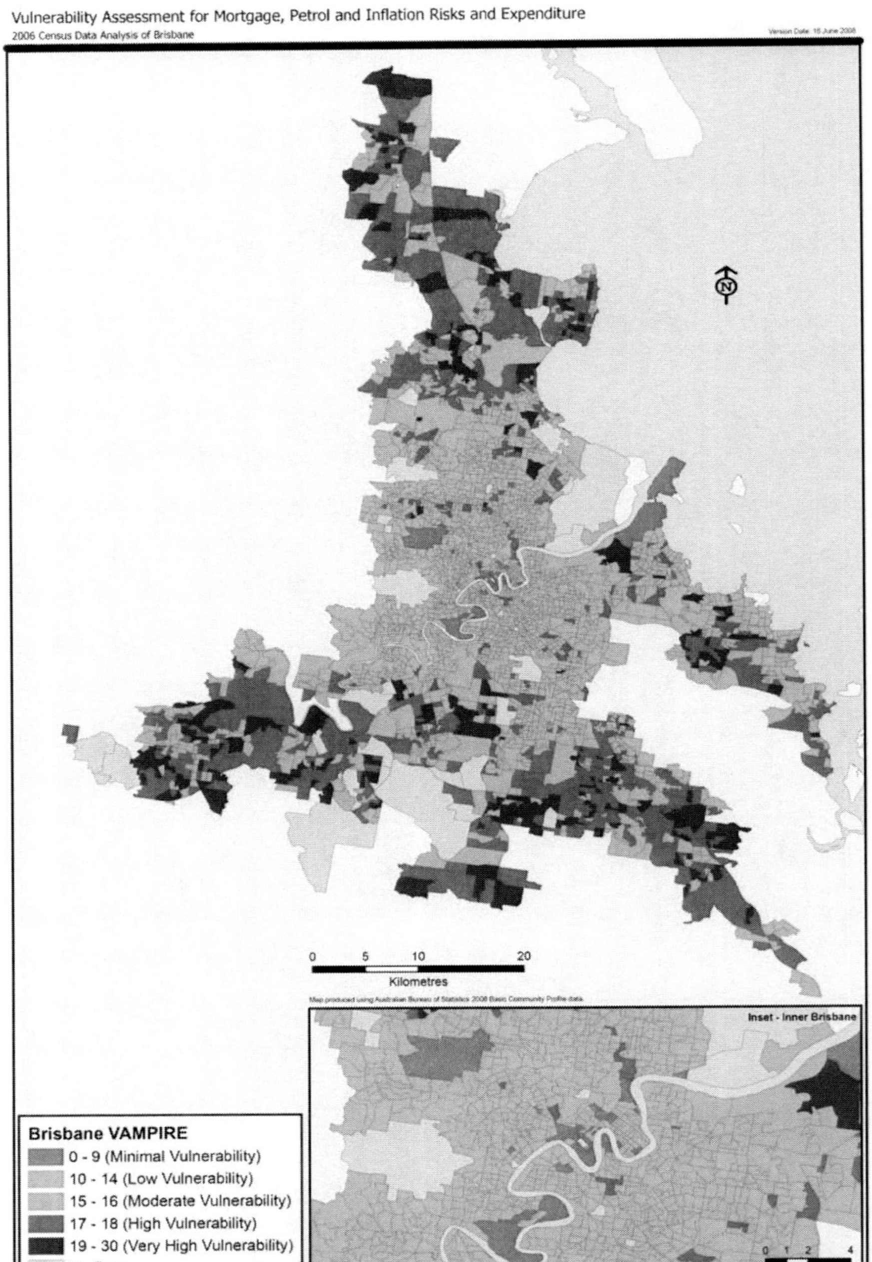

FIGURE 11.3 Oil and mortgage vulnerability in Brisbane, 2006

Source: Dodson and Sipe (2008b: 15).

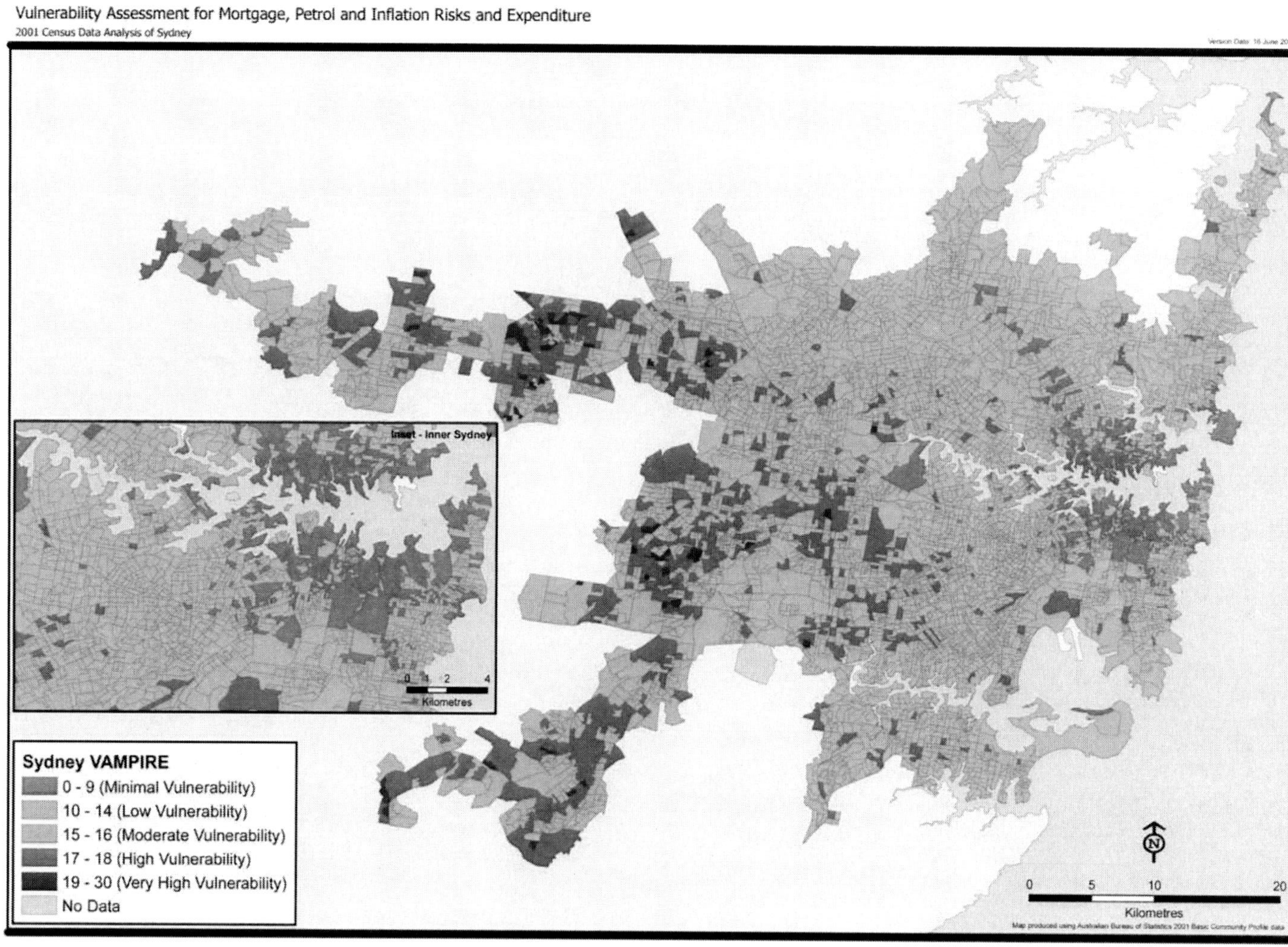

FIGURE 11.4 Oil and mortgage vulnerability in Sydney, 2001

Source: Dodson and Sipe (2008b: 16).

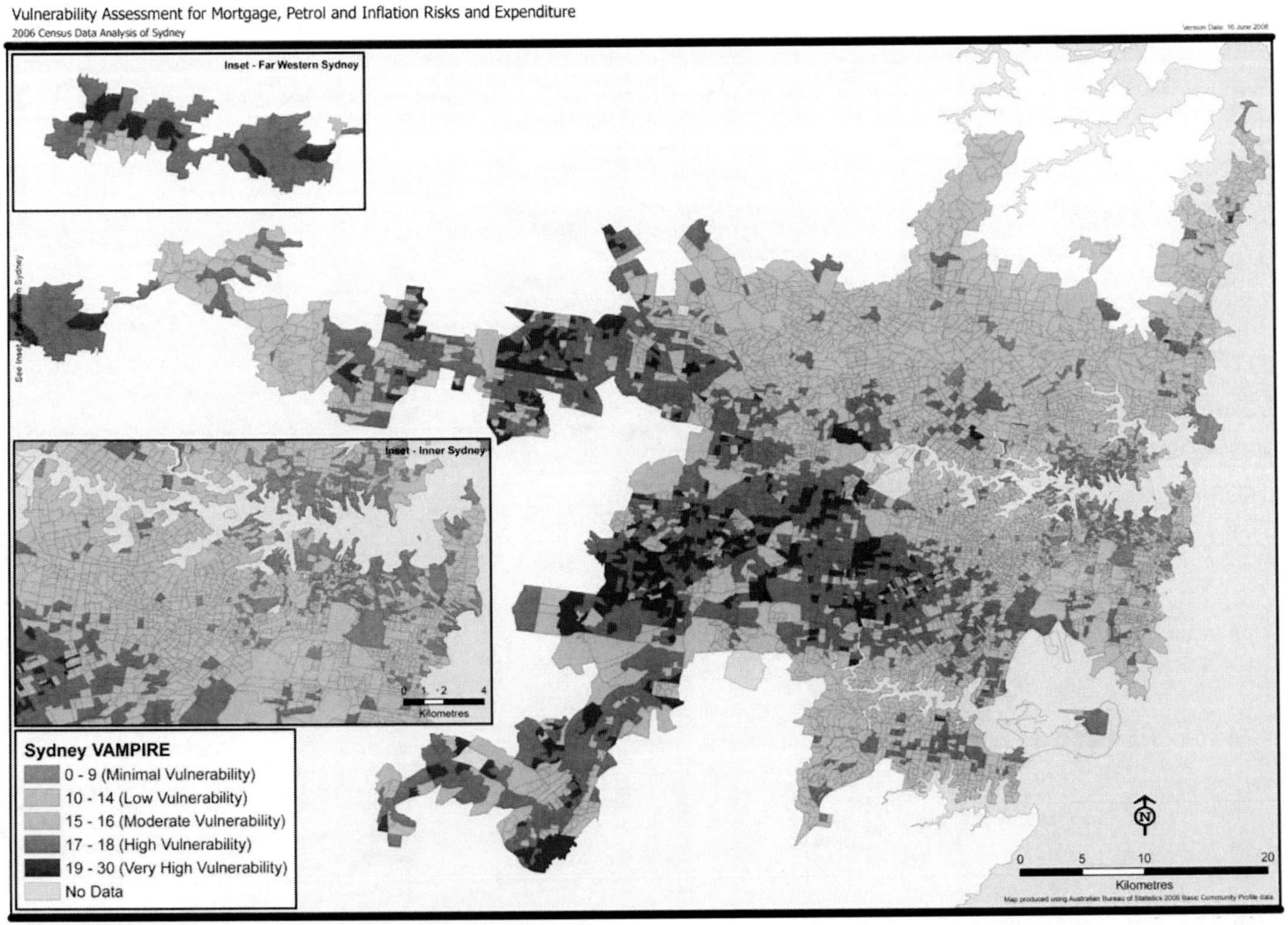

FIGURE 11.5 Oil and mortgage vulnerability in Sydney, 2006

Source: Dodson and Sipe (2008b: 18).

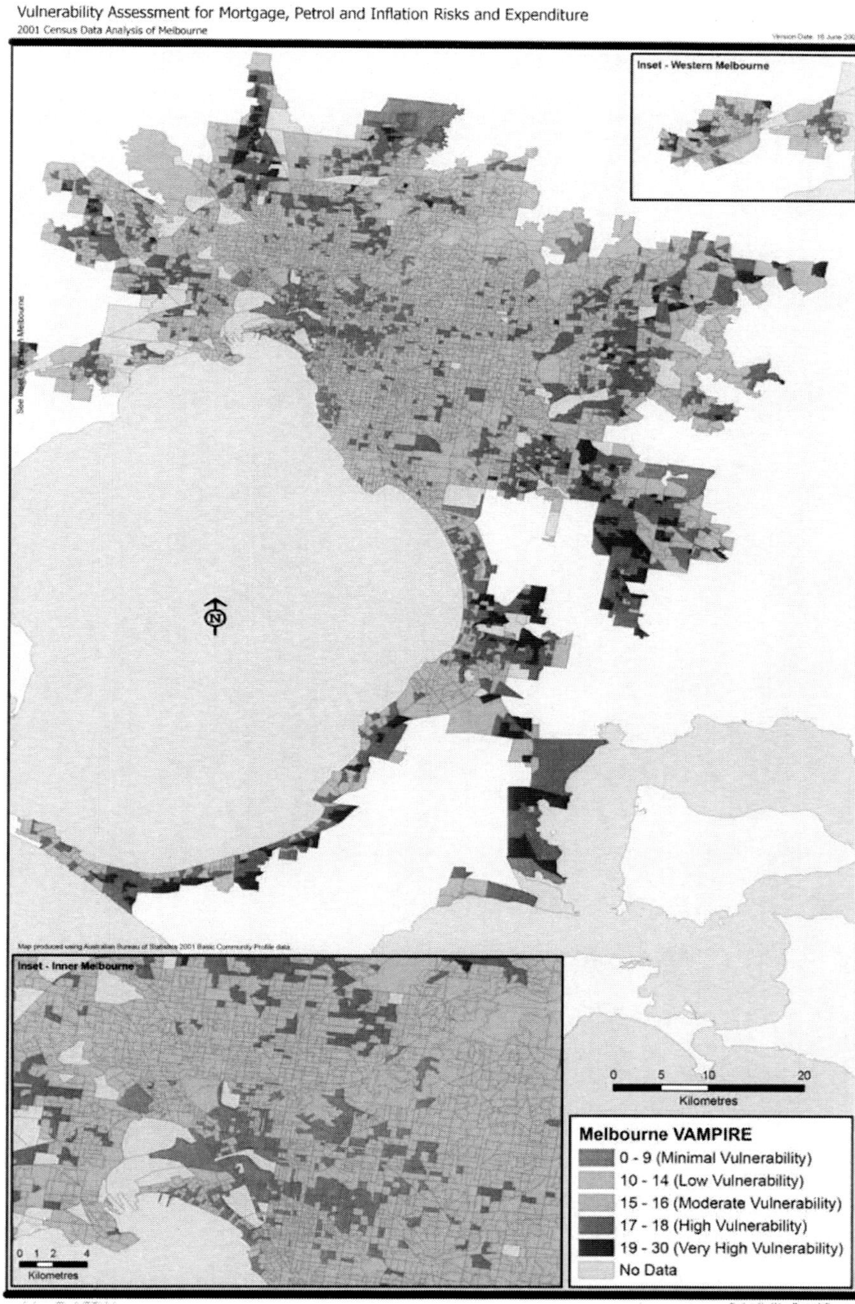

FIGURE 11.6 Oil and mortgage vulnerability in Melbourne, 2001

Source: Dodson and Sipe (2008b: 17).

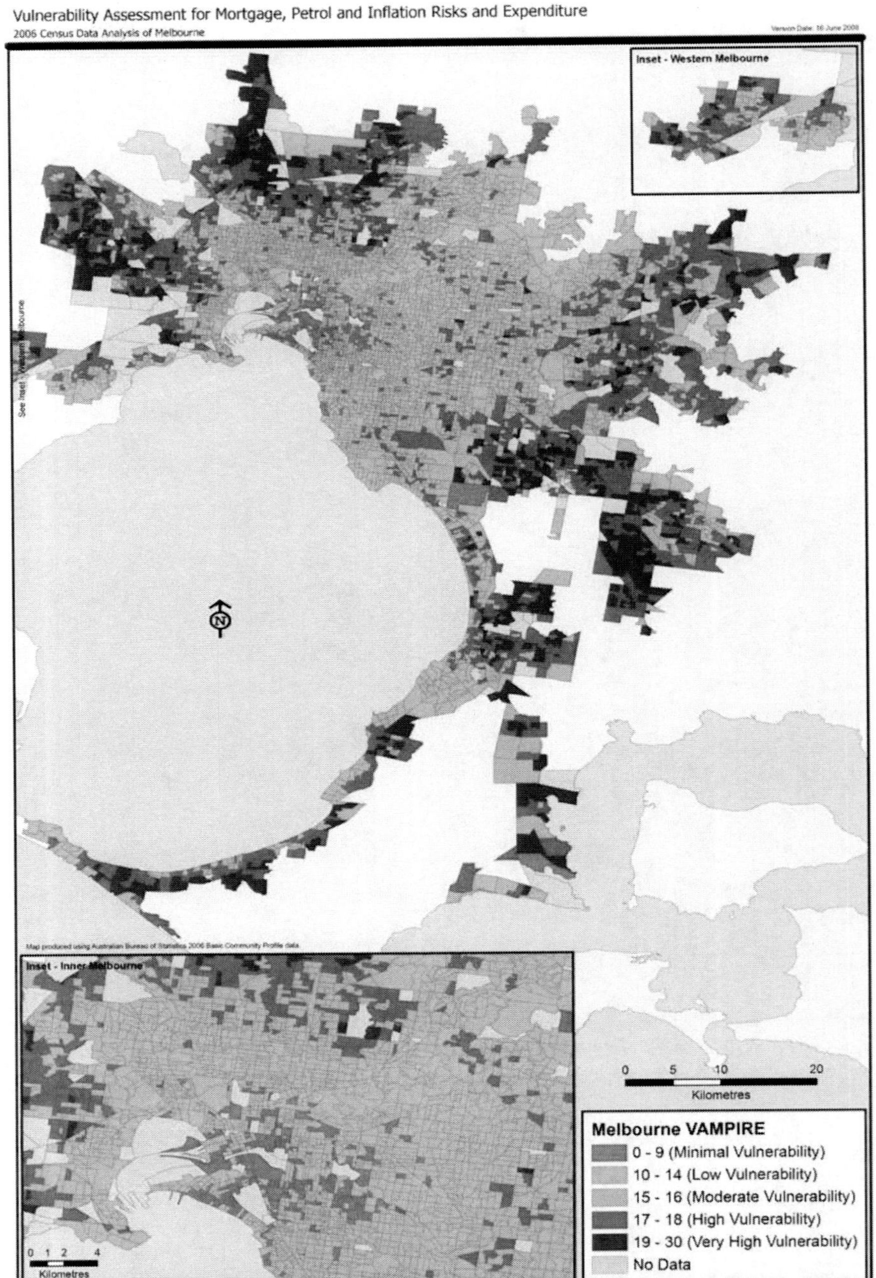

FIGURE 11.7 Oil and mortgage vulnerability in Melbourne, 2006

Source: Dodson and Sipe (2008b: 19).

TABLE 11.2 Changes in VAMPIRE indicators in Brisbane, Sydney and Melbourne CDs, 2001–2006

Change 2001–2006	*Brisbane*	*Sydney*	*Melbourne*
−7 or better	15	18	128
−4 to −6	182	297	502
−2 to −3	428	695	675
1 to −1	1,091	2,039	1,755
2 to 3	425	1,445	1,110
4 to 6	182	966	843
7 or worse	15	163	284

Brisbane

In Brisbane, a majority of CDs experienced minimal change in their level of vulnerability during the 2001–2006 period. The number of areas where oil and mortgage vulnerability declined offset, almost exactly, the number of areas where oil and mortgage vulnerability increased. Slightly more than a quarter of Brisbane's CDs became more vulnerable to oil and mortgage risks during the period and a similar proportion saw their oil and mortgage vulnerability improve. While in a few areas the VAMPIRE score increased more than fourfold, such increases were largely offset by similar numbers of areas where oil and mortgage vulnerability declined. As a whole Brisbane's oil vulnerability remained largely static over the most recent intercensal period.

Table 11.2 shows clearly that areas that became more vulnerable outweighed those where vulnerability had declined. While the distribution of increased oil and mortgage vulnerability within in Brisbane was quite dispersed, some unevenness within these patterns is discernible. However, the extent of clustering is limited, and almost every subregion of Brisbane includes areas that became more oil and mortgage vulnerable during 2001–2006.

Sydney

In Sydney, about 36 percent of CDs experienced minimal change in their level of oil and mortgage vulnerability between 2001 and 2006. The number of areas in which vulnerability declined was far outweighed by the number of areas where vulnerability increased. Approximately 18 percent of Sydney CDs became less vulnerable to oil and mortgage risks during 2001–2006, whereas 41 percent saw their oil and mortgage vulnerability worsen. This shift in the oil and mortgage vulnerability balance in Sydney was most pronounced among areas where their VAMPIRE score worsened by 2 or 3 points, with 20 percent of CDs experiencing this level of change. At the extremes, the number of Sydney CDs where the VAMPIRE Index worsened by at least 7 points (2.9 percent) far outweighed those that saw their fortunes improve by a similar scale (0.3 percent). In summary, Sydney as a whole became much more oil and mortgage vulnerable over the 2001–2006 period.

Melbourne

Many of Melbourne's CDs saw minimal changes in the extent of their oil and mortgage vulnerability during the period 2001–2006 (Table 11.2). Some 33.1 percent of CDs increased or decreased their VAMPIRE score by no more than 1 point. Meanwhile, the number of CDs that became less oil and mortgage vulnerable was substantially outweighed by the numerous CDs where vulnerability increased. Just under one-quarter (24.6 percent) of Melbourne CDs became less vulnerable to oil and mortgage risks between 2001 and 2006, whereas 42.3 percent saw their vulnerability worsen. This shift in the vulnerability balance was most pronounced among the 36.7 percent of areas where VAMPIRE scores fell into the two categories of worsening either by 2–3 points or by 4–6 points. At the extremes of the VAMPIRE Index, the numbers of Melbourne CDs that had seen oil vulnerability worsen by at least 7 points (5.4 percent) was more than double those whose fortunes had improved to a similar extent (2.4 percent). Melbourne, therefore, experienced a marked rise in the oil and mortgage vulnerable during the 2001–2006 period.

Discussion

Both the "snapshot" view of urban oil vulnerability, which was presented first, and the "dynamic" perspective, presented second, showed the major overall differences in spatial distribution of oil and mortgage vulnerability within Australia's three most populous cities. A consistent pattern was revealed by comparing the 2001 and 2006 VAMPIRE maps: inner areas tended to be less vulnerable to higher oil and mortgage costs than middle suburban zones; these middle areas, in turn, faced less exposure to such pressures than outer and fringe suburban areas. This differentiation in exposure to the socio-economic impacts of higher fuel costs and mortgage stress appears to be a structural feature of Australia's cities for the reasons described both earlier in this report and in previous studies.

The analysis demonstrates that the number of areas in which oil and mortgage vulnerability increased over the period 2001–2006 far outweighed those in which oil and mortgage vulnerability declined. This weakening of urban oil and mortgage resilience within these three large capital cities was, however, unevenly distributed. Brisbane appears to have remained relatively balanced in terms of areas of increased or reduced oil and mortgage vulnerability. Despite some areas becoming less oil and mortgage vulnerable, Sydney and Melbourne displayed a greater shift toward oil vulnerability at the local scale.

More analysis and evaluation of urban oil and mortgage vulnerability is needed to better understand the dynamic patterns and processes that contribute to spatial change in household fuel and housing purchase stress. This research represented a first attempt to evaluate longitudinal spatial changes in urban oil vulnerability in Australian cities. Further effort, drawing on this study, is necessary to both improve the evaluative capacity of the index and expand its measures to incorporate a wider set of transport and social variables and data.

Conclusions: The policy challenge of urban oil and mortgage vulnerability

The research we undertook demonstrates that there is a broad distribution of socio-economic exposure to higher fuel prices and rising mortgage interest rates within Australian cities. In general, higher and lower vulnerability are concentrated in different subregions of our cities. Despite some local variation, higher vulnerability tends to be found in outer suburban areas where cheaper housing attracts home purchasers with modest incomes and where transport systems are highly dependent on automobile travel. By comparison, households in inner suburban locations typically experience the advantages, from an oil vulnerability perspective, of higher incomes and lower reliance on automobiles for transport than those in outer suburban zones. These patterns are not transitory but a durable structural feature of the Australian metropolis. Urban structure and the local conditions of resilience and adaptability that urban structure engenders will be critical factors shaping household socio-economic circumstances under conditions of higher petroleum prices or mortgage interest rates.

The regressive city

The distribution of oil and mortgage vulnerability is highly inequitable because the impacts of higher fuel and home purchase costs are borne most by those in outer suburban tracts. As a result, the households that will face the greatest adaptive task in coping with higher transport and housing costs are among those with the least resources and weakest access to local infrastructure that could assist them through the adjustment process. Under current conditions, in which largely unhindered fuel and housing markets intersect with highly differentiated local infrastructure deficits, higher fuel prices and higher mortgage interest rates are a highly regressive phenomenon within Australian cities. Government has a responsibility to redress spatial failures on features of the suburban landscape over which they exert the greatest control, such as the distribution of high-quality public transport infrastructure and services.

The planning imperative

Since WWII, Australian cities have been planned on the presumption that automobiles running on low-priced fuel would be the dominant means of transportation. That dominance is now looking less secure. If Australian cities are to remain socio-economically resilient in the coming decades, our urban planning will need to give much greater emphasis to less oil dependent modes of travel, such as public transport, walking and cycling.

There are many reasons why governments should begin to take the problem of petroleum security and the oil dependence of Australia's urban transport systems more seriously. Shifting away from the business-as-usual approach to transport planning to a less petroleum dependent approach will not be easy for either elected representatives or government officials. A broad and comprehensive plan to reconstruct Australian suburbs to reduce their petroleum reliance is urgently needed.

Missing the train

Some government planning responses risk misallocating resources when addressing the effects of oil and mortgage vulnerability. There has been considerable discussion concerning pressures on public transport systems – partly due to the pressure of higher fuel prices – and appropriate measures for successfully addressing such pressures. In Brisbane, feasibility and planning studies have been conducted into the viability of an underground central business district rail loop. A metro rail line was proposed for Sydney, while a similar plan for Melbourne is currently in the planning and pre-construction phase. These schemes share worrying similarities from an oil vulnerability perspective; such infrastructure may, at great cost, perversely perpetuate the oil vulnerability of Australia's outer suburban regions.

First, such schemes are likely to mainly direct new investment to central and middle suburban areas, yet these zones are already well served by high-quality public transport infrastructure – reflected in the high use of such services by residents. For example, residents in Sydney's suburbs of the inner northwest are already among the least oil vulnerable in the city, but they are being provided with a new metro line. By comparison, many of the highly oil and mortgage vulnerable parts of Australian cities identified by the VAMPIRE Index will miss out on such investment even though lack of infrastructure is a main reason for their present vulnerability.

Second, the scale of finance that these schemes would require is extraordinary. Brisbane's rail tunnel comes at a price of A$7 billion, Sydney's metro at A$12 billion and Melbourne's metro at A$10 billion. An oil vulnerability perspective suggests that rather than using this A$29 billion for public transport improvements across these three cities to replicate existing inner-city infrastructure, it ought to be directed to ameliorating the exposure of suburban households to higher transport costs. This could be done via new outer suburban rail extensions and through improvements to interconnecting local bus services. Newman (cited in Campion 2008) has cautioned about the risk of "suburban abandonment" of oil dependent suburbs and has argued in favor of new rail lines to outer suburban areas.

Public planning, public transport

A more effective response to suburban oil vulnerability than current planning approaches would be to improve public transport services to match the quality found in inner and middle suburban areas. Such services are already providing a good level of resilience to these zones. There is a clear equity imperative to ensure that this level of public benefit is shared by those in the outer areas of Australian cities. This would not require large expenditures. Significant gains could be achieved through better management of suburban public transport, including more efficient institutions and improved integration of bus and rail services. In areas where new public transport infrastructure and services are required, these need not be grossly expensive. Modest extensions to existing suburban rail networks, combined with comprehensively planned and high-quality local suburban bus services, would be a cheaper and more sustainable option for redressing suburban oil vulnerability than costly underground inner-city services. The VAMPIRE maps we have provided can assist in the planning of new suburban public transport services by identifying where vulnerability is greatest.

At time of writing (early 2016), Australian cities remain highly oil vulnerable to adverse price rises in a changing and insecure global petroleum environment. Although the direct price of oil now sits at around $30 per barrel, this level is dependent on continued security in the volatile Middle East and the continued expansion of US tight oil production. Neither of these factors is likely to persist over the long term. Moreover, with the effects of the GFC lingering, official interest rates in Australia are at historic lows of 2.5 percent. This leaves us with limited macro-economic capability to respond to a further fuel price shock, should it occur. If our cities are to remain resilient in the face of declining global petroleum security – and its diabolical cousin, climate change – we will need comprehensive government planning of our suburban public transport networks. The legacy of the past half-century of suburban infrastructure neglect and weakened institutional capacity in service provision means that the task of planning for oil resilient suburbs has been made much more difficult. The historical personal and community security provided by Australia's suburbs is too great an achievement to let fail under the threat of declining petroleum security. We must begin planning now, so that the challenges we face do not end up unsettling suburbia.

References

ABS (2006) *Household Expenditure Survey, Australia: Detailed Expenditure Items, 2003–2004, Cat. 6535.0.55.001*. Canberra, Australian Bureau of Statistics.

AFP (2004) "50-Dollar Crude Oil Becomes Reality, and Some Look Higher," *Agence-France Presse*, 28 September.

Australian Senate (2007) *Australia's Future Oil Supply and Alternative Transport Fuels: Final Report*, Canberra: Australian Senate Standing Committee on Rural and Regional Affairs and Transport.

Burnley, I., Murphy, P. and A. Jenner (1997) "Selecting Suburbia: Residential Relocation to Outer Sydney," *Urban Studies* 34(7): 1109–127.

Campion, V. (2008) "Transport Crisis Turns West into Wasteland," *Daily Telegraph*, 1 May.

Currie, G. and Z. Senbergs (2007) "Exploring Forced Car Ownership in Metropolitan Melbourne," *Proceedings of 30th Australasian Transportation Research Forum*, Melbourne.

DIPNR (2003) *Regional Transport Indicators for Sydney*, Sydney: Transport and Population Data Centre, NSW Government.

Dodson, J. and N. Sipe (2008a) "Shocking the Suburbs: Urban Location, Home Ownership and Oil Vulnerability in the Australian City," *Housing Studies* 23(3): 377–401.

——— (2008b) *Unsettling Suburbia: The New Landscape of Oil and Mortgage Vulnerability in Australia's Cities*, Research Paper 18, Brisbane: Urban Research Program, Griffith University.

——— (2007) "Oil Vulnerability in the Australian City: Assessing Socio-economic Risks from Higher Urban Fuel Prices," *Urban Studies* 44(1): 37–62.

——— (2006) *Shocking the Suburbs: Urban Location, Housing Debt and Oil Vulnerability in the Australian City*, Research Paper 8, Brisbane: Urban Research Program, Griffith University.

Fitch Ratings (2008) *Australian Mortgage Delinquency by Postcode*, New York: Fitch Ratings.

Fournier, D.F. and E.T. Westervelt (2005) *Energy Trends and their Implications for US Army Installations*, Washington, DC: US Army Corps of Engineers.

Government Accountability Office (2007) *Crude Oil: Uncertainty about Future Oil Supply Makes It Important to Develop a Strategy for Addressing a Peak and Decline in Oil Production*, Washington, DC: US Government.

IEA (2006) *World Energy Outlook 2006*. Paris: International Energy Agency and Organization for Economic Cooperation and Development.

International Monetary Fund (2007) "World Economic Outlook Database, April 2007," accessed 28 January 2016 — www.imf.org/external/pubs/ft/weo/2007/01/data/download.aspx

La Cava, G. and J. Simon (2005) "Household Debt and Financial Constraints in Australia," *Australian Economic Review* 38(1): 40–60.

Maher, C., Whitelaw, J., McAllister, A., Francis, R., Palmer, J., Chee, E. and P. Taylor (1992) *Mobility and Locational Disadvantage within Australian Cities*, Canberra: Department of Prime Minister and Cabinet Social Justice Research Program into Locational Disadvantage.

Mees, P. (2008) *Does Melbourne Need Another Rail Tunnel?* Melbourne: Environment and Planning Program, RMIT University.

NRMA (2007) *Private Whole of Vehicle Operating Costs – June 2007*, Sydney: NRMA.

Productivity Commission (2004) *First Home Ownership: Productivity Commission Inquiry Report*, Canberra: Productivity Commission.

Queensland Government (2007) *Queensland's Vulnerability to Rising Oil Prices*, Brisbane: Queensland Oil Vulnerability Taskforce.

RailCorp (2007) *RailCorp Annual Report – 2006/2007*, Sydney: RailCorp.

——— (2006) *RailCorp Annual Report – 2005/2006*, Sydney: RailCorp.

RBA (2008a) *Household Finances – Selected Ratios (D21)*, Canberra: Reserve Bank of Australia.

——— (2008b) *Lending Commitments – All Lenders (D06)*, Canberra: Reserve Bank of Australia.

Saulwick, J. (2008) "Huge Rise in Home Evictions," *Sydney Morning Herald*, 23 January: 1.

US Energy Information Administration (2016) "Cushing, OK WTI Spot Price FOB (Dollars per Barrel)" in *Petroleum and Other Liquids*, accessed 15 April 2016 — www.eia.gov/dnav/pet/pet_pri_spt_s1_d.htm

12

OUTER SUBURBS, CAR DEPENDENCE AND RESIDENTIAL CHOICE IN FRANCE

Benjamin Motte-Baumvol and Leslie Belton-Chevallier

The expected decline in petroleum resources and concomitant rise in petrol prices for private transport raise questions about the future of car dependent areas, such as the outer suburban fringes of cities. Car ownership is high among those in urban populations who drive many kilometers every day. Suburbanization is reliant on the intensive use of the car as the main, even exclusive, means of transport for people who live or work in the outer suburbs. Transport costs may account for up to a quarter of budgets in some locations in the outer suburbs of Paris (Berri 2007), and costs are heavily dependent on increases in petrol prices (Dodson and Sipe 2008).

In large French cities, residents of outer suburbs are more under threat because they host a growing proportion of low-income households drawn in by more affordable housing than those downtown (Cavailhès and Selod 2003). Strong and sustained increases in petrol prices would cause a decline in living standards and adversely affect the activities of households without sufficient provision for transport in their budgets. The number of households impacted is likely to vary depending on the extent of increases in petrol prices. In France, policy makers seem unprepared for permanency in such developments. Stopgap emergency measures are introduced whenever petrol prices flare. In 2008, transport vouchers were distributed on a case-by-case basis by welfare agencies. In 2012, fuel taxes were cut back for several weeks.

In a discussion of the findings of extensive research on everyday household practices, this chapter suggests effective policy responses to increasingly more expensive petrol prices. Our research involved interviews, observations and analyses of low-income households adapting their mobility practices and residential locations within highly car dependent urban environments. The chapter examines two main types of practices in car dependent outer suburbs: either a household opts "to stay," forgoing sustained mobility and a varied range of activities, or they choose "to go," relocating in areas where travel costs are less of a burden on their budgets. Our research analyzed statistics in the French National Census (FNC) 2008 and the Household Travel Survey (HTS) 2003 relating to the Paris region, and conducted two qualitative surveys based on interviews with sixty households in 2010 and 2011. This research was funded by the French Ministry of Sustainable Development and Transport.

Public policy measures were considered for each of these to-stay or to-go practices. Overall, the results presented show that, on the one hand, no single comprehensive solution, such as massive provision of public transport or subsidies to car mobility, is likely to be socially or economically effective.

On the other hand, the solution to rising petrol prices cannot be limited to transport. Any solution must necessarily and closely tie in with housing and urban planning policies. For example, a complementary policy of supplying cheap housing in the centers of suburbs could help lower-income households by a measure that is more sustainable and less expensive than by addressing transport issues exclusively.

Our conclusion briefly contextualizes these French findings in an international research context and identifies which findings and policy solutions presented for France might be applicable in other countries. The specific demographics and location of households and recommended measures would all need to be adjusted according to the levels of car dependence by area and to leeway for policy intervention by public authorities in the areas of transport, social housing and urban planning.

Staying put at any cost

In all types of areas, mobility declines as income declines (Orfeuil 2004; Pucher and Renne 2003). In the outer suburbs, differences in mobility between the poorest and the wealthiest households are even greater (Morency et al. 2011), whether measured by number of trips or kilometers traveled. Low-income households make sparing use of cars, especially in such areas, to keep costs down. Even so, the low mobility of lower-income households does not mean that they are trapped within a limited living space. By making the most of the local environment, such households deploy a wide range of alternative practices to car mobility.

The interview responses suggested alternatives arranged around four pillars. The first is to limit what are considered to be superfluous car trips. The second is to mobilize local social networks. The third relates to use of public transport. The fourth involves increasing patronage of local resources. Close scrutiny of these practices with respect to daily mobility results in an initial set of proposed measures for enabling low-income households to cope better with higher fuel costs.

By moving into car dependent municipalities, householders highlighted the differences in their traveling patterns. Most of the French households surveyed in this research engaged in few, if any, leisure activities either outside the home or that required payment. Having limited levels of income, they did not consider such spending a priority, except for children who were often the main beneficiaries of leisure activities. Quantitative analyses of French HTS data display the lowest diversity of reasons for trips among low-income households.

Working-class, suburban, low-income households regularly engage in leisure activities that are directed more toward the home, such as do-it-yourself hobbies and gardening – entailing less travel (Coulangeon 2004). Generally, the priority given to activities enjoyed at home combined with tight budgetary constraints explains why the households that we surveyed did not go away on holiday often and engaged in few leisure activities that required payment:

> Holidays? We can't afford holidays, not with tax bills and everything . . . We don't go on holiday. Or it's with relatives, at a push, if they can put us up and everyone manages to have meals at the same price as we do . . . And, with holidays, you come back home and what's left of them? Memories? Memories are all very well, but I'd rather my daughter have her PlayStation, her toys, and play here all year round.
>
> *(Floriane, 37, married, housewife, one 11-year-old daughter)*

Leisure activities are often avoided because of previous wearing experiences when needing to travel to work and for shopping. Exhausting travel makes leisure activities at home or close to home more attractive. The proximity of the countryside allows for free leisure activities involving little car travel, such as going for walks or sporting activities on foot or by bike:

> Yes, because it is a bind to have to travel 10 km to go out on purpose, I like my house. I'm happy there, I have things to do whether my housework or my Internet job. So I don't feel like going out every 5 minutes.
>
> *(Colette, 48, single, alternates between unemployment and temporary work, homeowner)*

> I force myself a bit because, well, I don't want to give up sports, so I go running here . . . I take my trainers and I go running here, or get on my indoor trainer . . . I don't waste time travelling. But as a result I do a lot of sport all alone.
>
> *(Christophe, 49, married, after-sales technician, two daughters, homeowner)*

But some trips prove unavoidable. Financial constraints of home mortgages force households to have a regular occupation, or even overtime or second jobs for at least one member of the household. Accordingly, commuting occupies a central position in household travel and the budget allocated to traveling. However, the direct costs of such travels are offset in part financially. Spending on fuel and vehicle maintenance can be partly deducted from income tax. For workers who live and/or work in the same place, with the same working hours, car-sharing is often used to travel in congenial company and split the associated costs:

> I never drove alone for just about 15 years. I knew there were . . . people, youngsters from my municipality who found temporary work in my company so they came to see me . . . Well, at one time I travelled with a guy who lived in my municipality. We were on shift work at the time. I drove with him for 9 years. It's worth it given the cost of fuel. I think it's the answer. It's good, you take your car one week in three. I fill it up once a month instead of twice.
>
> *(Vincent, 54, married, supervisor, homeowner)*

Many women with children, especially babies and toddlers, decided to stop working because of the expense of travel and, above all, child care. However, given the very nature of the households interviewed, having just one person going out to work was not viable in the long term. When they joined the labor market again, the women interviewed tended to opt for work near home or at home – home help, nursery school help, self-employed – so they could manage the domestic sphere and be there for their children.

Child minders and people available to travel with other members were in demand in the family entourage. Usually relatives live close to low-income households (Bonvalet et al. 1999), as was the case with many of the householders we interviewed. In addition to relatives, the social network generally offered more help than just with the children and was involved in aspects of daily transport for low-income households. The social network provided assistance by offers of rides, lending vehicles or through advice- and information-sharing. The social network meant saving on certain trips by swapping and combining services for dropping off and picking up children, for lessons or other activities. The social network was a major provider of leisure activities, doing away with expenditure on paying for leisure and long-distance travel by combining leisure trips with visits to see family or friends. In exchange, low-income households gave up much free time for family and friends, rewarding or paying them to ensure such services continued. Options were "created, arranged, and planned," demonstrating and demanding the organizational capacity of householders (Clifton 2004).

Public transport is another alternative to car-based mobility used by low-income households, even in car dependent areas with less public transport. Some move to municipalities just with train stations – and no other services – so that they can travel without using the car. Among the interviewees, several were highly mobile, traveling long distances almost daily by public transport. For them, the cost of mobility was not an issue because their public transport subscription was covered by their employers. Asked why he "chose to go by train rather than by car," one interviewee responded:

> I made the choice for several reasons actually. Because public transport is more economical, because my employer refunds half. And, if I went by car, they wouldn't refund anything at all. And by car, it's a long drive. And when you get to Paris, it's hard to find anywhere to park. There are several good reasons, actually. It's longer and more tiring than by train.
>
> *(Akif, 40, married, dustman, tenant)*

Examination of the HTS confirms a higher propensity among lower-income households to use public transport for commuting. Given that in many of the outer suburbs provision is minimal, with limited services, concentrated around the morning and evening rush hours, it is difficult to use public transport for other purposes.

For shopping, proximity is sought – using local shops and those of nearby secondary centers – to the detriment of the major shopping areas in the immediate periphery of the central cities. The aim is to avoid pointless and tiring trips. Trips are often clustered "to avoid going for nothing" and require serious planning, once a week or once a month, depending on the days available, and are made without the children to prevent excess spending. Such trips are doubly tiring, both because of the traveling and the activities engaged in upon arrival. To avoid such chores, households consider less wearing alternatives, including home-growing, such as having a kitchen garden or orchard; picking, say mushrooms, and gathering, say wood; relying on bulk storage, such as freezing and preserving; use of the Internet for online banking, e-shopping, and click and collect systems; and sharing trips with others. Maximizing local resources went hand in hand with the job-seeking strategies of low-income households. Local work was preferred, these households having recognized the low gains and high costs of mobility given their skill and wage levels (Chapple 2001).

All told, low-income households living in the outer suburbs tend to drastically reduce their car mobility by avoiding what they see as superfluous travel; by relying on social networks for car-sharing, for other services or for leisure activities; by using public transport; and by replacing long trips by local ones. While these practices reveal the economic vulnerability of the households interviewed, they show that these households manage to minimize their fuel spending while maintaining a varied program of activities through local travel practices. To promote such practices, governments can consider two types of measures.

The first are traditional actions in transport, such as improving the public transport network coverage. However, because flows are very diffuse and densities low in these areas, such measures are expensive or have limited effectiveness. A system of subsidies for car mobility would also be very expensive, as shown by Fol et al. (2007), and would necessarily threaten public transport, which would lose its target market. An innovative answer might come from mobility management (Gärling and Schuitema 2007) and local actors participating in organizing carpooling or car-sharing arrangements. Even if they concern only a limited population and limited users, and even if they are currently based mainly in city centers, such mobility solutions might also offer alternatives to "traditional" car-based mobility in suburban areas.

Low-income households have shown their propensity to seize varied opportunities to make savings in their transport budgets, while no single comprehensive solution seems capable of solving the problem.

Apart from transport, a second type of intervention by government might improve the local commercial supplies. Thus the development of e-shopping, of local networks of collection points, where delivery costs would be lower than for home delivery, might be a sustainable solution in suburban areas assisting less car use by resident populations. Obviously, the costs of delivery and service in these areas remain a challenge. Who will pay the bill – customers, sellers, or transporters? Authorities have their part to play in diffusing and generalizing intermediate points, perhaps by public service delegation, as with the postal network or local utility networks.

Moving out and moving on

Some low-income households seem unable to cope with the costs of car-based mobility or constraints on their daily mobility. They leave the outermost suburbs for municipalities with more shops, services and public transport (Motte-Baumvol et al. 2010). This type of residential trajectory is not just anecdotal. The population census clearly identifies the growth of low-income households leaving the outer suburbs of French cities. Yet this trend is a minor phenomenon that probably only allows for marginal regulation of the numbers of low-income households in suburban areas. The proportion of low-income houses there has not dropped.

Low-income households leave the outer suburbs when the cost of mobility of living there is compounded by changes in occupation or personal circumstances, such as young people leaving home, people setting up home together as couples, deaths (of spouses or parents), divorce and separation, job postings and unemployment. In most cases, these events are not planned or expected by households when they move to suburban areas. The census data do not provide such a detailed view of household circumstances and changes, but do identify single-parent families and unemployed people as the most likely to leave the outer suburbs. The census data can be used to spot the main trends in the residential trajectories of these households because the place of residence of householders at the time of the previous census is shown. However, there is neither information about characteristics of any previous dwelling nor details about any change in an individual's job or marital status. This affects the findings to a degree that is hard to quantify. However, the scale of the observed phenomenon is such that it is not in doubt; only its scope is hard to determine.

While single-parent families are more likely to leave a municipality where car dependence is high, they often move to other municipalities where dependence is just as high. These areas continue to be attractive, probably because housing costs are similar and because they do not want to compound their difficult life episode by uprooting both the parent and children from the area central to their social networks of family and friends. The propensity of such households to leave municipalities where car dependence is high mainly seems to reflect circumstances associated with a separation. There might be a requirement to move out of accommodation occupied before the separation. The household might need to find more suitable housing for their new situation close to the social network on which households with children, and single-parent families especially, largely rely.

After her divorce, Claudie left her municipality for a village nearby so as to stay close to her ex-husband (they shared custody) and her parents who helped her with child care when she worked nights. Now she is living with a new partner but is keen to stay in the same area because "my children have grown up here, so . . . well, their lives are here. Their friends are here. They have everything, everything . . . And I wouldn't go back to the city."

Things are different for the unemployed. Those who move away from a highly car dependent municipality forgo lower housing costs for their knowledge and experience of the local employment market and its opportunities. Conversely, unemployed householders whose previous residence was not in car dependent areas are more likely to move. Presumably, the tradeoff between housing prices and job opportunities is not judged in the same way by these households, perhaps because of specific residential experiences.

While lifecycle stage and changes are decisive in explaining why low-income households leave highly car dependent areas, they do not always shed light on the choice of destination, beyond the major trends described earlier for single-parent families and the unemployed. We need to look at other factors for a clearer understanding of the residential trajectories of these households, and especially their access to social housing, their more or less traumatic experience of car dependence and, lastly, their immediate past experience and social network.

Access to social housing is a significant factor in explaining why people move from highly car dependent municipalities. Reduced income in addition to separation or job loss gives greater access to social housing for which the rent is lower. So, whatever the earlier occupation status, the probability of moving is higher for households with access to social housing at their place of destination. But access to social housing also determines the places to which people move.

Social housing is generally situated in a more central location that has more shops and services, and provides access to a larger labor market. Yet the availability of social housing leads some low-income households to favor destinations that are just as car dependent as the municipality they have left. This counterintuitive finding is explained by the fact that the outer suburbs have an attractive supply of social housing. Furthermore, this choice of social housing in car dependent municipalities can probably be explained by a number of factors. Access to inexpensive housing eases the strain on the household budget and offers more leeway for mobility. Social housing in more central areas is often concentrated in districts of ill repute. Certain choices enable some households, especially single-parent families, to remain in the heart of their social networks.

Where people move to depends on another factor, namely on whether their experience of car dependence has been more or less traumatic. The cost of mobility is as important a factor as how tiring it might be. Long commuting times, congestion, infrequent public transport at inconvenient times, fuel budgets and vehicle wear and tear are all financial and psychological costs that make living in the outer suburbs intolerable:

> The real problem for me was the distance. It took me about three-quarters of an hour. It was hard going. In the evenings, I was just about fresh. But, in the mornings, when I had to drive, it was tiring. [Using public transport would have been] less tiring. [The problem is that it takes longer and is hard to arrange], it's extra stress . . . Whereas by car you can travel door-to-door. It's more direct . . . But [public transport] would have been far more economical because I had a free pass and it would have meant I could have saved up quite a bit.
>
> [My car budget] was huge. I filled the tank about every two weeks, about €150 or even a bit more per month just for short trips and work . . . So it was substantial.
>
> *(Réda, 27, single, police officer in Paris, no children, living in Paris in social housing)*

Apart from actual trips, several respondents had great fears of the consequences of being located in the outer suburbs at some time in the future. The fear of themselves or their children being isolated and the fear of accidents were reasons that prompted households to relocate to more densely populated areas.

This isolation was felt much more when households were confronted with a near total absence of shops and services, which they were not necessarily used to:

> And since I was shut up indoors all day long, when he came home in the evening all I wanted to do was to get out. We stayed out at the shops until 8 pm, until closing time, because it was the only social time in the day. And then, one day, I said: "Look, it's all very well, but there's not much more choice about it, we have to move, we have to move because it's not going to work out." . . . And we moved to a flat that is very bright because there are big windows everywhere. It is very big, and great for me, in the city centre, with, well, two bus services, because I don't drive. Which means I feel, well, free to go out in the middle of the day if I have a mind to. To pick up my children, to take the bus and come and go like I want to.
>
> *(Justine, 33, married, desktop publishing operator, two boys aged 9 and 8 years, in social housing in an urban municipality)*

When they have lived in urban areas or in suburban areas with more amenities, people find it difficult to adapt to areas with few amenities and prefer to move away. Often the low-income households who give up trying to be homeowners in the outer suburbs are those with special ties to urban areas, especially through work, networks of family and friends or personal and family history. Conversely, households with the same kinds of ties with suburban areas look to stay, despite the costs of concessions in terms of mobility and standard of living:

> It was funny because I lived in Noisiel for 20 years and so I was used to having everything to hand, a small shopping center just two underground stops away, getting out and about a lot . . . There's always something nearby, whereas at Gouvernes, you get here, and there's practically nothing, a baker's, a garage, there isn't much. So it was a bit difficult and then, even if you had a car . . . well, he had the car, we were a bit further away from my grandmother who was still there and from his family, who still lived in Noisiel. We were in Noisiel more than we were at our place. We only came home to sleep, or pretty much so.
>
> *(Coralie, 34, married, police officer in Paris, one son aged 13 years, in social housing in an urban municipality)*

While the growing awareness of the consequences of car dependence, such as cost and isolation, varies from one person to another, it is materialized by a return to urban areas sooner or later. The absence of a local social network is decisive to explain this choice of relocation. Unlike households still living in car dependent areas, the social network of those who have moved away is often far less immediate. Without a social network, it is difficult to offset any difficulties related to traveling and to car dependence.

Once relocated in less car dependent areas, households change their traveling habits and see a marked fall in transport costs. They do not necessarily give up on driving – many keep their cars. Other outgoings offset the financial saving made: more activities and greater temptations to spend. However, they turn more often to public transport or to other means, such as walking or cycling to carry out their activities. Accordingly, the scope of their daily activities is smaller. However, many admit underestimating the savings made, given the rise in fuel prices and that they still keep their cars. Although the costs associated with car-based mobility fall, households still have the fixed costs associated with their use, such as purchase and insurance. However, to adapt to the constraints of their new area of residence, some tend to make changes by keeping just one car if they used to have two, buying smaller vehicles, or switching from cars to other individual motorized transport, such as motorbikes or scooters:

> Out there, there was more spending on cars, I had to fill up nearly every week. Here it's every two weeks. Apart from when I go down for the weekend, but logically, a full tank lasts me longer here because I have no more commuting. I use the car less than I did out there. It was always the car because there aren't many trains.
>
> *(Lylie, 26, single, temporary secretary, one 14-month-old son, lives in social housing in an urban municipality)*

> I had a big petrol budget at the time. And now, paradoxically, I have a bus-pass and I use the car even so. I come by car twice a week because on Wednesdays I bring my son to the leisure center. So I use the car because it's a little out of the way.
>
> *(Catherine, 44, divorced, medical secretary, two children aged 9 and 14 years, tenant in an urban municipality)*

All told, going back to more densely populated areas involves giving up cars or at the very least some fuel costs. Such returns are possible through subsidized housing, by paying rents below the market price. Whether from government or from private firms, housing support makes it possible for these households to move to places that they could not have afforded otherwise. So, by promoting the return of households to urban areas through support for rents and to home buyers, government has a role to play in reducing car use.

However, a return to inner urban areas is impossible unless housing characteristics correspond to people's values. Having lived in the outer suburbs, households tend to prefer mixed sectors, away from what is seen as the negative image of dormitory suburbs. Similarly, many people find it difficult to give up the size of their housing or its detached character. For this reason, and because some ties continue to hold there, many low-income households have opted for social housing in areas that are highly dependent on cars close by the municipalities they lived in formerly. In these cases, moving house may partially or temporarily alleviate transport budgets by proximity to their place of work, social networks, family and friends, which remain under threat from higher energy costs.

Access to social housing stock allows households to substantially reduce their housing budgets to provide extra leeway for mobility. Steering new building projects toward municipalities that are only weakly car dependent is a policy option: transport costs may remain a problem while allowing some households to remain in familiar areas, close to their social networks. Favoring more secondary centers can strike a better balance between aspirations to a life in the "countryside" and a level of amenities and services that is in keeping with their expectations and urban habits. Moreover, the cost of this social housing for the welfare state is lower and can be built faster than in more densely populated central areas where there is less land available to build on and where costs of land and building are higher.

International comparisons with French ways of escaping car dependency

Are the circumstances and practices of low-income households in the outer suburbs and prospects in terms of public policy outlined in this chapter specific to France, or do they concern other European countries or other continents? The answer is necessarily nuanced given marked social, economic, institutional and political differences. An analysis of international literature about the different questions addressed in this chapter provided many responses.

With respect to the initial analysis of the alternative practices of low-income households relating to expensive car mobility, there are marked similarities with the state of affairs described by Lucas (2006) for the UK. There are many similarities with findings reported by Morency et al. (2011) for Canada, and Delbosc and Currie (2011) for Australian cities. In the US, where car use is even greater, alternatives to generalized car use are fewer, but there are similarities with the mobility of the poorest households, especially immigrant families, described by Lovejoy and Handy (2011). The measures proposed with regard to this analysis, such as mobility management or local offers for carpooling and car-sharing, are already implemented in the countries mentioned to some degree.

Differences among countries seem far greater with respect to concerns addressed in the second part of the chapter. First, there is no research that shows that low-income households have better exit strategies from highly car dependent municipalities. Nor are there any divergent results, precisely because the question does not seem to have been investigated elsewhere as far as we are aware. The scale of the phenomenon may well vary with the social structures of the cities and the local and national real-estate markets, where there are significant differences. Moreover, full residential trajectories and the public policy measures contemplated may vary greatly by country, given that both rely on social housing. But there is less welfare-state provision for housing in many other countries than in France, even if it is found in varying degrees throughout Europe.

References

Berri, A. (2007) *Residential Location and Household Expenditures on Transport and Housing: The Example of the Greater Paris Region*, Berkeley: 11th World Conference on Transport Research.

Bonvalet, C. Gotman, A. and Y. Grafmeyer (1999) *La famille et ses Proches: L'Aménagement des Territoires*. Paris: INED/PUF.

Cavailhès, J. and H. Selod (2003) "Ségrégation Sociale et Périurbanisation," *Inra Sciences Sociales* 1–2(3): 1–4.

Chapple, K. (2001) "Time to Work: Job Search Strategies and Commute Time for Women on Welfare in San Francisco," *Journal of Urban Affairs* 23(2): 155–73.

Clifton, K.J. (2004) "Mobility Strategies and Food Shopping for Low-Income Families," *Journal of Planning Education and Research* 23: 402–13.

Coulangeon, P. (2004) "Classes Sociales, Pratiques Culturelles Et Styles De Vie," *Sociologie et Sociétés* 36(1): 59–85.

Delbosc, A. and G. Currie (2011) "The Spatial Context of Transport Disadvantage, Social Exclusion and Well-being," *Journal of Transport Geography* 19(6): 1130–37.

Dodson, J. and N. Sipe (2008) "Shocking the Suburbs: Urban Location, Homeownership and Oil Vulnerability in the Australian City," *Housing Studies* 23(3): 377–401.

Fol, S., Dupuy, G. and O. Coutard (2007) "Transport Policy and the Car Divide in the UK, the US and France: Beyond the Environmental Debate," International Journal of Urban and Regional Research 31(4): 802–18.

Gärling, T. and G. Schuitema (2007) "Travel Demand Management Targeting Reduced Private Car Use: Effectiveness, Public Acceptability and Political Feasibility," *Journal of Social Issues* 63(1): 139–53.

Lovejoy, K. and S. Handy (2011) "Social Networks as a Source of Private-Vehicle Transportation: The Practice of Getting Rides and Borrowing Vehicles among Mexican Immigrants in California," *Transportation Research Part A* 45: 248–57.

Lucas, K. (2006) "Providing Transport for Social Inclusion within a Framework for Environmental Justice in the UK," *Transportation Research Part A* 40: 801–9.

Morency, C., Paez, A., Roorda, M.J., Mercado, R. and S. Farber (2011) "Distance Traveled in Three Canadian Cities: Spatial Analysis from the Perspective of Vulnerable Population Segments," *Journal of Transport Geography* 19: 39–50.

Motte-Baumvol, B., Massot, M.-H. and A. Byrd (2010) "Escaping Car Dependence in the Outer Suburbs of Paris," *Urban Studies* 47(3): 604–19.

Orfeuil, J.-P. (2004) "Accessibilité, Mobilité, Inégalités: Regards sur la Question en France Aujourd'hui," in J.-P. Orfeuil (ed.) *Transports, Pauvretés, Exclusions: Pouvoir Bouger pour s'en Sortir*, La Tour d'Aigues: Editions de l'Aube, 27–47.

Pucher, J. and J. Renne (2003) "Socioeconomics of Urban Travel: Evidence from the 2001 NHTS," *Transportation Quarterly* 57(3): 49–77.

13

GREENSPACE AFTER PETROLEUM

From freeways to greenways

Jason Byrne

George Seddon (1972: 230) once observed that "parkland would probably . . . succumb . . . to the motor car [because] . . . parklands . . . have a fatal charm for freeway engineers." In the car-dominated and oil-fueled twentieth century, this notion seemed inevitable, but in a post-oil future, the converse may well be true. Although it might not be the first suggestion that comes to mind when considering the urgent imperative imposed by peak oil planning, greenspace will be a vitally important concern this century. Five reasons for this follow.

First, parks, greenways and other urban greenspaces are highly modified landscapes and require constant maintenance. The lawn and vegetation in these greenspaces is highly dependent on petroleum derivatives due to the fuel and chemicals used, for instance, in maintaining fences and firebreaks, pruning and planting trees, controlling weeds and pests, cleaning facilities and landscaping (Robbins and Birkenholtz 2003). Many of these activities use petroleum products. Even naturalistic greenspaces require ongoing management.

Second, greenspace confers a wide range of benefits on built environments – arguably far greater than petroleum ever has. Beyond social and economic benefits, greenspaces provide an array of "ecosystem services" necessary for planetary life support (Bolund and Hunhammar 1999). Decades of automobile dependent development have fragmented habitats, fouled air and water, ravaged ecosystems and undermined the productive capacity of urban ecologies (O'Rourke and Connolly 2003). Peak oil creates an opportunity to reverse these harms.

Third, car dependence is implicated in the obesity epidemic now gripping cities within both the Global North and Global South. Researchers are finding close correlations between car-oriented built environments, mental and physical illness and premature death (Badland and Schofield 2005; Saelens et al. 2003). It seems the car dependent landscapes created in the twentieth century may now be killing us. But the good news is that active lifestyles fostered by greenspace use can counter many of these problems (Bedimo-Rung et al. 2005; Giles-Corti et al. 2005).

Fourth, rapacious car-oriented sprawl of the twentieth century has left many built environments bereft of greenspace. A decade of environmental justice research has shown that green and open spaces are not uniformly distributed throughout cities (Jennings et al. 2012). Places occupied by wealthier, white residents tend to have better access to greenspaces, and are less exposed to major roads and associated noise and pollution. Not only is this unfair, but also may be condemning people of color, such as African Americans and Latino/as, to shorter and insalubrious lives (Abercrombie et al. 2008; Bedimo-Rung et al. 2005). Converting some roads to parks could redress this inequity.

Fifth, the urban consolidation (densification) policies of many governments in the Global North and Global South are concentrating residents in inner-city areas, which typically have fewer greenspaces than outer suburban areas (Byrne et al. 2010). This trend is likely to continue as cheap oil declines. How can we ensure that regional and national parks and greenspaces designed in the automobile age remain accessible in a post-petroleum world, and that those residents of denser urban environments will have access to a variety of urban greenspaces, such as pocket parks?

This chapter focuses on the vexing question of how we will design, access, use and maintain parks and other greenspaces post-petroleum. We begin by defining what is meant by "greenspace" and identify its many benefits. Next, we examine how access to parks has been shaped by cheap oil. We consider a case of re-greening a built environment and ask if it provides cues for greenspace planning in the post-oil future. We conclude by suggesting some policy implications that will require attention in the future.

This chapter is framed around five assumptions, as follows. First, oil scarcity and associated social vulnerability will manifest before mass-produced, affordable and socially and environmentally acceptable substitutes for petroleum products (such as algae-based biofuels) can be widely adopted. Second, the inertia locked into built environments will mean that people will tend to remain where they currently live, albeit with some major adjustments to their travel behavior and work patterns. Third, peak oil will not result in widespread social revolution, and capitalism will remain the dominant economic mode of production, although adapted to new conditions – in other words, the future will look much like today. Fourth, existing modes of mass transporting goods, including food, over vast distances may no longer be viable or feasible, necessitating local and regional production and distribution. Fifth, remaining oil will become increasingly expensive to discover, mine, process and distribute. Scarce oil supplies are likely to be rationed for essential services and infrastructure maintenance while society transitions to alternatives, such as biogas, natural gas and electric vehicles (Newman 2013).

Of course we cannot accurately predict the future, and many of these assumptions are arguable. In the limited space of a chapter they cannot be addressed in more than a cursory fashion, but they do serve to help us imagine alternatives.

Defining and prioritizing greenspaces

Greenspaces are common elements in built environments. They include formal areas, such as a variety of parks – from pocket parks through to national parks and recreation areas, such as playing fields and ovals, botanic gardens and golf courses – the sides of streams and riverbanks, and other riparian areas. Greenspaces include street trees and urban forests, cemeteries, road verges, community gardens, school grounds and publicly accessible campuses. Among elements usually included as greenspace are rock walls, rooftop gardens, greenways, utility easements and drainage corridors – provided they contain vegetation. Private areas, such as yards and gardens, common property around apartment buildings and corporate and commercial green areas, can also constitute greenspace (Byrne 2013).

Greenspaces confer a wide range of benefits upon both users and non-users. There is insufficient space here to discuss their myriad benefits as others, such as Byrne and Wolch (2009), have done. Suffice to say, greenspaces provide "free" ecosystem service benefits to built environments, including economic, social and environmental benefits (Bolund and Hunhammar 1999). Economic benefits include promoting tourism, reducing pollution, providing food, decreasing health care expenses, increasing property values and reducing energy expenditure. Social benefits include relieving stress and anxiety, promoting mental restoration, hastening disease recovery, fostering active living, encouraging social interaction and conviviality, moderating incivility, promoting child development and cultivating environmental values. Environmental benefits include regulating ambient temperatures, filtering air, reducing noise, lowering wind speeds, sequestering carbon, attenuating storm water, providing habitats and preserving biodiversity. Many of these benefits have direct relevance to post-petroleum planning, and planners need to consider how they can be protected and managed following the end of cheap oil. This is because synergistic effects between greenspace and transportation have been poorly studied to date and they warrant closer attention.

For example, safe, attractive and well-maintained greenspace has been demonstrated to foster increased walking and cycling (Bedimo-Rung et al. 2005; Giles-Corti et al. 2005). This is especially where parks are large, attractive, offer a range of facilities and are close to home (Giles-Corti et al. 2005). But lack of access to public transportation is a significant impediment to park use (Mowen et al. 2005). This is a problem. Increased physical activity provides dividends through lower health care costs and higher levels of productivity – thus freeing up finances that can be used for retrofitting built environments and urban infrastructure (like public transportation). Integrating public transportation and greenspace can have flow-on benefits, such as increasing levels of transit patronage, boosting multipurpose trips (shopping and recreation) and increasing property values (Crompton 2005; Kenworthy 2007). Retrofitting greenspace to built environments can provide a mechanism for value capture or "betterment." Providing new greenspaces increases nearby property values. Higher land values and thus more property taxes can be used to pay for further greening and public transport investment.

More greenspace can reduce traffic speeds, lower levels of stress and anxiety, promote improved concentration and recovery from disease, and reduce incivility (Byrne et al. 2010). Research has shown that greener neighborhoods have higher levels of social interaction and lower levels of crime (Kuo and Sullivan 2001a, 2001b). Extra greenspace can cool urban environments too, an important consideration with climate change and heat island impacts (Byrne and Wolch 2009). In a post-oil world, we can assume that private automobile use will decline. With decreased car use, land presently occupied by roads could become available for other purposes. If the primary modes of transportation in a post-oil future are walking, cycling and public transport, then land currently used for roads could be converted to urban forests, community gardens, storm water management, power generation, wildlife habitat, community facilities, and parks and recreation areas (Schroepfer and Hee 2008). It could also be used for biofuel production. This begs the question: "Why hasn't more attention been given to the relationship between transportation and greenspace?" Surprisingly, parks and automobiles do have intersecting histories.

How oil and automotive transport shaped park development

The first urban greenspaces, such as commons, squares, plazas and parade grounds, developed long before the automobile. The world's first public parks in the UK and US were oriented toward walking and horse-drawn carriages, not cars (Cranz 1982). Indeed, many early parks, like Central Park in New York City, featured "promenades" where people could stroll and watch the scenery. However, these early parks were criticized for being outside "walking distance" and too far away from public transport (Rosenzweig and Blackmar 1992). As new parks were developed in cities, park reformers devised the widely adopted standard that residents should have to walk no further than one-quarter of a mile to access a park (Rosenzweig and Blackmar 1992). This notion of "walking distance" has since become enshrined in park planning. In the US, UK and Australia, park planning standards all require parks within one-quarter to one-half of a mile (400–800 m) of most urban residents (Byrne 2013). But there are also strong connections between parks and public transportation.

Many parks in cities such as Los Angeles were developed as elements of property development and land speculation schemes. Land developers would often build a park at the terminus of a tramline, and advertise it as a destination for family picnics and entertainment (Hall 1977). Early amusement parks developed this way (Shaw 1986). While people were enjoying the sights, they could also learn about the adjacent property development, and would be encouraged to purchase land. Henry Huntington used this strategy effectively in Los Angeles, and was known to build tramlines to associates' properties in return for company shares, landholdings and a right-of-way for the railway line (Byrne et al. 2007). Many parks also developed in lockstep with metropolitan subway systems. Subway lines had stops at public parks that were already popular destinations, and new parks developed closer to subways had higher levels of patronage (Cranz 1982). Some early national parks were also accessible by train, such as Yosemite National Park (US) or Sydney's Royal National Park (Australia).

But, as automobiles became more affordable, new interactions between parks and transport emerged. The first freeways in New York City and Los Angeles for example, were imagined as "parkways" where motorists could drive along expansive green corridors for both rapid transport and leisure (Hise and Deverell 2000). With increasing reliance on cheap oil and private automobile use, planners developed alternative types of parks. No longer was "walking distance" a constraint. Larger regional parks were provided on the outskirts of cities, designed only for car accessibility. Many new national parks were also developed with the car in mind (Byrne and Wolch 2009). Not surprisingly, private automobile use undermined walking-oriented park standards.

Recent studies of metropolitan park systems in the US have found that many municipalities seldom achieved the park standards articulated in their planning instruments; most have failed to provide parks within walking distance (Harnik and Simms 2004). Public health researchers have found that walking distance is a problematic concept because residents are often unable to accurately judge their distance from a park (Macintyre et al. 2008). Even parks within prescribed walking distances may be beyond the time, physical or motivational capabilities of many residents (Giles-Corti et al. 2005). Will peak oil and pressures associated with climate change alter these relationships? If so, what issues will planners and design professionals need to consider when planning for greenspace?

Factors in planning for greenspace after oil

Parks and recreation areas are not the only types of greenspace in built environments. Many other types of greenspaces and open spaces characterize cities, towns and villages. They include drains, stream and riverbanks, railway and power transmission easements, road verges, vacant lots and derelict brownfields, such as former industrial areas. Enormous effort is currently expended in maintaining such spaces, even if they may appear unkempt. Maintenance regimes currently depend on petroleum products, such as fuel, herbicides and lubricants. What will happen to such places when oil is a scarce commodity? Will oil substitutes become readily available, such as algae-based fuels, lubricants and chemicals? If not, how will greenspaces be maintained and by whom – if at all? What are the implications if such spaces are left unmanaged?

Greenspace maintenance activities currently rely intensively on petroleum products. Petrol- or diesel-fueled vehicles and equipment – chainsaws, leaf-blowers, tractors, brush-cutters, line-trimmers, lawn-movers and mulchers – need oil. Park workers need to be able to travel to and around parks: currently many rely on cars, trucks and even boats to do so. Plastic pots are widely used in plant propagation; without oil we'll need substitutes. Indeed, without oil, or a cheap and readily substitutable, competitive non-food alternative, such as algae-based biodiesel fuel or electric vehicles, in the future park maintenance will need to be radically different. How we plan, use and manage parks and greenspaces will have to be rethought.

In a post-oil future, we may have to redesign parks for less energy-intensive maintenance. It is likely that we will need to change the type of greenery in our parks and greenspaces. We may have to cultivate community acceptance of less manicured greenspaces. Without petrochemicals – such as pesticides, herbicides and petroleum-based fuels used in current practices (Figure 13.1 and Figure 13.2) – changed practices may include the type and frequency of lawn mowing, species of lawn cultivated, use of

FIGURE 13.1 Worker spraying insecticide on park trees in Singapore's Gardens by the Bay

Photographer: Jason Byrne.

FIGURE 13.2 Workers trimming park grass in Singapore's Gardens by the Bay

Photographer: Jason Byrne.

pest-resistant plants and use of companion planting. Landscapers may also need to reconsider the expense and difficulty of moving large quantities of material.

If greenspaces are left unmanaged, crime rates might increase, including violent crimes against women, and parks could become liabilities rather than treasured assets (Brownlow 2005). Without an army of low-paid workers or alternative fuels for machinery, landscape design may need to become more modest. Park-based living spaces for rangers and park workers could be required, as was once common in the West and still is in China. We will probably need different types of cleaning equipment too, as well as different cleaners, solvents and air-fresheners; petroleum-based plastics will be prohibitively expensive and an alternative is yet to become available.

Emma Marris (2011) suggests that the future of post-petroleum greenspace might already be found in many "neglected" landscapes within cities, where vegetation generates spontaneously, both as so-called weeds and natives. Here, sometimes, animals prowl, children play and people walk their dogs or grow vegetables. Such places are gaining increasing recognition as important "green infrastructure" within cities. Existing parks that have evolved out of post-petroleum landscapes provide cues for the future. The Kenneth Hahn State Park in Los Angeles is an example. That park was once an operational oilfield but is now a treasured urban greenspace (Byrne et al. 2007).

We can look to abandoned transport infrastructure for ideas too – infrastructure that has been repurposed. The Highline Park in New York City is an innovative example. Built atop an old railway line, this urban greenspace is cherished by residents and visitors. It is not manicured. Pieces of relict rail-line are interspersed among flowers. Yet it attracts international tourists and locals. Jennifer Wolch and her collaborators have recently shown how back alleys across the US are being converted from asphalt eyesores to multipurpose spaces, if not verdant oases (Newell et al. 2013). But how feasible is it to convert large areas of surplus transport infrastructure to greenspaces? And, is it worth the cost? The Cheonggyecheon Stream restoration project in Korea provides some insights.

From freeways to greenways: The Cheonggyecheon Stream restoration experience

The Cheonggyecheon Stream is a tributary of the Han River and flows through downtown Seoul, a metropolis of approximately twenty-six million residents and the capital city of South Korea. Following the end of the Korean War, when Seoul burgeoned, the banks of the stream became cluttered with housing, factories, warehouses and other land uses. According to the Landscape Architecture Foundation (2011), beginning in 1958, the stream was progressively channelized and gradually enclosed in concrete; by 1976 a length of 5.6 km had been capped by an elevated freeway. But, after thirty years of use, the elevated expressway was rapidly deteriorating. Methane and nitrogen dioxide gases around the structure exceeded acceptable levels, concrete pylons and supports had deteriorated, and the structure had become unsafe (Seoul Metropolitan Facilities Management Corporation 2013). The Seoul Metropolitan Government faced a hefty repair bill (Landscape Architecture Foundation 2011). In 2003, under the management of then city mayor and future president Lee Myung-bak, a radical idea was proposed – to remove the freeway and replace it with a landscaped greenspace – see Figure 13.3.

Removing the freeway initially met with some resistance due to its cost (ultimately US$280 million) and the fact that the freeway carried 169,000 vehicles a day. Seoul is a large, densely populated megacity, and transport engineers feared that the project would create traffic congestion. However, the municipal government acknowledged that repairing the structure would cost around US$90 million, and wider benefits were at stake. The official objectives of the restoration project included improving resident

FIGURE 13.3 Transformation of Cheonggyecheon Stream: as a drain, before channelization (top left); as a freeway, c. 2000 (top right); as a site under restoration (bottom left); once restored, c. 2006 (bottom right)

Source: Seoul Metropolitan Government.

wellbeing, stimulating economic development, and improving environmental quality, under the broader rubric of making Seoul a sustainable city. Performance indicators measuring implementation of these objectives appear to have been met or exceeded (Landscape Architecture Foundation 2011).

According to the Seoul Metropolitan Government (Seoul Metropolitan Facilities Management Corporation 2013), restoration of the Cheonggyecheon Stream stimulated around US$2 billion in tourism and investment income. Biodiversity in the corridor also increased markedly from 2003 to 2008: plant species increased (from 62 to 308 species), as did aquatic invertebrates (from 5 to 53) and terrestrial insect species (from 15 to 192), fish species (from 4 to 25), amphibians (from 4 to 8), bird species (from 6 to 36), and mammals from (2 to 4) (Landscape Architecture Foundation 2011). Over the same period, temperatures decreased notably, between 3 and 6 degrees – partly due to an increase in wind speed, up around 8 percent – and air pollution decreased, with a 34 percent reduction in nitrogen dioxide and a 19 percent decline in big or coarse particulate matter (PM10) (Seoul Metropolitan Facilities Management Corporation 2013).

Moreover, the project increased surrounding property values by up to 50 percent in 2008 compared with 2003, and markedly improved use of public transportation. Much of the freeway was recycled in the project: 100 percent of scrap iron and about 95 percent of other materials (Landscape Architecture Foundation 2011). The project was not without its detractors, however, and there were local demonstrations. The Seoul Metropolitan Government eventually held over 4,200 meetings with residents, community groups and businesses to develop consensus.

The site now hosts museum and exhibition spaces, street performances, and cultural, sporting and musical events. It has become a tourism destination and is functioning as an incubator for spin-off beneficial impacts. Unfortunately, health data is not yet available, but the lower level of air pollutants has likely decreased respiratory diseases, such as asthma. Higher levels of physical activity fostered by the new greenspace are evident, and are likely to have concomitant health benefits. Significantly, this restoration project differs from other well-known restoration projects, such as the Madrid Rio, because it did not relocate or bury an existing main arterial road, but instead removed one altogether (Kimmelman 2011).

Policy recommendations

Greenspaces, such as the Cheonggyecheon Stream restoration project, are becoming increasingly common. Indeed, both the Sydney and London Olympics featured parks created through ecological restoration of former brownfields. Repurposed green infrastructure is evolving in cities across the globe. Los Angeles for example, is currently revitalizing the downtown section of its dilapidated river, converting surplus rail marshaling yards into a large urban park. The City of Sydney is contemplating removing the Cahill Expressway and replacing it with a park and apartment buildings. In Hangzhou (China), surplus land adjacent the main rail line to Shanghai was combined with land, by demolishing obsolete factories, to create the now vibrant Chengdong Park. The Amsterdam Westerpark, another rail-side park, was once a gasworks. The Greenwich Ecology Park, on the banks of the Thames River and accessible via the city's subway, was once "wasteland" (Byrne 2013).

However, these types of greenspace projects that point in the post-oil direction pose some policy dilemmas and challenges. We need to ask some important questions about how such greenspaces align with the transition to a post-petroleum future. If we plan to redevelop road infrastructure into greenspaces, which roads should be converted? When should this be done? How will we pay for the costs? Who will do the work?

Demolishing freeways and factories, and restoring brownfields, are energy intensive and relatively expensive (although long-term dividends can amortize costs). In a post-petroleum world, where will the energy and finances come from to undertake such projects? Is a betterment tax feasible? Where will the fuel come from to run the heavy machinery needed in demolition and reconstruction?

If we remake post-petroleum landscapes into greenspaces and housing, how can we ensure that rising property values will not displace poor and vulnerable people? Environmental justice research (Wolch et al. 2014) demonstrates that gentrification often accompanies green-infrastructure development, and can displace vulnerable populations. We may need to re-evaluate our park standards in post-petroleum cities, reasserting the importance of greenspaces that can be accessed by walking and cycling. But this raises further questions: How many "local" parks are needed? What size should the parks be? Where should they be located? Who will maintain them, and how? What facilities should they include, if any? How can we ensure that they will meet residents' needs?

Another important question is: How can we better plan for modal integration in greenspaces? Post-petroleum parks will need to facilitate walking and cycling and be accessible by public transport. But we currently know very little about the percentage of urban greenspaces within cities that are accessible via public transport. This research is urgently needed. Are all populations covered? If not, this is another potential social and environmental justice issue.

Post-petroleum greenspaces will also need to be multifunctional. They will have to provide important recreational functions, but they will also need to be spaces for growing food, rearing animals, urban wildlife and nature conservation, and cultural pursuits. How can we maximize the use of these spaces but avoid conflicts between users? How can we plan for flexible designs and park programs?

Some of the answers to these questions are obvious. People living in higher densities must have good access to parks and open space. Repurposed greenspaces must be within easy access to existing and planned public transport. We might begin by identifying those areas that currently have poor access to greenspace.

This may necessitate developing a park access vulnerability index, much like the VAMPIRE Index for oil vulnerability developed by Dodson and Sipe (2008) or the "park congestion" index of Sister et al. (2010).

Redevelopment of urban greenspaces cannot, and should not, solely occur on a massive scale, such as with the Cheonggyecheon Stream restoration project. Such projects need not be top-down, either. Indeed, transformations must occur across a variety of scales, but especially at the local scale, if they are to be effective. The trend in San Francisco and New York City for turning small sections of road and sidewalk into "parklets" is a good example (Brozen and Loukaitou-Sideris 2013). This initiative is being driven from the community level and has the power to transform built environments. An Australian model is Melbourne Water's retrofitting of green areas to roads, at the metropolitan scale, as part of water-sensitive urban design.

Conclusion

In the post-petroleum landscapes of the future we are still likely to see widespread private automobile use, such as electric vehicles and various types of hybrid vehicles. But the cost of transitioning to new vehicle fleets, and the time required to develop commercially viable petroleum replacements, such as algae-based biofuels, may inhibit the level of car use seen in the twentieth century. Provided that public transportation systems are developed to meet the needs of residents with personal mobility constraints, this will not prove negative.

Less reliance on the private automobile in the future could well mean using streets for a variety of new purposes. If we do not need roads for cars, roads can be converted into water catchments, communal food gardens, cycle ways, wildlife habitat, urban forests, power generation areas and perhaps even places for growing biofuels. As petroleum use declines there are enormous opportunities to re-green cities and improve not only human health and wellbeing, but also to heal an ailing planet. The future looks verdant.

References

Abercrombie, L.C., Sallis, J.F., Conway, T.L., Frank, L.D., Saelens, B.E. and J.E. Chapman (2008) "Income and Racial Disparities in Access to Public Parks and Private Recreation Facilities," *American Journal of Preventive Medicine* 34(1): 9–15.

Badland, H. and G. Schofield (2005) "Transport, Urban Design, and Physical Activity: An Evidence-Based Update," *Transportation Research Part D: Transport and Environment* 10(3): 177–96.

Bedimo-Rung, A.L., Mowen, A.J. and D.A. Cohen (2005) "The Significance of Parks to Physical Activity and Public health: A Conceptual Model," *American Journal of Preventive Medicine* 28(2): 159–68.

Bolund, P. and S. Hunhammar (1999) "Ecosystem Services in Urban Areas," *Ecological Economics* 29(2): 293–301.

Brownlow, A. (2005) "Inherited Fragmentations and Narratives of Environmental Control in Entrepreneurial Philadelphia," in N. Heynen, E. Swyngedouw and M. Kaika (eds) *In the Nature of Cities*, New York: Routledge, 208–25.

Brozen, M. and A. Loukaitou-Sideris (2013) "Reclaiming the Right-of-Way: Best Practices for Implementing and Designing Parklets," *Transportation Research Board 92nd Annual Meeting*, Washington, DC, 1–13.

Byrne, J. (2013) "Greenspace Planning: Problems with Standards, Lessons from Research, and Best Practices," *Citygreen* (6): 50–55.

Byrne, J., Kendrick, M. and D. Sroaf (2007) "The Park Made of Oil: Towards a Historical Political Ecology of the Kenneth Hahn State Recreation Area," *Local Environment* 12(2): 153–81.

Byrne, J., Sipe, N. and G. Searle (2010) "Green Around the Gills? The Challenge of Density for Urban Greenspace Planning in SEQ," *Australian Planner* 47(3): 162–77.

Byrne, J. and J. Wolch (2009) "Nature, Race, and Parks: Past Research and Future Directions for Geographic Research," *Progress in Human Geography* 33(6): 743–65.

Cranz, G. (1982) *The Politics of Park Design: A History of Urban Parks*, Cambridge, MA: Harvard University Press.

Crompton, J.L. (2005) "The Impact of Parks on Property Values: Empirical Evidence from the Past Two Decades in the United States," *Managing Leisure* 10(4): 203–18.

Dodson, J. and N. Sipe (2008) "Shocking the Suburbs: Urban Location, Homeownership and Oil Vulnerability in the Australian city," *Housing Studies* 23(3): 377–401.

Giles-Corti, B., Broomhall, M.H., Knuiman, M., Collins, C., Douglas, K., Ng, K., Lange, A. and R.J. Donovan (2005) "Increasing Walking: How Important is Distance to, Attractiveness, and Size of Public Open Space?" *American Journal of Preventive Medicine* 28(2, Supplement 2): 169–76.

Hall, M. (1977) "The Park at the End of the Trolley," *Landscape* 22(1): 11–18.

Harnik, P. and J. Simms (2004) "Parks: How Far is Too Far?" *Planning Magazine* 70(11): 8–11.

Hise, G. and W. Deverell (2000) *Eden by Design, The 1930 Olmstead-Bartholomew Plan for the Los Angeles Region*, Berkeley: University of California Press.

Jennings, V., Johnson Gaither, C. and Gragg, R.S. (2012) "Promoting Environmental Justice Through Urban Green Space Access: A Synopsis," *Environmental Justice* 5(1): 1–7.

Kenworthy, J. (2007) "Urban Planning and Transport Paradigm Shifts for Cities of the Post-Petroleum Age," *Journal of Urban Technology* 14(2): 47–70.

Kimmelman, M. (2011) "In Madrid's Heart, Park Blooms where a Freeway once Blighted," 26 December, *New York Times*, (Art and Design, Critic's notebook), accessed 7 October 2015 — www.nytimes.com/2011/12/27/arts/design/in-madrid-even-maybe-the-bronx-parks-replace-freeways.html?_r=1

Kuo, F.E. and W.C. Sullivan (2001a) "Aggression and Violence in the Inner City Effects of Environment via Mental Fatigue," *Environment and Behavior* 33(4): 543–71.

——— (2001b) "Environment and Crime in the Inner City: Does Vegetation Reduce Crime?" *Environment and Behavior* 33(3): 343–67.

Landscape Architecture Foundation (2011) *Cheonggyecheon Stream Restoration Project*, Washington, DC: Landscape Architecture Foundation, accessed 7 October 2015 — www.lafoundation.org/research/landscape-performance-series/case-studies/case-study/382

Macintyre, S., Macdonald, L. and Ellaway, A. (2008) "Lack of Agreement Between Measured and Self-Reported Distance from Public Green Parks in Glasgow, Scotland," *International Journal of Behavioral Nutrition and Physical Activity* 5(26), accessed 7 October 2015 — www.ijbnpa.org/content/5/1/26

Marris, E. (2011) *Rambunctious Garden: Saving Nature in a Post-Wild World*, New York: Bloomsbury USA.

Mowen, A.J., Payne, L.L. and D. Scott (2005) "Change and Stability in Park Visitation: Constraints Revisited," *Leisure Sciences* 27: 191–204.

Newell, J.P., Seymour, M., Yee, T., Renteria, J., Longcore, T., Wolch, J.R. and A. Shishkovsky, (2013) "Green Alley Programs: Planning for a Sustainable Urban Infrastructure?" *Cities* 31: 144–55.

Newman, P. (2013) "Imagining a Future Without Oil for Car-Dependent Cities and Regions," in J.L. Renne and B. Fields (eds) *Transport Beyond Oil*, Dordrecht: Springer/Island Press, 203–25.

O'Rourke, D. and S. Connolly (2003) "Just Oil? The Distribution of Environmental and Social Impacts of Oil Production and Consumption," *Annual Review of Environment and Resources* 28(1): 587–617.

Robbins, P. and T. Birkenholtz (2003) "Turfgrass Revolution: Measuring the Expansion of the American Lawn," *Land Use Policy* 20(2): 181–94.

Rosenzweig, R. and E. Blackmar (1992) *The Park and the People*, Ithaca: Cornell University Press.

Saelens, B.E., Sallis, J.F. and L.D. Frank (2003) "Environmental Correlates of Walking and Cycling: Findings from the Transportation, Urban Design, and Planning Literatures," *Annals of Behavioral Medicine* 25(2): 80–91.

Schroepfer, T. and L. Hee (2008) "Emerging Forms of Sustainable Urbanism: Case Studies of Vauban Freiburg and SolarCity Linz," *Journal of Green Building* 3(2): 65–76.

Seddon, G. (1972) *Sense of Place: A Response to an Environment, the Swan Coastal Plan, Western Australia*, Perth: University of Western Australia Press.

Seoul Metropolitan Facilities Management Corporation (2013) *Welcome to Seoul's Pristine Stream: Cheong Gye Cheon*, Seoul: Seoul Metropolitan Government. http://english.sisul.or.kr/grobal/cheonggye/eng/WebContent/index.html

Shaw, D.V. (1986) "Making Leisure Pay: Street Railway Owned Amusement Parks in the United States, 1900–1925," *Journal of Cultural Economics* 10(2): 67–79.

Sister, C., Wolch, J. and J. Wilson (2010) "Got Green? Addressing Environmental Justice in Park Provision," *GeoJournal* 75(3): 229–48.

Wolch, J.R., Byrne, J. and J.P. Newell (2014) "Urban Green Space, Public Health and Environmental Justice: The Challenge of Making Cities 'Just Green Enough,'" *Landscape and Urban Planning* 125: 234–44.

PART III

Urban systems

14

LOCAL ENERGY PLANS FOR TRANSITIONS TO A LOW CARBON FUTURE

Brendan F.D. Barrett and Ralph Horne

Energy transitions in the face of climate change and peak oil are emerging across various scales of government, often with widely varying and sometimes disappointing effects for their electorates. Civil society organizations are also engaged at international, national and local levels, finding willing coalitions in communities disaffected by the lack of decarbonization actions by governmental actors. The private sector is increasingly profiting from the new "green" economy, via the rhetoric of "green growth" and associated grants and incentives from governments. These somewhat stereotypical and simplistic statements about energy transition "stakeholders" conceal a rich – and getting richer – range of organizations and actors working in between and across the decarbonization and clean energy fields, ranging from policy and advocacy to active experiments and large-scale adoption of socio-technological programs.

Ideas of ecological modernization (Barrett 2005; Mol 2002; Mol et al. 2009; Toke 2011) still dominate policy discourse, with indomitable confidence that technology and markets can steer us toward the achievement of a decarbonized, cleaner world, starting with reconfiguring pipes and wires. Meanwhile, scholars across multiple disciplines such as law, politics, economics, sociology and geography are engaging in energy transition debates. Studies in governance and power, on social-material change, and in technology and society are of particular relevance to this chapter, where our focus is at a local scale (the local authority, city or subregion).

When promoting an energy transition, one challenge that local governments and cities across the world face is that often energy supply lies outside their direct control. While local governments may appear to retain control over urban planning, buildings, transport, water and waste (Mohanty 2012), the wave of post 1970s privatizations has resulted in splintered ownership of the urban project (Graham and Marvin 2001). As part-players in regulating privately run energy infrastructures, local governments need to negotiate with private sector counterparts. Increasingly locked into global economic markets that promote consumption, municipalities face multiple conflicts and constraints, and limited control in attempts to curb consumption, including of energy, within their boundaries. Even if they do embark on purposive attempts to shift energy-using practices, "the outcomes of actions are unknowable, the system unsteerable, and the effects of deliberate intervention inherently unpredictable" (Shove and Walker 2007: 768).

An urban planning perspective has highlighted the importance of limited local government influence over energy consumption with respect to the vulnerability of residents in suburbia to oil price spikes (Dodson and Sipe 2008). "Peak oil" – when world production of oil reaches its maximum, only to gradually decline thereafter – raises a similar dilemma. A UK report has argued that: "Peak Oil is not well recognized in spatial planning although some direct effects of Peak Oil will be experienced in the UK within time horizon of national and local plans" (RTPI 2011: 16). Yet, in most instances, urban planners' priority remains the promotion and accommodation of population and economic growth (Fainstein 2010).

Given the "stickiness" of energy in cities, markets and social life, it is unsurprising that the interplay of governance, technology and social change is recognized by advocates and scholars alike as critical to the energy transition. The notion of transition has become increasingly central to futures-oriented thinking (Moloney et al. 2010) and draws on contributions from fields of urban studies and technological transitions to develop conceptual frameworks and empirical research on how we might understand urban transitions and the multiple scales and actors involved. The idea of governed, socio-technical transitions has been consistently proposed, including "transition management" (Loorbach 2007) and a "multi-level perspective" (Geels 2010). We argue that the latter is a useful way to think about alignment (or otherwise) of the existing energy regime, how it is affected by larger global framing landscape factors, and how decarbonization niches may be created in protected spaces that may at some point challenge or displace the existing energy regime.

These transitional strategies have been criticized by sociologists, who note the lack of evidence for successful steering by government in such ways, and who emphasize, instead, the roles of city and climate governance, social and political structures (Bulkeley 2010; Bulkeley et al. 2011) and social practices (Shove 2010; Shove and Walker 2007). We accept these criticisms and argue that social change needs to be framed by practices rather than solely through a cognitive "behavioral" frame. Furthermore, urban scholars point out the spatial and temporal dynamics of the city, and the critical role of intermediaries who operate between civil society, government and the private sector in activating or resisting decarbonization (Guy et al. 2011). We have also found the idea of intermediaries highly relevant to the study of local energy transitions (Horne and Dalton 2014). Accordingly, using the three ontological framings of the multi-level perspective, social practices and intermediaries, this chapter examines how local authorities are responding to the proposition of an energy transition. It explores some of the plans, policies and targets that they have developed.

To frame this discussion, we provide a brief overview of the global context of peak oil and decarbonization policies. We discuss four distinct scenarios inferred from current proponents of action on peak oil and decarbonization: do nothing, mitigation, adaptation and systemic/regime/revolutionary change. Then we turn to the plans of cities and local authorities in such activities. Specifically, we report on a survey of eighteen cities that are signatories to the UN Global Compact Cities Programme. The main argument is that, after an initial upsurge in local authority action from 2007 onward – primarily in the form of mitigation measures that included both peak oil response plans and oil emergency preparedness plans – there has been a significant shift in how local governments engage with such issues. Essentially their approach has morphed into measures designed to facilitate climate change mitigation and adaptation, while at the same time promoting energy security and resilience. We conclude by questioning the prospects for success of local energy transitions, using a multi-level perspective, social practices and intermediaries as means to assess these prospects.

Peak oil and decarbonization: International debates

Peak oil has been debated for more than half a century. In 1956, M. King Hubbert predicted that, for the US, peak oil would be reached between 1965 and 1970. The Hirsch Report by the US Department of Energy (Hirsch et al. 2005) predicted the peak within twenty years, and argued for an early response, which promoted the issue up the international policy agenda. In Australia, steep declines in oil production from 2016 onward are expected (Australian Government 2009). Similarly, a UK Energy Research Centre report (Sorrell et al. 2009) predicting peak oil before 2020 prompted the UK chief scientist to canvass opinion, finding that many believed that 2020–30 would be a critical decade for peak oil (Department of Energy and Climate Change 2011). Subsequently, the Royal Society *Future of Oil Supply* report (Miller and Sorrell 2014: 17) echoed these findings.

The fact that the peak oil "date" has moved closer to the present since the 1950s, and that global oil prices have fluctuated significantly from time to time, has provided cause for reflection on the need for preparation for peak oil. Bearing in mind the uncertainty around the timing of the peaking of oil production, some commentators such as the UK Energy Research Centre (2010) prefer the term "oil depletion." This recognizes that oil is a finite resource and that oil reserves are being depleted at a rate far faster than any new discoveries. However, reserve prediction techniques have improved. One key explanation for the dramatic oil price increase in July 2008 (to US$148 per barrel) was that conventional oil production effectively reached a plateau in 2005 with an apparent production cap since then (King and Murray 2012). Subsequent price falls, including to US$40 in 2015, have not negated the assertion that peak oil is approaching. Rather, the gap in supply has been filled by unconventional oil (Mushalik 2014), reduced global demand has been due to slow economic growth, and there has been a glut in supply as Saudi Arabia has maintained high levels of oil production (perhaps with the aim of undermining unconventional oil production in the US) (Raval and Hume 2015).

Since the early 2000s, the decarbonization debate has paralleled the peak oil debate even though they have different underlying concerns and trajectories (Friedrichs 2013). Researchers, activists and non-governmental organizations have advocated for various "power down" scenarios (Heinberg 2004; Holmgren 2009; Jacobson and Delucchi 2009) in response to increasingly concerning data and modeling from the Intergovernmental Panel on Climate Change (IPCC), which has pointed to the need for greenhouse gas (GHG) emission reductions of 40–50 percent by 2050 to achieve a near-zero emission level by 2100 (IPCC 2014). Rationing or "carbon budgets" are increasingly mentioned and advocated by the 2000 Watt Society (Notter et al. 2013). The IPCC *Fifth Assessment Report* (IPCC 2014) indicates that, if emissions could be stabilized at 1,000 Gt CO_2 from 2011 onward, we would have a greater than 66 percent probability of limiting human-induced warming to less than 2 degrees Celsius. However, other reports are more pessimistic (Carbon Tracker Initiative 2011; Meinshausen et al. 2009).

In addition to the obvious problem of free-market theory and rationing not being easy bedfellows, there are two further challenges at the core of global debates on decarbonization. The first challenge is whether or not there is enough global political will to leave fossil fuels in the ground. One-third of oil reserves, half of gas reserves and over four-fifths of current coal reserves should remain unused from 2010 to 2050 to cap global temperature increase below 2 degrees Celsius (McGlade and Ekins 2015). The second challenge is the cost and feasibility of required socio-technical changes to national economies while preserving globally competitiveness. In a review of seventeen decarbonization scenarios, Loftus et al. (2015) noted a requisite 50–90 percent reduction in GHG emissions by 2050, and on a case-by-case basis, transformations of energy systems that included shifts up to 100 percent renewable energy and expansion of nuclear energy. Another report (SDSN 2014) has showed that it is technically feasible to reduce total CO_2 energy emissions but only at significant cost.

In Japan, the Low Carbon Society has established the technical feasibility of 70 percent GHG reduction and developed a technology-driven scenario and another based on a decentralized society (National Institute for Environmental Studies 2008). The central finding for both scenarios was that the transition would only require investing the equivalent of 1 percent of Japan's annual GDP each year to 2050. This in-depth research project had a major impact on Japanese political leadership at the time, which subsequently adopted ambitious targets to reduce Japan's carbon emissions. At the 2008 Tokyo G8 Summit, then Prime Minister Yasuo Fukuda announced that Japan would set a goal of reducing emissions from current levels by 60–80 percent by 2050. Since then, the government has reneged on this commitment. According to Climate Action Tracker (2015), Japan's Intended Nationally Determined Contribution included an emissions reduction target of 26 percent below 2013 emission levels by 2030 (equivalent to 18 percent below 1990 levels by 2030). This case illustrates the often fragile and reversible nature of national government moves around decarbonization and raises decentralization/localization proposals treated later in detail.

Four energy transition scenarios

Drawing on global debates, we identify four key scenarios for a local energy transition (presented in Table 14.1) derived from a diverse range of literature concerning how society can respond to the energy and climate challenge (Barrett 2012). The do-nothing scenario is not described in detail since it does not involve change from a current business-as-usual approach; the do-nothing scenario represents the

TABLE 14.1 Four energy transition scenarios

Scenario	*Do nothing*	*Mitigation/emergency measures*	*Adaptation*	*Systemic/regime change*
Assumptions	Assumes that peak oil and climate change are not transformative issues and that we will find alternative fuels or a technological fix.	Assumes that it is too late to adapt and, therefore, we need emergency measures. It is unclear how this will play out in the long term, since this is mainly a bridging scenario.	Works within existing political, economic and social structures to try to gradually shift them to energy security future, while solving climate change at the same time.	Promotes a long-term, sustainable solution, requiring systemic social and economic change that will face resistance from those who favor the status quo.
Proponents	Advocates of business as usual.	Robert Hirsch, Chatham House, UK Industry Peak Oil Taskforce, International Energy Agency.	David MacKay (Cambridge University), Breakthrough Institute (*An Ecomodernist Manifesto*), IPCC, Low Carbon Society. Mark Lynas (Nuclear 2.0), Federal Institute of Technology in Zurich – ETHZ (2,000-Watt Society).	Jacobson and Delucchi (Stanford University), Greenpeace, WWF. Amory Lovins (Rocky Mountain Institute). Jeremy Rifkin (3rd Industrial Revolution).

Scenario	*Do nothing*	*Mitigation/emergency measures*	*Adaptation*	*Systemic/regime change*
Measures proposed	Continue with the current energy mix based entirely on pricing and let the market decide (situation of most local governments today). Technological change (unspecified carbon capture/storage) is expected as peak oil affects prices, but climate change remains a major externality.	Promote administrative measures such as carpooling, telecommuting and fuel rationing. Promote physical measures like fuel-efficient transportation. As this is a peak oil plan, it includes Enhanced Oil Recovery in existing oil fields, liquid fuels from tar sands and coal-to-liquid/gas-to-liquid operations. Ignores impact on the environment, as the exploitation of existing energy reserves remains the priority in order to minimize potential negative impacts on the economy and society.	Promote renewable (and some include nuclear) energy, but reduce fossil fuel use (or use of carbon capture and sequestration) as rapidly as economically and technically feasible. Massive technological innovation is expected to address climate and energy security challenges at the same time. Maintain existing lifestyles, convenience and services. Promote equitable solutions for developed and developing countries (recognizing the right of developing countries to meet their development and energy needs). Assume technology risks will be minimized (including nuclear where relevant).	100% renewable by specific date, phase out fossil fuels and nuclear power (the latter due to externalities, costs and risks). Reduce greenhouse gas emissions to zero by a specific date, at latest by 2100. Reduce energy consumption through greater efficiencies and savings, mainly because current energy levels cannot be met by renewables. Introduce smart grid, electric vehicle fleet and off-peak power storage (e.g. hydrogen fuel, batteries). Long-term, sustainable solution, but massive social change. Resistance highly likely from vested interests tied to current energy paradigm.

Source: Data drawn from Barrett (2012).

existing energy regime. In some respects, do-nothing thinking is embodied in the report by Harvard Kennedy School's Leonardo Maugeri (2012), which argued that oil production would climb from 2011 levels by around seventeen million barrels per day by 2020, predicting an oil glut and pressure on oil prices (which has been realized), without any reference to the potential impact on climate change. These projections reflected an optimistic perspective on the future of oil supplies, and Sorrell (2012) has described his long-term assumptions as inconsistent with available evidence.

The "mitigation" scenario draws on the work of Robert Hirsch et al. (2010), who identified various emergency measures that were designed to mitigate the impact of peak oil. In this long emergency – which already may have started (Kunstler 2007) – two kinds of measures, administrative and physical, are advocated. According to the calculations of Hirsch et al. (2010), if implemented globally, such measures would save the equivalent of thirty million barrels of oil per day. Effectively, this would provide a buffer or period of transition for the economy to adjust to a world with less oil. The effectiveness of these mitigation measures depends upon the speed at which oil production might decline, which is projected to be around 4–6 percent per annum (p.a.). The results of the survey in the *World Energy Outlook* by IEA (2008a), a survey of 800 oil fields observed a post-peak decline rate, averaged across all fields and weighted by their production over their whole lives, of 5.1 percent; decline rates are lowest for the biggest fields, averaging 3.4 percent for super-giant fields, 6.5 percent for giant fields and 10.4 percent for large fields.

Apart from the concern that the mitigation scenario focuses on managing oil depletion to the neglect of impacts on the environment, its focus on current prices and markets means that biomass-to-liquids do not feature in the solution (mainly because they currently require subsidies). In addition to the work of Hirsch, other bodies that promote scenarios along similar lines and that focus on a potential oil supply emergency or oil crunch are the International Energy Agency (IEA 2008b; Young 2014), Chatham House (Stevens 2008) and the UK Industry Peak Oil Taskforce (ITPOES 2008).

The "adaptation" scenario promotes renewable energy and some argue for nuclear energy, alongside a reduction in dependency on fossil fuels. Researchers at the US Breakthrough Institute, *An Ecomodernist Manifesto* (Asafu-Adjaye et al. 2015) and Nuclear 2.0 (Lynas 2014) support the pro-nuclear stance. This perspective also resonates closely with the IPCC, the 2,000-Watt Society at the Federal Institute of Technology (Zurich, Switzerland – ETHZ) and the Japan-based Low Carbon Society (of the National Institute for Environmental Studies).

"Systemic/regime change" is the most ambitious scenario and requires that the entire world shift to a 100 percent renewable energy supply by a specific date. A Stanford University study proposes ways to sustainable energy by 2030 (Jacobson and Delucchi 2009) and related proponents include Jeremy Rifkin (2013) and Amory Lovins (Lovins et al. 2011) as well as various proposals from non-governmental organizations, such as Greenpeace (2015) and WWF (WWF Global 2011). Proponents call for a reduction of energy consumption, in the order of 50 percent compared to current levels, through energy efficiency measures. This scenario includes measures such as the introduction of smart grids, electric vehicle fleets and off-peak power storage (for example, conversion to hydrogen fuel or storage in batteries). Anticipated near-future scenarios typically include home and car energy storage, and widespread transformations in markets and services to upscale the renewables sector. For instance, electronic appliance manufacturers become electric car manufacturers, computer companies build electric cars, car companies build energy efficient homes and shopping centers become electric car charging points. Stability of supply and production of sufficient liquid energy fuels remain key technology concerns. On the "up" side, advocates anticipate new green jobs, enhanced energy security, no more nuclear waste, reduced impacts on the climate and less likelihood of resource-based conflicts.

There has been a significant groundswell of bottom-up, community-led initiatives; for example, the Transition Towns Network started in 2005 in Totnes (Feola and Nunes 2013) has now spread to communities around the world. Focused on peak oil and climate change, this initiative guides each community in developing and implementing an Energy Descent Action plan within twenty years. This activity is at the scale of local government, the subject of the next section.

Local urban scale plans

How have local governments responded to the dual pressures of an impending peak in oil production and the need to decarbonize in order to avoid potentially dangerous climate change? Local urban-scale plans can be classified into three scenarios, as described in this section: a mitigation scenario, an adaptation scenario, and a systemic/regime change scenario.

Mitigation scenario

Plans prepared both before the July 2008 oil price shock and just after the global financial crisis (GFC) in 2008 and 2009 broadly fit the mitigation scenario, as shown in Table 14.2, including examples from the US, UK and Australia. The plans tended to take the 2005 Hirsch Report as their starting point and included measures to deal with impacts of peak oil, particularly concerning how to maintain vital public services (such as fire, police and health) once liquid fuel availability becomes severely restricted, or declines year on year. The focus of these plans was on long-term disruptions and confronting a new reality of a world with less oil. The plans tended to identify various local vulnerabilities in relation to land use and transportation, food and agriculture, public safety and social services, and proposed a range of mitigation measures.

TABLE 14.2 The mitigation plans of five US, UK and Australian city councils

MITIGATION PLANS					
City government	*City of Portland, US*	*City of Oakland, US*	*City and County of San Francisco, US*	*Bristol City Council, UK*	*Maribyrnong City Council, Australia*
Area	376 km^2	202 km^2	600 km^2	110 km^2	31.2 km^2
Population	583,700	406,000	852,000	432,500	79,300
Title of plan	Descending the Oil Peak: Navigating the Transition from Oil and Natural Gas	Oil Independent Oakland Action Plan	San Francisco Peak Oil Preparedness Task Force Report	Building a Positive Future for Bristol After Peak Oil	Peak Oil Contingency Plan
Date	March 2007	February 2008	March 2009	2009	June 2009
Time period	To 2030			To 2030	To 2034
Target	Reduce total oil and natural gas consumption by 50% by 2033	3% reduction in oil usage per year	Policies to reduce reliance on fossil fuels, introduce more renewable energy, expand public mass transit, and retrofit buildings for energy conservation	Reduce oil consumption by 50% by 2020	Reduce total oil and natural gas consumption by 50% over the next 25 years

Source: Based on data drawn from City of Portland (2007), City of Oakland (2008), San Francisco Peak Oil Preparedness Task Force (2009), Bristol City Council (2009), Maribyrnong City Council (2009).

TABLE 14.3 Adaptation plans for four Australian cities and one US city

City government	*City of Whitehorse, Australia*	*Sunshine Coast Council, Australia*	*City of Lawrence, US*	*Manningham Council, Australia*	*Meander Valley Council, Australia*
Area	2,365 km²	2,291 km²	88.7 km²	1,031 km²	3,821 km²
Population	151,300	282,800	87,600	117,500	18,900
Title of plan	Whitehorse Peak Oil Action Plan 2011	Sunshine Coast Energy Transition Plan	Solutions to Peak Oil Vulnerabilities – A Response Plan for Lawrence, Kansas	Securing the Future: Responding to Climate Change, Peak Oil and Food Security	Meander Valley Oil Vulnerability Action Plan
Date	May 2011	December 2010	September 2011	July 2012	October 2013
Time period covered	To 2021	To 2020			To 2023

Source: Based on data drawn from Sunshine City Council (2010), City of Whitehorse Council (2011), City of Lawrence (2011), Manningham Council (2012), Meander Valley Council (2013).

Adaptation scenario

Plans prepared from the 2010 economic recovery period onward included consideration of oil price decline and sector divestment (Table 14.3). This planning was more complex, given that the US shale gas revolution was already influential and was superseding the previously core peak oil argument. In some cases, such as the Australian City of Whitehorse Council and Meander Valley Council, two scenarios were explored. The first was a short-term disruption, whereby oil supply was restricted for four to six weeks, perhaps as a result of a prolonged conflict in the Middle East that caused serious interruptions to global supply, as simulated for the US in "Oil Shockwave" since 2005 (SAFE 2015). The second scenario was a long-term disruption that resulted in a 50 percent reduction in global oil production (2010–30) and a consequent major increase in oil prices from 2015 onward. Maribyrnong City Council included a third scenario called "disintegration," where oil depletes rapidly from 2010 onward, causing high prices, economic recession, political conflicts, and health and food crises. The focus was very much on risk management in relation to local government services, the residents and local business and industry. Climate change was not generally considered.

Systemic/regime change scenario

More radical plans are implied where carbon-free or deep decarbonization is set as a goal. A Stanford University study (Jacobson et al. 2015) ran "100% renewable by 2050" energy scenarios for all fifty US states (Solutions Project 2015a). In a similar vein, the IPCC *Fifth Assessment Report* summarized the GHG emission reduction targets of local governments across the globe, to show that some local authorities had targets to reduce emissions by 100 percent (such as Copenhagen by 2030, and Stockholm and Oslo by 2050). Oslo has developed a new climate and energy strategy to reduce fossil emissions by 50 percent by 2030 and become fossil fuel free by 2050. Meanwhile, another signatory city, Berlin, has produced a feasibility study on how the city could become climate neutral by 2050 (Table 14.4).

The City of San Francisco aims to reduce emissions by 80 percent by 2050, compared to the 1990 level. This would involve measures to ensure that 50 percent of all trips would be sustainable (walk, bike, public transport) and that electricity generated would be 100 percent renewable (from a baseline of 17 percent).

TABLE 14.4 Plans for systemic scenarios in two US and two European cities

City government	*Oslo City*	*Berlin*	*City and County of San Francisco*	*New York City*
Area	139.5 km^2	891.8 km^2	600 km^2	1,214 km^2
Population	647,676	3.5 million	852,000	8.5 million
Title of plan	The Green Shift – Climate and Energy Strategy for Oslo	Climate-Neutral Berlin 2050 — Results of a Feasibility Study	Climate Action Strategy	One New York – The Plan for a Strong and Just City
Date	January 2015 (Draft Plan)	March 2014	2013	2015
Targets	Reduce GHG emissions by 50% by 2030, become fossil fuel free by 2050	Reduce GHG emissions by 85% by 2050 relative to 1990	100% renewable electricity by 2030; 40% reduction in GHG emissions by 2030 relative to 1990; 80% reduction by 2050	Reduce GHG emissions by 80% by 2050 relative to 2005 levels

Note: Oslo, Berlin and San Francisco are signatories to the UN Global Compact.

Source: Data drawn from City and County of San Francisco (2013), New York City (2015), City of Berlin (2014), Oslo Municipality (2015).

This important evolution, instigated in a relatively short period of time since the San Francisco Peak Oil Preparedness Report (SFPOPTF 2009), is indicative of how rapidly things are changing at the local level. The 2015 New York "Plan for a Strong and Just City" contains a target to reduce GHG emissions by 80 percent by 2050 compared to 2005, to be achieved by promoting cleaner power generation, fossil fuel free modes of transportation, reducing solid waste and improvements to the energy efficiency of buildings.

Prospects for local energy transitions

Although the rhetoric on social and behavior change has grown over recent years, overwhelmingly the implied and explicit focus of both peak oil and decarbonization plans is technology substitution. This is most evident in the IPCC Fifth Assessment Report, which places considerable hope in the effectiveness of unproven carbon sequestration technologies, and makes little or no mention of social or behavioral change. Beyond this, across each local scenario presented earlier, there is little sophistication across plans on the "how" of transitioning beyond the aspiration that change will follow the setting of targets. The central normative assumption is that measures such as promoting renewable technology, resource efficiency and waste reduction are solutions only if they are economically feasible or job-creating. Feasibility considerations range from whether the technology can meet expected capacity or efficiency "requirements" to whether treasuries and finance departments might deem the technologies to be cost-efficient. Both carry inherent assumptions about future demand and market economic settings based on continuing existing regime settings, although progress is shown in recent works, such as the New Climate Economy (GCEC 2015), a major international initiative to analyze and communicate the economic benefits and costs of acting on climate change. Yet, even in cases where behavior change programs are advocated or implemented, normativity persists as policy settings imply a need to "just try harder" for the energy transition to unfold (Moloney et al. 2010).

Given this situation, it is unsurprising that the dominant justifications for peak oil and decarbonization planning are twofold: cost-effectiveness over the long term and to ensure energy security. For example, the ReFresh Milwaukee climate and energy plan notes that US$12.5 billion p. a. flows out of Wisconsin's economy to import fossil fuels, including $4.3 billion for natural gas and coal (City of Milwaukee 2013). This suggests a clear economic argument for the redirection of these funds from fossil fuel to investment in low carbon energy. In the UK, a study of Leeds claimed that, if 125 energy-efficient small-scale renewable and low carbon measures were implemented, the city region could cut its energy bills by £1.2 billion (1.6 percent of Leeds' GDP) and carbon emissions by 36 percent a year (Gouldson et al. 2012). A recent report by Klenas et al. (2015) argues that, in the long-run, renewables are safer bets than fossil fuels, suggesting that renewables projects can effectively lock in the cost of energy for twenty years or more, whereas oil prices will continue to be volatile, creating uncertainty and risk.

The World Bank argues for policy packages that extend beyond pricing and cost efficiency to accompany long-term plans; to trigger changes in investment patterns, technologies and behaviors; and to guide the political economy to ensure a "smooth transition" (Fay et al. 2015). The technical and economic challenges of powering major population centers, like Berlin and New York City (NYC), completely on renewable energy are obvious, but nothing compared to the difficulties of navigating transition pathways "from here to there" if one takes into account the social engineering implied by a predicted drop in demand across NYC from 91.8 GW, predominately met by fossil fuels and nuclear, to 54.9 GW in 2050 (Solutions Project 2015b). As a city epitomizing capitalism, ideas of rationing are anathema in NYC, and the idea of "limits" runs counter to the American Dream.

In conclusion, while the dominant narratives of decarbonization and peak oil in local energy plans are about adjusting economic and technology settings, they contain more emergent and divergent currents. The Transition Towns movement has been influential in energy transition planning and forms part of a bottom-up community-based movement of experiments, within niches more or less protected from the market by rebates and subsidies and featuring countercultures of downscaling. There is a role here for thinking about how the multi-level perspective of Geels (2010) provides for ways of thinking about socio-technical change in designing policy and practices for local energy transitioning. The local or city scale is an appropriate one, we argue, for tapping into local movements and promoting regime change through new narratives, governance and social practices in local energy transitions.

We also appreciate that the examples presented in this chapter are in cities of the Global North and that there is considerable diversity of settings and challenges for local energy transitions. ICLEI (2009) and others have been arguing that "Sustainable Urban Energy" has the potential to bring benefits for cities of the Global South, in terms of improvements in air quality, financial savings, new jobs and economic developments. It is also clear that the next phase on the local energy transition will occur when the decarbonization and energy shift becomes part of mainstream city development, which is pretty much what we find in the 2015 One New York – The Plan for a Strong and Just City.

Local energy transition plans need to be cognizant of theories of social and socio-technical change, to be open to the fact that such change is not often steerable and to the serendipity and opportunities that arise with unintended changes. For example, in Australia the number of rooftop solar photovoltaic (PV) systems grew from a mere 8,000 in 2007 to over one million in 2013 (Flannery and Sahajwalla 2013: 6). While this technical observation is remarkable and visually apparent on rooftops,

> what is not apparent is what might emerge given those millions of customers who may soon have access to storage technology to manage "their" new resource harvest locally, nor the potential impact of other businesses seeking to take advantage of new opportunities to provide off-grid services.
>
> *(Horne and Fudge 2014: 283)*

Even though not explicitly recognized by policy makers, it seems inevitable that daily domestic routines will shift as a result of these new material arrangements, and combine with changing meanings and understandings of renewable energy, and the skills and knowledge required to exploit it.

We speculate that local energy plans should begin by overtly promoting protected niches where local scale experiments, networks, co-learning and innovations can occur, and should aim to be flexible enough to promote local energy transitioning by taking advantage of unfolding social and technical conditions. The politics and priorities of organizations and individuals must be explored to understand the implications for local energy transitions. The development, nurturing and proliferation of active intermediaries who can work within and between civil society, government and the private sector is essential here, "to constitute a space outside of the obduracy of both existing urban governance regimes and existing socio-technical regimes" (Hodson and Marvin 2010: 482).

Returning to the Australian domestic PV case, one set of intermediaries active in promoting this technical transition is the Victorian Climate Change Alliances. An analysis of roles and relationships between these interstitial organizations of different levels of government, community and environmental organizations, and other actors enables some assessment of the progress of the local energy transition (Moloney and Horne 2015). Such plans rely, of course, only on coalitions of the willing around them. At some point, there has to be a qualitative leap from niche to mainstreaming. We anticipate that this will be soon in coming, based on how local energy and climate plans have evolved in the past decade. Ending obduracy of fossil energy regimes is a necessarily local as well as multi-stakeholder, social and cultural projects as well as being an opportunity for technology providers.

References

Asafu-Adjaye, J., Blomqvist, L., Brand, S., Brook, B., Defries, R., Ellis, E., Foreman, C., Keith, D., Lewis, M., Lynas, M., Nordhaus, T., Pielke, R., Pritzker, R., Roy, J., Sagoff, M., Shellenberger, M., Stone, R. and P. Teague (2015) *An Ecomodernist Manifesto*, accessed 5 October — www.ecomodernism.org/manifesto-english/

Australian Government (2009) *Transport Energy Futures: Long-Term Oil Supply Trends and Projections*, Report 117, Department of Infrastructure, Transport, Regional Development and Local Government.

Barrett, B.F.D. (2012) "Big Picture Policy Responses to Peak Oil," *Our World*, 18 June, accessed 5 October 2015 — http://ourworld.unu.edu/en/big-picture-policy-responses-to-peak-oil

——— (2005) *Ecological Modernisation and Japan*, London/New York: Routledge.

Bristol City Council (2009) Building a Positive Future for Bristol After Peak Oil, accessed 11 October 2015 — www.bristol.gov.uk/page/environment/peak-oil-and-transition-bristol

Bulkeley, H. (2010) "Cities and Governing of Climate Change," *Annual Review of Environmental Resources* 35: 229–53.

Bulkeley, H., Castan-Broto, V., Hodson, M. and S. Marvin (eds) (2011) *Cities and Low Carbon Transitions*, Hoboken, NJ: Routledge.

Carbon Tracker Initiative (2011) "Unburnable Carbon: Are the World's Financial Markets Carrying a Carbon Bubble?" accessed 5 October 2015 — www.carbontracker.org/wp-content/uploads/2014/09/Unburnable-Carbon-Full-rev2–1.pdf

City and County of San Francisco (2013) *San Francisco Climate Action Strategy*, accessed 11 October 2015 — www.sfenvironment.org/climate-change

City of Berlin (2014) *Climate Neutral Berlin: Results of a Feasibility Study*, Senate Department of Urban Development and the Environment, accessed 11 October 2015 — www.stadtentwicklung.berlin.de/index_en.shtml

City of Lawrence (2011) *Solution to Peak Oil Vulnerabilities – A Response Plan for Lawrence, Kansas*, The Lawrence Mayor's Peak Oil Task Force, 29 September, accessed 11 October 2015 — www.lawrenceks.org/boards/task-forces/archives/peak-oil-task-force

City of Milwaukee (2013) *ReFresh Milwaukee – Fresh Coast, Fresh Ideas, A Vision for Community Sustainability, City of Milwaukee Sustainability Plan 2013–2023*, accessed 11 October 2015 — www.refreshmke.com

City of Oakland (2008) *Oil Independent Oakland Action Plan*, accessed 11 October 2015 — www.ecocitybuilders.org/wp-content/uploads/2010/11/OIO-ActionPlan.pdf

City of Portland (2007) *Descending the Oil Peak: Navigating the Transition from Oil and Natural Gas*, Report of the City of Portland Peak Oil Task Force, accessed 11 October — www.portlandoregon.gov/bps/42894

City of Whitehorse (2011) *Whitehorse Peak Oil Action Plan*, Nunawading: City of Whitehorse, accessed 11 October 2015 — www.whitehorse.vic.gov.au/Peak-Oil.html

Climate Action Tracker (2015) "Japan," Countries page, accessed 5 October 2015 — http://climateactiontracker.org/countries/japan.html

Department of Energy and Climate Change (2011) *DECC Call for Evidence – Summary of Response: Prospects for Crude Oil Supply and Demand*, June, London, Department of Energy and Climate Change, accessed on 11 October 2015 — www.gov.uk/government/consultations/call-for-evidence-prospects-for-crude-oil-supply-and-demand

Dodson, J. and N. Sipe (2008) "Energy Security, Oil Vulnerability and the Future of Australia's Cities," in P. Newton (ed) *Transitions: Pathways Towards Sustainable Urban Development in Australia*, Melbourne: CSIRO, 57–73.

Fainstein, S. (2010) *The Just City*, New York: Cornell University Press.

Fay, M., Hallegatte, S., Vogt-Schilb, A., Rozenberg, J., Narloch, U. and T. Kerr (2015) *Decarbonizing Development: Three Steps to a Zero-Carbon Future*, Washington, DC: World Bank.

Feola G. and R.J. Nunes (2013) *Failure and Success of Transition Initiatives: A Study of the International Replication of the Transition Movement*, Research Note 4 (August), Walker Institute for Climate System Research. Reading: University of Reading, accessed 1 July 2016 — www.walker-institute.ac.uk/publications/research_notes/WalkerInResNote4.pdf

Flannery, T. and V. Sahajwalla (2013) *The Critical Decade: Australia's Future – Solar Energy*, Canberra: Commonwealth of Australia – Department of Industry, Innovation, Climate Change, Science, Research and Tertiary Education.

Friedrichs, J. (2013) *The Future is Not What It Used to Be: Climate Change and Energy Security*, Cambridge: Massachusetts Institute of Technology Press.

GCEC (2015) *New Climate Economy*, Global Commission on the Economy and Climate, accessed 5 October 2005 — http://newclimateeconomy.net

Geels, F. (2010) "Ontologies, Socio-technical Transitions (to Sustainability), and the Multi-level Perspective," *Research Policy* 39(4): 495–510.

Gouldson, A., Kerr, N., Topi, C., Dawkins, E., Kuylenstierna, J. and R. Pearce (2012) *The Economics of Low Carbon Cities: A Mini-Stern Review for the Leeds City Region*, Leeds: Centre for Low Carbon Futures.

Graham, S. and S. Marvin (2001) *Splintering Urbanism: Networked Infrastructures, Technological Mobilities and the Urban Condition*, London: Routledge.

Greenpeace (2015) *Energy [R]evolution 2015, A Sustainable World Energy Outlook*, Amsterdam: Greenpeace.

Guy, S., Marvin, S., Medd, W. and T. Moss (2011) *Shaping Urban Infrastructures: Intermediaries and the Governance of Socio-Technical Networks*, London: Earthscan.

Heinberg, R. (2004) *Power Down: Options and Actions for a Post-Carbon World*, Gabriola Island, British Columbia: New Society.

Hirsch, R.L., Bezdek, R. and R. Wendling (2010) *The Impending World Energy Mess: What It Is and What It Means to You*, Burlington: Apogee Prime.

——— (2005) *Peaking of World Oil Production: Impacts, Mitigation and Risk Management*, Pittsburgh: US Department of Energy.

Hodson, M. and S. Marvin (2010) "Can Cities Shape Socio-Technical Transitions and How Would We Know If They Were?" *Research Policy* 39: 377–485.

Holmgren, D. (2009) *Future Scenarios: How Communities Can Adapt to Peak Oil and Climate Change*, White River Junction, VT: Chelsea Green.

Horne, R. and T. Dalton (2014) "Transition to Low Carbon? An Analysis of Socio-Technical Change in Housing Renovation," *Urban Studies* 51(16): 3445–58.

Horne, R. and C. Fudge (2014) "Low Carbon Urban Australia in a Time of Transition," in C. Miller and L. Orchard, *Australian Public Policy: Progressive Ideas in the Neo-Liberal Ascendency*, Bristol: Policy Press, 279–95.

ICLEI (2009) *Sustainable Urban Energy Planning: A Handbook for Cities and Towns in Developing Countries*, International Council for Local Environmental Initiatives, Nairobi: UN Habitat.

IEA (2008a) *IEA Response System for Oil Supply Emergencies*, Paris: International Energy Agency.

——— (2008b) *World Energy Outlook*, Paris: International Energy Agency.

IPCC (2014) *Climate Change 2014: Synthesis Report. Contribution of Working Groups I, II and III to the Fifth Assessment Report of the Intergovernmental Panel on Climate Change*, Core Writing Team – R.K. Pachauri and L.A. Meyer (eds) Geneva: Intergovernmental Panel on Climate Change.

ITPOES (2008) *The Oil Crunch: Securing the UK's Energy Future: First Report*, UK Industry Taskforce on Peak Oil and Energy Security accessed 5 October 2015 — http://aie.org.au/AIE/Documents/The_Oil_Crunch.pdf

Jacobson, M.Z. and M.A. Delucchi (2009) "A Path to Sustainable Energy by 2030," *Scientific American* November: 58–65.

Jacobson, M.Z., Delucchi, M.A., Bazouin, G., Bauer, Z.A.F., Heavey, C.C., Fisher, E., Morris, S.B., Piekutowski, D.J.Y., Vencilla, T.A. and T.W. Yeskooa (2015) "100% Clean and Renewable Wind, Water, and Sunlight (WWS) All-Sector Energy Roadmaps for the 50 United States," *Energy Environmental Sciences* 8: 2093.

King, D. and J. Murray (2012) "Climate Policy: Oil's Tipping Point has Passed," *Nature* 481: 433–35, accessed 5 October 2015 — www.nature.com/nature/journal/v481/n7382/full/481433a.html

Klenas, P., Stern, N. and J. Frejova (2015) *Oil Prices and the New Climate Economy*, Global Commission on the Economy and Climate briefing paper, May, accessed 14 October — http://newclimateeconomy.report

Kunstler, J.H. (2007) *The Long Emergency: Surviving the End of Oil, Climate Change, and Other Converging Catastrophes of the Twenty-First Century*, New York: Grove Press.

Loftus, P.J., Cohan, A.M., Long, J.C.S and J.D. Jenkins (2015) "A Critical Review of Global Decarbonization Scenarios: What Do They Tell us about Feasibility?" *WIRE's Climate Change* 6(1): 93–112.

Loorbach, D. (2007) *Transition Management: New Mode of Governance for Sustainable Development*, Utrecht: International Books.

Lovins, A. Odum, M. and J.W. Rowe (2011) *Reinventing Fire: Bold Business Solutions for the New Energy Era*, White River Junction, VT: Chelsea Green.

Lynas, M. (2014), *Nuclear 2.0: Why a Green Future Needs Nuclear Power*, Seattle: Amazon Digital Services.

Manningham City Council (2012) "Securing the Future - Responding to climate change, peak oil and food scarcity," accessed 1 July 2016 — www.meander.tas.gov.au/webdata/resources/files/Meander%20Valley%20Oil%20Vulnerability%20Action%20Plan%20Final.pdf

Maugeri, L. (2012) "Oil: The Next Revolution: The Unprecedented Upsurge of Oil Production Capacity and What it Means for the World," Belfer Center for Science and International Affairs, Harvard Kennedy School, accessed 5 October 2015 — http://belfercenter.ksg.harvard.edu/files/Oil-%20The%20Next%20Revolution.pdf

McGlade, C. and P. Ekins (2015) "The Geographical Distribution of Fossil Fuels Unused When Limiting Global Warming to 2°C," *Nature* 517: 187–90.

Meander Valley Council (2013) "Meander Valley Oil Vulnerability Action Plan, Melbourne: AECOM in Association with Griffith University," accessed 1 July 2016 — www.meander.tas.gov.au/webdata/resources/files/Meander%20Valley%20Oil%20Vulnerability%20Action%20Plan%20Final.pdf

Meinshausen, M., Meinshausen, N., Hare, W., Sarah, C.B., Raper, K. F., Knutti, R., Frame, D.J. and R. Allen Myles (2009) "Greenhouse-Gas Emission Targets for Limiting Global Warming to 2°C," *Nature* 458: 1158–62.

Miller, R.G. and S.R. Sorrell (2014) "The Future of Oil Supply, theme issue," *Philosophical Transactions of the Royal Society* 372(2006): 1–27, accessed 10 October 2015 — http://rsta.royalsocietypublishing.org/content/future-oil-supply

Mohanty, B. (2012) *Sustainable Urban Energy: A Sourcebook for Asia*, Nairobi: UN Habitat.

Mol, A.P.J. (2002) "Ecological Modernization and the Global Economy," *Global Environmental Politics* 2: 92–115.

Mol, A.P.J., Sonnenfield, D.A. and G. Spaargaren (2009) *The Ecological Modernisation Reader: Environmental Reform in Theory and Practice*, London/New York: Routledge.

Moloney, S. and R. Horne (2015) "Low Carbon Urban Transitioning: From Local Experimentation to Urban Transformation?" *Sustainability* 7: 2437–53.

Moloney, S., Horne, R. and J. Fien (2010) "Transitioning to Low Carbon Communities – From Behaviour Change to Systemic Change: Lessons from Australia," *Energy Policy* 38: 7614–23.

Mushalik, M. (2014) "IEA Report Implies US Crude Production May Start to Peak 2016," 14 August, accessed 5 October 2015 — http://crudeoilpeak.info/iea-report-implies-us-crude-production-may-start-to-peak-2016

National Institute for Environmental Studies (2008) *Japan Scenarios and Actions Towards Low-Carbon Societies*, accessed 11 October 2015 — http://2050.nies.go.jp/report/file/lcs_japan/2050_LCS_Scenarios_Actions_English_080715.pdf

New York City (2015) *One New York: The Plan for a Strong and Just City*, accessed 11 October 2015 — www1.nyc.gov/html/onenyc/index.html

Notter, D.A., Meyer, R. and H.J. Althaus (2013) "The Western Lifestyle and its Long Way to Sustainability," *Environmental Science & Technology* 47(9): 4014–21.

Oslo Municipality (2015) *The Green Shift: Climate and Energy Strategy for Oslo*, June, accessed 11 October 2015 — www.oslo.kommune.no/politikk-og-administrasjon/miljo/miljopolitikk/

Raval, A. and N. Hume (2015) "OPEC Agrees to Keep Oil Output High in Battle for Market," *Financial Times*, 5 June, accessed 5 October 2015 — www.ft.com/cms/s/0/afcf2658–0b5b-11e5–994d-00144feabdc0.html#axzz3jyuVo6ZN

Rifkin, J. (2013) *The Third Industrial Revolution: How Lateral Power is Transforming Energy, the Economy and the World*, Basingstoke/New York: Palgrave McMillan.

RTPI (2011) *Peak Oil: The Implications for Planning*, Discussion paper, Royal Town Planning Institute (Transport Planning Network), accessed 5 October 2015 — www.rtpi.org.uk/media/6326/Report-Final.pdf

SAFE (2015) "Oil shock wave," Securing America's Future Energy, accessed 5 October 2015 — www.secureenergy.org/projects/oil-shockwave

SDSN (2014) "Pathways to Deep Decarbonization. France/New York: Sustainable Development Solutions Network and Institute for Sustainable Development and International Relations," accessed 1 July 2016 — http://unsdsn.org/wp-content/uploads/2014/09/DDPP_Digit_updated.pdf

SFPOPTF (2009) "San Francisco Peak Oil Preparedness Task Force Report," Report to the Board of Supervisors, City and County of San Francisco, accessed 11 October 2015 — http://sfenvironment.org/sites/default/files/policy/sfe_en_peakoil_final_report.pdf

Shove, E. (2010) "Beyond the ABC: Climate Change Policy and Theories of Social Change," *Environment and Planning A* 42: 1272–85.

Shove, E. and G. Walker (2007) "Caution! Transitions Ahead: Politics, Practice, and Sustainable Transition Management," *Environment and Planning A* 39(4): 763–70.

Solutions Project (2015a) "100% Renewables Energy Vision," accessed 5 October 2015 — http://thesolutionsproject.org

——— (2015b) "Projected Energy Supply and Demand, New York," accessed 5 October 2015 — http://thesolutionsproject.org/wp-content/uploads/2015/07/solutions-ramp-NY-20150615.pdf

Sorrell, S. (2012) "Response to Leonardo Maugeri's Decline Rate Assumptions in 'Oil: The Next Revolution'," *The Oil Drum*, 11 July, accessed 5 October 2015 — www.theoildrum.com/node/9327

Sorrell, S., Speirs, J., Bentley, R., Brandt, A. and R. Miller (2009) *Global Oil Depletion Report: An Assessment of the Evidence for a Near-Term Peak in Global Oil* Production, London: UK Energy Research Centre.

Stevens, P. (2008) *The Coming Oil Supply Crunch*, A Chatham House Report, London: Royal Institute of International Affairs.

Sunshine Coast Council (2010) *Sunshine Coast Energy Transition Plan 2010–2020*, accessed 11 October 2015 — www.sunshinecoast.qld.gov.au/sitePage.cfm?code=energy-trans-plan

Toke, D. (2011) *Ecological Modernisation and Renewable Energy*, Basingstoke/New York: Palgrave Macmillan.

WWF Global (2011) *The Energy Report: 100% Renewable Energy by 2050*, Gland: WWF, accessed 5 October 2015 — http://wwf.panda.org/what_we_do/footprint/climate_carbon_energy/energy_solutions22/renewable_energy/sustainable_energy_report/

Young, M. (2014) "Oil That Limits Supply Disruptions: Emergency Stocks," *IEA Energy* 7: 16–17.

15

MOTOR VEHICLE FLEETS IN OIL VULNERABLE SUBURBS

A prospect of technology innovations

Tiebei Li, Neil Sipe and Jago Dodson

Australia is among the most petroleum dependent countries in the world. Continued growth in car travel and traffic congestion has resulted in increasing levels of fuel consumption and greenhouse gas emissions. In 2010, there were 12.3 million private vehicles registered in Australia, representing an increase of 12.6 percent from 2005 (ABS 2010). In addition, the increase in automobile dependence has placed Australian cities at a greater risk for potential adverse social and economic outcomes arising from increasing petrol prices. A series of studies has shown that many lower-income and car dependent households in Australia will experience high levels of risk and financial stress if fuel prices increase substantially beyond their current levels (Dodson and Sipe 2007, 2008, 2010).

Because of their high car dependency, oil prices have placed Australian cities at greater risk of energy security and climate mitigation challenges. The automotive transport sector faces a considerable transformation in order to respond to the challenges of petroleum depletion and climate mitigation. Commentators and policy makers have identified technological innovation as a central component of reducing the energy and carbon intensity of the Australian transport system, especially the private motor vehicle fleet (Romm 2006). Around the world, research into energy efficient vehicles have shown that fuel economy improvements have accelerated rapidly in many countries (Kuhnimhof et al. 2013; Millard-Ball and Schipper 2011; Van Dender and Clever 2013). The US alone showed an annual decline of 4.6 percent in fuel consumption per vehicle between 2001 and 2010, reflecting the contribution of improved vehicle fuel economy (EIA 2012).

Australia has had a modest record on vehicle fuel economy improvements, with average fleet efficiency remaining effectively unchanged, at around 11.4 L/100km, since the early 1960s through to the latter half of the 2000s. Contrast this limited change in Australia, contrasted to that observed for cars in the US over that period, which saw fuel economy improve from 17.6 L/100km, in 1973, to 10.5 L/100km, in 2006 (Sivak and Tsimhoni 2009). Consequently, Australia committed to reduce the energy and carbon intensity of its transport system, especially private motor vehicle fleets, during the second half of the 2000s.

Australian government policy for reducing the energy dependence and greenhouse gas impact of motor vehicle use focused on two components. The first strategy was to reduce motor vehicle emissions by improving vehicle fuel efficiency (VFE) via a new drivetrain. The drivetrain is the system connecting the transmission to the drive axles. New drivetrains include hybrid petrol electric or plug-in electric motors. At the federal level this policy shift was signaled by the government's A$6.2 billion Green Vehicle Plan to encourage innovation within the domestic automotive sector, motivated by a combination of "global warming, the emergence of low-cost competitors, and rising fuel prices" (Carr and Rudd 2008). The two largest components of the plan were the A$3.4 billion automotive industry structural adjustment package and the A$1.3 billion Green Car Innovation Fund to assist the sector to design and sell "environmentally friendly" cars.

A second policy objective was to shift the composition of fuels consumed to alternative forms, such as biofuels and electricity. The main energy form being pursued by motor vehicle manufacturers globally was battery-powered electric vehicles. The Australian government's automotive industry review anticipated an array of electric and hybrid petrol-electric vehicles coming onto the market in the near future (Bracks 2008). In subsequent years, there was no mass production of new energy vehicles on sale in Australia. However, there were moves toward introducing new energy vehicle infrastructure (e.g. electric vehicle recharging stations) in anticipation of such vehicles entering widespread use (Andersen et al. 2009), along with an expectation that implementation of these strategies would have noticeable fuel economy gains and offer opportunities to reduce carbon emissions.

It was in this policy context, around the early 2010s, that we embarked on a study of oil vulnerability and vehicle use across Brisbane, the state capital of Queensland – a study that is the focus of this chapter. While it was well recognized that incorporating improved vehicle technologies (for example, more efficient vehicle engines and new vehicle energy use) saved vehicle fuel intensity and household fuel costs, most discussion had centered on discovering and supplying a new technology and fuel combination that could deliver to the level of current patterns of mobility. Little research had investigated the nature of the technological dimension of the motor vehicle fleet and its impact on household fuel consumption and vulnerability. There had been almost no assessment of the differential socio-economic demand of households for low carbon and new vehicle fuel types.

Consequently, a cluster of questions remained unanswered, or unsatisfactorily answered. Did households receive the benefits of vehicle technology and energy innovations equally? Did households in the lower income groups have the financial capacity to afford new and fuel-efficient vehicles? Could policies that relied on new vehicle technology meet greenhouse gas and energy objectives in a socially equitable way? Based on our study, this chapter examines the links between urban social structure and the energy efficiency of motor vehicles to examine the implications of transport policies that rely on vehicle and fuel innovation.

This research drew on datasets from the Australian motor vehicle registration records and from the Australian Green Vehicle Guide, the provider of information on vehicle energy and environmental performance. These datasets show how efficient vehicles and new energy vehicles are spatially distributed. Using further spatial analytical methods, these vehicle and efficiency datasets are linked to social and spatial variables that measure the distribution of oil vulnerability in Brisbane, thus delivering a picture of the relationship between innovative vehicles and socio-economic patterns in urban areas. We demonstrate that both these datasets and analytical methods offer the potential to illuminate social-spatial patterns of new motor vehicle technologies in Australia.

This chapter is structured as follows: first, the data and methods of analysis are outlined; second, the results are discussed; third, we provide a discussion of the results, along with limitations, and an outline of future research.

Data construction

For this analysis, the spatial distribution of VFE was modeled through a combination of private motor vehicle registration data and the Australian government's green VFE guide (Green Vehicle Guide – https://www.greenvehicleguide.gov.au/), to enable an investigation of oil vulnerability across Brisbane. This section briefly outlines our research method.

Motor vehicle energy and fuel efficiency

When this research was conducted, the Australian government's Green Vehicle Guide provided data on the environmental performance of 14,996 vehicle makes and models sold in Australia between 1986 and 2003 and manufactured between 2005 and 2009. The guide included information on air pollution, CO_2 emissions, noise and fuel consumption by vehicle type. Vehicle registration data was obtained for 2009 and included 520,576 private passenger vehicle records for Brisbane. Each record contained the make, model, year of production, body shape, engine capacity, energy type and suburb location – using Australian Bureau of Statistics (ABS) "State Suburbs," that is, their approximation of localities gazetted by the geographical place name agency in each state and territory. For this study, the fuel consumption rate (measured in L/100km) was extracted and used for the VFE analysis, as it provided accurate information on standard fuel consumption for each vehicle make and model. Motor vehicle records associated with alternative energy types (diesel, electricity, gas and hybrid energy) were also extracted from the fleet datasets.

Fuel consumption rates by make and model specified by the Green Vehicle Guide were matched to the individual vehicles in the vehicle registration database. The two datasets were developed independently and each contained a large number of motor vehicle types. The main issue with matching records between the two datasets was that there were many vehicle makes and models in the vehicle registration database that were not found in the Green Vehicle database. Therefore, a recognition procedure was used to identify the best possible match in the name of vehicle makes and models. If the vehicle make and model was not found in the Green Vehicle Guide database, the information for the closest vehicle match was allocated. For example, if the fuel consumption rate for the registered Alfa Romeo 156 was not available in the Green Vehicle Guide database, the fuel consumption rate for Alfa Romeo 159 – the closest match in terms of vehicle make and model – was used. Nevertheless, in some cases a match was not found, which resulted in approximately 20 percent of vehicle makes and models in the registered vehicle database being omitted from the analysis. Fortunately, these vehicle makes and models represent a small number (less than 1 percent) of vehicles in the overall dataset. Therefore, the impact of their omission was judged as not statistically significant. Once the fuel efficiency information was allocated to the vehicle registration data, the data was then aggregated at the suburb level and the average VFE was calculated for each suburb.

Oil vulnerability

We examined the energy and efficiency of private vehicles and their spatial relationships with socio-economic variables that measured household oil vulnerability in Brisbane. The purpose was to understand motor vehicle characteristics specifically in areas where people were car dependent and faced pressures due to vehicle fuel costs. The VAMPIRE Index for "vulnerability assessment for mortgage, petroleum, and inflation risks and expenditure" (Dodson and Sipe 2007; see also Dodson and Sipe, Chapter 11 in this book) was used to measure of household oil vulnerability in Queensland's capital city. The VAMPIRE Index, developed using ABS Census data to examine oil and mortgage vulnerability in Australian cities, was constructed from four socio-economic factors: household income, number of motor vehicles owned, method of journey to work and whether or not the household had a mortgage. These variables were applied to measure relative household oil vulnerability. However, the index did not identify vehicle types or their specific petrol usage.

The VAMPIRE Index used for this study was calculated using suburb level ABS census data for 2011 and involved four steps: first, taking each variable and determining its range from low to high and, then, dividing the range into six categories based on the following percentages: 0–9.9, 10–24.99, 25–49.99, 50–74.99, 75–89.99 and 90–100; second, assigning a value (0–5), depending on which category the data value fell into, from "0" as the least vulnerable to "5" as the most vulnerable; third, performing the same procedure for the remaining variables; and fourth, calculating the VAMPIRE Index by summing the values for the four variables assuming the following weightings:

VAMPIRE Index = (proportion of households owning two or more vehicles) + (proportion of people traveling to work by car) + (household weekly income × 2) + (proportion of homes purchased with a mortgage × 2)

The VAMPIRE Index was used to measure the 2011 household oil vulnerability in Brisbane. Households in a highly vulnerable suburb, as indicated by a high VAMPIRE score, were more exposed to potential adverse impacts from rising fuel costs than households in a suburb with a lower VAMPIRE score.

Findings

The findings of our Brisbane study concentrated on measuring the distribution of VFE, the distribution of alternative energy vehicles (AEVs) and the spatial distribution of oil vulnerability throughout the suburbs of Brisbane. Then, we compared the distribution of VFE and AEVs with household oil vulnerability with the following results.

Distribution of vehicle fuel efficiency

The distribution of average VFE for Brisbane is shown in Figure 15.1. Overall, the average VFE tends to be higher in inner urban areas surrounding the city center. These areas are surrounded by suburbs

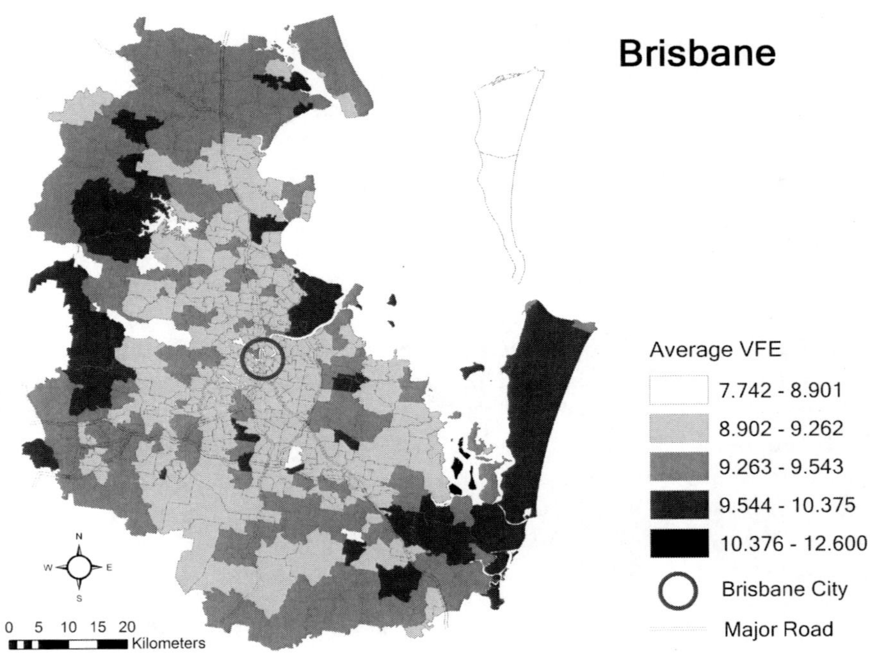

FIGURE 15.1 Distribution of average VFE (L/100km) for Brisbane suburbs

Source: Li et al. (2015, Fig. 2).

with average energy efficiencies. A household's preference for vehicle efficiency in those areas appears to be influenced by residential location and travel patterns. This is observable in high density areas that tend to have fewer and smaller parking spaces, narrower streets and more congested traffic. In addition, households in these areas tend to be closer to employment and public transport services and, thus, make fewer trips by motor vehicle travel and use less energy for transport. These conditions all work in favor of choosing smaller, more maneuverable and often fuel-efficient vehicles (Bhat et al. 2009; Choo and Mokhtarian 2004; Eluru et al. 2010; Kitamura et al. 2001).

In contrast, there are many outer suburbs in Brisbane that have poor vehicle efficiencies. Households located in the dispersed outer suburban areas typically have inadequate access to employment and public transport, so they are car dependent. In addition, households in low density areas have fewer constraints on car transport operations (e.g. more parking and less road traffic), resulting in higher frequencies of large vehicles. For these areas, this results in correspondingly larger and less efficiency vehicles (Cao et al. 2006; Kockelman and Zhao 2000). Some low vehicle efficiencies were observed in some middle suburbs, most likely due to a higher proportion of large or high performance vehicles used in some higher-income suburbs. The analysis also suggests that occupation may affect household vehicle choice. For example, suburbs with a high proportion of blue collar workers tend to have more light trucks and minivans.

Distribution of alternative energy vehicles

The proportion of AEVs within a suburb served as the second indicator of any technological transformation of the Brisbane vehicle fleet. This is because, generally, these vehicles have lower carbon emissions and therefore offer an alternative to current vehicle technologies for reducing vehicle fuel consumption. The proportion of AEVs by suburbs is presented in Figure 15.2. A variation in the distribution of AEVs

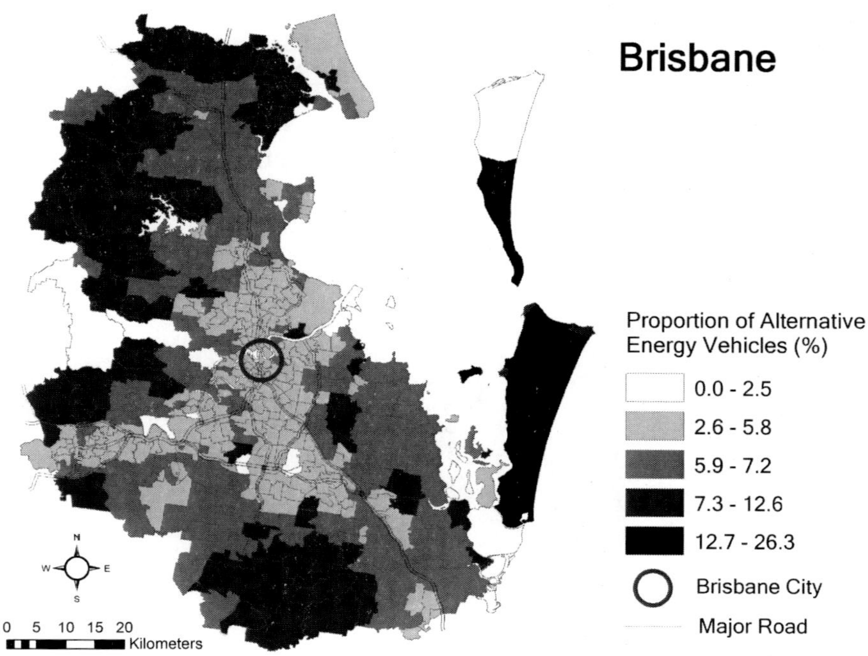

FIGURE 15.2 Distribution of alternative energy vehicles in Brisbane

can be observed between inner urban areas and middle and outer urban areas of Brisbane. An array of suburbs with very low proportions of AEVs are observable to the inner suburbs of Brisbane, stretching to the west, south and north along the major road networks. High rates of AEVs per suburb are observed among suburbs in the northwest, far north, and far south of Brisbane, particularly in suburbs beyond approximately 20 km from the central business district (CBD).

Oil vulnerability in Brisbane suburbs

The VAMPIRE results, as shown in Figure 15.3, demonstrate a wide variation in mortgage oil vulnerability levels between suburbs within Brisbane. Brisbane CBD exhibits the highest concentration of low vulnerability localities. Other areas with low or moderate VAMPIRE scores are dispersed throughout the middle suburbs, mostly within 15 km of the CBD. The middle suburbs of Brisbane also display some variation in vulnerability with a mix of moderate scores.

High mortgage and oil vulnerability scores were found in many middle, outer and fringe suburban areas. The highest VAMPIRE levels were found in the outer growth corridors to the north, west and east. Each of these corridors contained many suburbs with high VAMPIRE scores, indicating high levels of vulnerability. There were some small pockets of moderate mortgage and oil vulnerability in outer areas at Redcliffe (in the northeast), Cleveland (in the south) and near central Ipswich. However, these pockets were rare relative to the broad tracts of highly vulnerable densely mortgaged residential areas within the growth corridors.

The overall picture provided by Brisbane is of a graduated spatial divide between the inner and middle areas, with lower car dependency and good public transport services exhibiting lower vulnerability

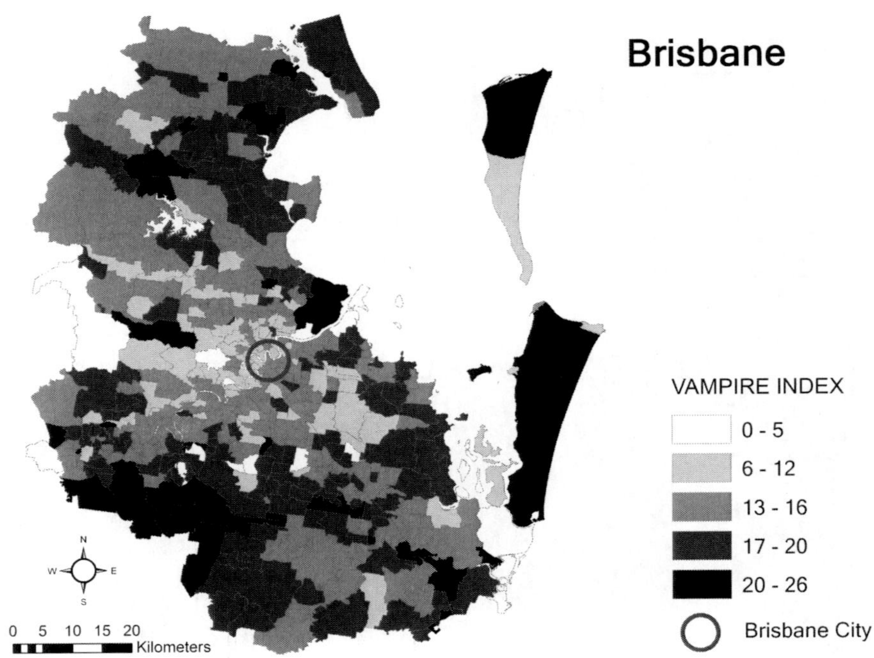

FIGURE 15.3 Oil vulnerability in Brisbane's suburbs

than those in the dispersed and outer suburban locations. Outer suburban and fringe areas that are highly car dependent are most vulnerable to both the impacts of rising fuel and mortgage interest costs.

Vehicle fuel efficiency and alternative energy vehicles in vulnerable suburbs

We compared the distribution ofVFE and AEVs with household oil vulnerability by overlaying the suburbs with the highestVFE and AEVs with those that were the most oil vulnerable, respectively, to identify the suburbs that were the most vulnerable and had high VFE or high proportions of AEVs. The suburbs with high VFE (or high proportions of AEVs) were identified as those suburbs with the average VFE (or AEVs) greater than one standard deviation from their mean. The most oil vulnerable suburbs were those with VAMPIRE scores in the highest decile. Figure 15.4 shows the distribution of the suburbs that are the most oil vulnerable and have high VFE or high proportion of alternative energy vehicles in the local fleet.

Overall, the level of spatial intersection between those areas of high VFE or AEVs and the most vulnerable suburbs is low. For example, only 6 out of the 55 most oil vulnerable suburbs in Brisbane had high levels ofVFE in the local fleet. These suburbs were identified at Griffin and Rush Creek (in the north) and Deebing Heights, Augustine Heights, Ripley and West Ipswich (in the west). The most vulnerable suburbs in Brisbane also had relatively low proportions of AEVs. Only 6 out of the 55 most oil vulnerable suburbs in Brisbane had high proportion of AEVs in the local fleet. Households in the most vulnerable suburbs showed a lower tendency to own and use more fuel-efficient vehicles.

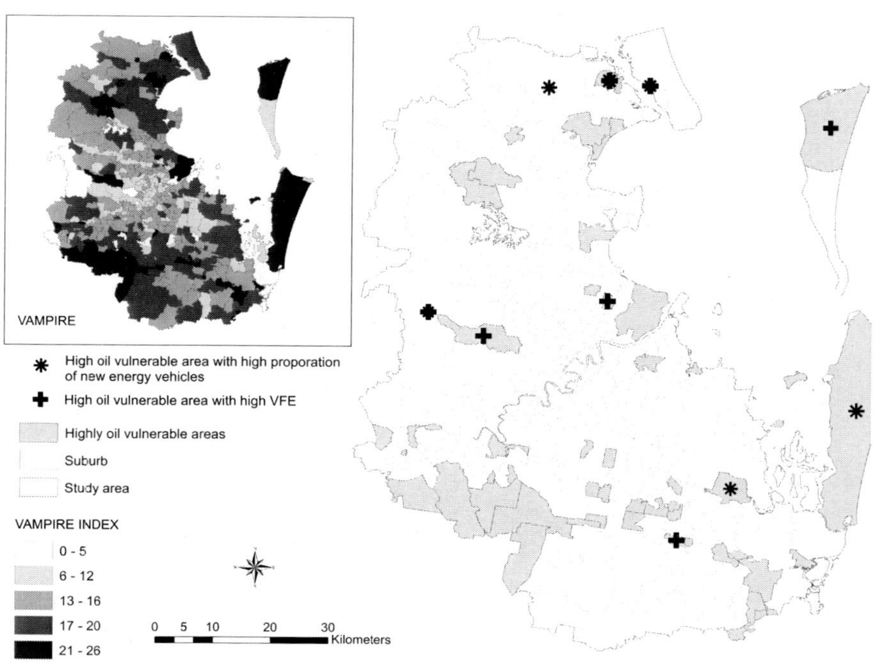

FIGURE 15.4 VFE change in Brisbane's most oil vulnerable suburbs

Discussion and conclusions

Australian cities face energy security and climate change mitigation challenges. The shifting energy environment will impact urban transport systems. Present policy settings assume that advances in motor vehicle technology combined with a shift to low carbon fuel sources will enable a relatively smooth transition to an energy secure, carbon neutral vehicle fleet.

While new technology holds considerable appeal among politicians and policy makers, there are a number of problems associated with a reliance on technology as a salve for urban transport problems. To address this issue, this chapter examined the distribution of VFE and AEVs in Brisbane in the early 2010s and how they intersected with social spatial variables of household oil vulnerability. The analysis showed a clear spatial variation in emerging fuel-efficient vehicles and alternative fuel technologies.

The overall VFE tended to be lower in outer suburbs than those in the middle and inner suburbs of Brisbane. The areas with low VFE were typically low density, distant to major employment and poorly served by public transport. AEVs tended to be found in outer suburban areas of Brisbane. Areas that were close to Brisbane CBD, or near the major transport corridors, show a lower proportion of AEVs. The results further illustrate that both AEVs and fuel-efficient vehicles are less likely to be found in areas where households have low incomes and are highly car dependent. This is due to the fact that people living in those areas do not have the capacity to afford vehicle upgrades to those with better VFE or AEVs.

Dodson et al. (2010) argued that policies that focus on vehicle technology alone face a number of social equity hurdles as measures to overcome urban transport fuel security problems. Policies need to account for the considerable social differences in household exposure to the costs of transport energy and the adaptability of households in altering their vehicle ownership patterns.

When links between household financial capacity and motor vehicle energy efficiencies are examined in combination, these patterns show that the government's carbon pollution reduction objectives will be frustrated because the areas with the highest levels of dependence on cars are also those with modest capacity to afford new vehicle types.

In short, contemporary efforts to reduce Australia's urban transport dependence on increasingly insecure petroleum and to reduce the carbon intensity of urban transport through the deployment of new vehicle technologies do not offer a comprehensive transformation of urban transport patterns. Households in the highly oil vulnerable areas would need higher financial support or significant fuel saving returns before it would become cost-effective to shift to new more efficient vehicles.

Further analysis should be done to explore the levels of household fuel savings that could be generated through adopting more efficient vehicles. This is because although the current uptake of high efficient and new energy vehicles in the oil vulnerable areas is relatively low, the potential benefit it generates for households is potentially considerable. Such a change could be driven by more dispersed vehicle travel patterns and higher net fuel energy consumption in those areas (Li et al. 2013). New technologies will provide an element within Australia's urban transport future. However, a much wider strategic approach to coping with declining energy security and reducing carbon emissions will be needed. As argued by Mees (2000) and Dodson and Sipe (2008), a holistic approach will include improving public transport networks and associated coordinating institutions.

References

ABS (2010) *Survey of Motor Vehicle Use* (Cat. No. 9208.0), Canberra: Australian Bureau of Statistics.

Andersen, P.H. Matthews, J.A. and M. Rask (2009) "Integrating Private Transport into Renewable Energy Policy: The Strategy of Creating Intelligent Recharging Grids for Electric Vehicles," *Energy Policy* 37(7): 2481–86.

Bhat, C.R., Sen, S. and N. Eluru (2009) "The Impact of Demographic, Built Environment Attributes, Vehicle Characteristics and Gasoline Prices on Household Vehicle Holdings and Use," *Transport Research Part B* 43(1): 1–18.

Bracks, S. (2008) *Review of Australia's Automotive Industry*, Canberra: Australian Government.

Cao, X., Mokhtarian, P.L. and S.L. Handy (2006) "Neighborhood Design and Vehicle Type Choice: Evidence from Northern California," *Transportation Research Part D* 11: 133–45.

Carr, K. and K. Rudd (2008) "A New Car Plan for a Greener Future," 10 November, Canberra: Joint Media Release by Prime Minister and Minister for Innovation, Industry, Science and Research, accessed 30 January 2016 at the Australian Government's Department of Industry, Innovation, Science, Research and Tertiary Education Website Archive — http://archive.industry.gov.au/ministerarchive2011/carr/MediaReleases/Pages/ANEWCARPLANFORAGREENERFUTURE.html

Choo, S. and P.L. Mokhtarian (2004) "What Type of Vehicle Do People Drive? The Role of Attitude and Lifestyle in Influencing Vehicle Type Choice," *Transportation Research Part A* 38(3): 201–22.

Dodson, J., Li, T. and N. Sipe (2010) "Urban Structure and Socio-economic Barriers to Consumer Adoption of Energy Efficient Automobile Technology in a Dispersed City: A Case Study of Brisbane, Australia," *Transportation Research Board* 2157(2): 111–18.

Dodson, J. and N. Sipe (2010) "Emerging Australian Planning Practice and Oil Vulnerability Responses," *Australian Planner* 47(4): 293–301.

——— (2008) "Shocking the Suburbs: Urban Location, Home Ownership and Oil Vulnerability in the Australian City," *Housing Studies* 23(3): 377–401.

——— (2007) "Oil Vulnerability in the Australian City: Assessing Socio-economic Risks from Higher Urban Fuel Prices," *Urban Studies* 44(1): 37–62.

EIA (2012) *Annual Energy Review 2011*, Pittsburgh, PA: US Energy Information Administration, accessed 30 January 2016 — www.eia.gov/totalenergy/data/annual

Eluru, N., Bhat, R., Pendyala, R.M. and K.C. Konduri (2010) "A Joint Flexible Econometric Model System of Household Residential Location and Vehicle Fleet Composition/Usage Choices," *Transportation* 37: 603–26.

Kitamura, R., Akiyama, T., Yamamoto, T. and T.F. Golob (2001) "Accessibility in a Metropolis: Towards a Better Understanding of Land Use and Travel," *Transportation Research Record* 1780: 64–75.

Kockelman, K.M. and Y. Zhao (2000) "Behavioral Distinctions: The Use of Light-Duty Trucks and Passenger Cars," *Journal of Transportation and Statistics* 3(3): 47–60.

Kuhnimhof, T., Zumkeller, D. and B. Chlond (2013) "Who Made Peak Car, and How? A Breakdown of Trends over Four Decades in Four Countries," *Transport Reviews* 33(3): 325–42.

Li, T., Sipe, N. and J. Dodson (2015) "Exploring Social and Spatial Patterns in Private Vehicle Fuel Efficiency: A Case Study of Brisbane and Sydney, Australia," *Australian Geographer* 46(2): 217–33 — www.tandfonline.com

——— (2013) "Investigating Private Motorized Travel and Vehicle Fleet Efficiency: Using New Data and Methods to Reveal Socio-Spatial Patterns in Brisbane, Australia," *Geographical Research* 51(3): 269–78.

Mees, P. (2000) *Rethinking Public Transport in Sydney*, Urban Futures Program Issues Paper 5, Sydney: Urban Frontiers Program, University of Western Sydney.

Millard-Ball, A. and L. Schipper (2011) "Are We Reaching Peak Travel? Trends in Passenger Transport in Eight Industrialized Countries," *Transport Reviews* 31(3): 357–78.

Romm, J. (2006) "The Car and Fuel of the Future," *Energy Policy* 34(17): 2609–14.

Sivak, M. and O. Tsimhoni (2009) "Fuel Efficiency of Vehicles on US Roads: 1923–2006," *Energy Policy* 37(8): 3168–70.

Van Dender, K. and M. Clever (2013) *Recent Trends in Car Usage in Advanced Economies-Slower Growth Ahead?* Discussion paper 2013–09, Paris: OECD International Transport Forum.

16

ENERGY FOR CITIES

Cheryl Desha and Angela Reeve

This chapter considers the altered dynamics of energy in cities where fossil fuels are no longer the dominant fuel sources. Using energy as a central planning paradigm, we discuss the need to take immediate action to build resilience for an energy future that fundamentally differs from current conditions. With so many rapidly emerging and diverse socio-technological innovations, it is imperative that these innovations are integrated within the energy matrix of cities in a systematic way that supports reliability and resilience to pressures such as increasing populations and climate change. This would include managing transition factors such as changing the amount and type of demand for energy in the built environment, and building capacity to implement systematic ways of meeting these demands with appropriate energy supply solutions.

We contextualize this challenge as one of "decoupling," where city growth and prosperity need to be separated from using fossil fuels. Precedents for such decoupling exist and provide exciting opportunities for reimagining the way that cities function. We reflect on the Dutch Sustainable Technological Development program of the 1990s as a seminal example that has informed global efforts to manage transitions of complex systems to sustainable alternatives. Then we focus on the more recent leadership shown by the Australian Townsville City Council, using energy management as a core concept to drive significant changes in energy flows, from infrastructure planning and maintenance through to industry, business and community behaviour.

Townsville City is the largest urban centre in North Queensland with a population of approximately 200,000 (including neighboring Magnetic Island). The city has a "dry tropics" savannah climate, with southeast trade winds resulting in reduced winter rainfall compared to other tropical locations. We highlight the non-linear and organic journey of this regional city that has prospered through a whole-of-system approach to energy and opportunities arising through day-to-day operations. Key factors of success include Townsville's involvement in the Australian Solar City program, and engagement with innovative initiatives such as Network Demand Management, Energy Sense Communities, Smart Infrastructure and Sustainable Development and Energy Transformation Townsville. A multifaceted, systemic approach with local authority leadership has resulted in multiple benefits, including increased community capacity to address energy issues and opportunities, a shift in the use of electricity in the community, and engagement with alternative transportation options.

We conclude this chapter by considering the need for immediate capacity building and decision making to facilitate the complex transition to life beyond petroleum as a dominant fuel source. Capacity building and decision making must span professions such as planning, design and engineering, and the community at large, as well as fostering small-scale pilot projects to take advantage of existing opportunities to build capacity within the system for new technologies. Lessons from Townsville City are particularly applicable to regional cities around the world facing energy challenges. In situations where existing generation or transmission capacities are being reached, or where a city or neighborhood can be partially isolated from the wider electricity grid, opportunities exist to explore alternative electricity generation and supply systems, and to focus on electricity demand reductions. Cities with relatively small populations – or sections of cities (comparable with Townsville) that can be partially isolated from the electricity grid – may have greater agility in terms of piloting and implementing new systems. Identifying such critical opportunities for testing and integrating new systems into urban environments lays foundations for broad-scale transitions in the future.

Energy complexities in planning after petroleum

Every aspect of the built environment depends on being able to access energy in some form or another. Reliable access to energy underpins our ability to create and live in the twenty-first century, whether the energy is used for producing infrastructure in the city's skeleton, whether it is flowing through the city's electricity network to power domestic, retail or commercial needs, pumping water or wastewater around, or being harnessed for human transportation. Globally, affordable and reliable energy is essential for improving living standards and alleviating poverty.

Over the last four decades, numerous authors – such as Burby and Flemming Bell (1979), Beaumont and Keys (1982), Owens (1986) and OECD (1995) – have considered the central role of energy in urban development planning, from contemplating impacts of the oil crises in the mid-1970s to considering the relationship between fossil fuel dependence and climate change. Regardless of the driver for considering energy issues, there is a common thread of argument for the need for more systemic approaches to addressing energy issues in the built environment. Such recommendations are also evident in recent publications by authors such as Gray and Gleeson (2007) and Gray et al. (2010), who consider the energy demands of urban living and the role for planning in Australia; Lerch (2007), who discusses the energy context for "post carbon" cities in an environment of energy and climate uncertainty; and Madlener and Sunak (2011), who discuss the impacts of urbanization on urban structures and energy demand. Clearly urban energy planning and urbanization management will be pivotal for creating the right framework conditions for a sustainable energy future.

Technological development, over the last 200 years in particular, has dramatically increased access to stationary and non-stationary energy. As shown in Figure 16.1, each wave of innovation through history has expanded opportunities for city developments and created additional energy demand in a variety of forms. With their prolific availability, fossil fuels such as coal, oil and gas have underpinned the creation of materials and electricity for city development and urban transport and, consequently, have become embedded in the functioning of most urban systems.

However, it is less well appreciated that the development of energy systems servicing urban demand for energy has been largely piecemeal and disconnected, resulting in systems that are vulnerable to physical (e.g. environment-related) and economic (e.g. pricing) shocks. Substantial time and effort is currently invested in caring for these systems to avoid potentially calamitous citywide and regional issues. For instance, the infamous 2003 Northeast blackout in the US and Canada resulted from overloaded power lines touching unpruned foliage and was compounded by an alarm software failure; the blackout affected 55 million people (NYISO 2003). Such fragility has been discussed in detail in Rocky Mountain Institute publications *Winning the Oil End-Game* (Lovins et al. 2004) and *Reinventing Fire* (Lovins et al. 2011).

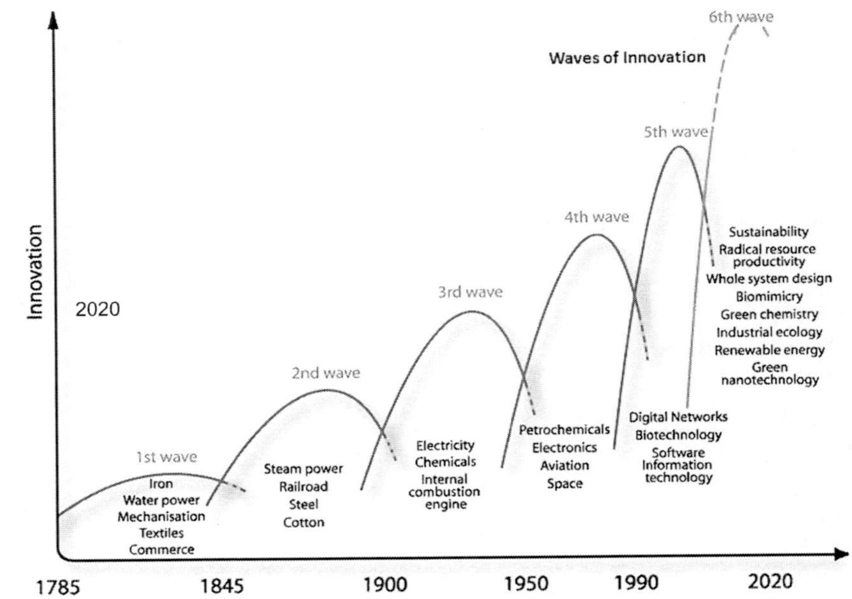

FIGURE 16.1 Waves of innovation diagram, highlighting associated leaps in energy needs

Source: Hargroves and Smith (2005: 17).

Existing energy infrastructure, spanning production, transmission and storage, needs to be altered to varying extents to use different emerging alternative energy options. These options range from alternative energy sources, such as distributed renewable energy systems that feed into the grid, to alternative technologies for energy consuming systems, such as electric plug-in vehicles or district networked chiller-plant systems for cooling, right through to energy efficiency and demand management interventions such as behaviour change programs and peak load shifting. Achieving a broad-scale transition to alternative energy systems will likely require implementation of many of these initiatives in concert. The complex interactions between such innovations, and the significant potential for achieving whole-system synergies through their concurrent adoption, requires early and integrated planning, considering whole-of-city opportunities and impacts.

Around the world, changing urban populations will have an impact on the type and scale of energy infrastructure in cities. Looking ahead, there is a pressing need for future energy systems to cater to higher-density urban environments, with an increasing proportion of aging populations and associated energy demands. An additional consideration for energy systems in these economies is the reduced spending capacity of aging populations and potentially smaller workforces to support the construction of new infrastructure. Depending on the rate of increase in the cost of fossil fuels, many economies may face complex challenges in providing care and support for a rapidly aging population and funding transitions to other energy sources.

Considering energy as a central planning paradigm, we are reminded of the point made by Albert Einstein (1905), that "energy can neither be created nor destroyed, it can only be changed from one form to another." The form and availability of energy will be critical to the success of twenty-first-century cities. Furthermore, in the mid-2010s, cities cover less than 2 percent of the earth's surface while consuming three-quarters of global energy and producing around two-thirds of all carbon dioxide (UN Habitat n.d.).

Our ability to access energy in a way that does not diminish our global environmental systems and, potentially, can restore degraded systems is also critical to success in mitigating further climate change.

Given the potential for sudden changes in energy markets, early efforts are necessary to begin a transition to diversified, low carbon energy sources, and to increase the efficiency and resilience of urban systems that rely on this energy. As highlighted by the United Nations Conference on Trade and Development Secretariat (UNCTAD 2010: 17):

> It is, therefore, imperative to introduce low-carbon energy sources in the national and global energy mix, while shedding the dependence on fossil fuels so that countries are able to attain energy security without jeopardizing the efforts made in attaining food security.

Key factors to consider in the transition include holistically matching demand to supply and addressing both energy security and resource security – as briefly addressed in the following paragraphs.

Managing our energy future requires a whole-of-system approach that integrates energy demand and supply considerations and opportunities across all urban systems. Historically, transportation energy and stationary energy systems have been relatively separate. However, several significant changes to electricity production and the context for transportation have resulted in much greater interdependence. Such changes include the increased use of gas for electricity production (particularly for peak generation), the linking of gas prices to oil prices over recent years and the increased interest in hydrogen cell and electric plug-in vehicles.

Looking ahead to future energy security with regard to sourcing petroleum, it is advantageous to diversify into alternative and domestically available fuels. This has the added benefit of stabilizing prices at the pumps and addressing carbon dioxide emission limits. Considering that twelve countries control nearly 80 percent of the world's oil reserves and nearly 50 percent of global oil production (Lovins et al. 2011), once petroleum is no longer the dominant fuel source, the planning environment will look substantially different. Indeed, in contrast to a planet vulnerable to oil supply shocks, there is potential for a vastly different array of global alliances and agreements around other energy sources. This is particularly evident for a primary consumer of crude oil, the transportation sector, whose energy matrix may comprise fuel options ranging from biofuels to fuel cells and battery-powered vehicles.

With the expansion of urban populations and processes of globalization and a concentration of the ownership and location of resource production systems, resources are increasingly coming from more distant places. Indeed, cities have become increasingly reliant on the import of resources and export of wastes. Rees (1992) has highlighted the reliance of urban areas on resources outside of the city limits, and similarly, transfers of ecological degradation through movement of waste products, pollution and other impacts of resource consumption. While Rees highlighted the need to recognize and address these ecological impacts caused by cities, more recently their vulnerability to resource shortages has become a critical focus in considering urban design and development (Gleeson 2008; Newman et al. 2009). The design of most cities in the Global North has locked in significant levels of resource consumption, through specific forms of housing stock, land-use mix, transportation systems and urban density, all of which determine reliance on cars, external sources of food and water, and electricity to maintain buildings, infrastructure and urban systems more generally.

Within the context of these energy factors and complexities, there is no silver bullet solution for meeting the energy needs of cities. Rather, the opportunity to plan for "after petroleum" is multifaceted, and requires planning for decoupling from fossil fuels as a dominant energy source (as we experience peaking and then tailing reliance on fossil fuels for energy) in addition to coupling, through increasing reliance on alternative fuel sources as energy supplies for cities. These mechanisms are illustrated in Figure 16.2, which draws on the decoupling theory described by Smith et al. (2010: 32).

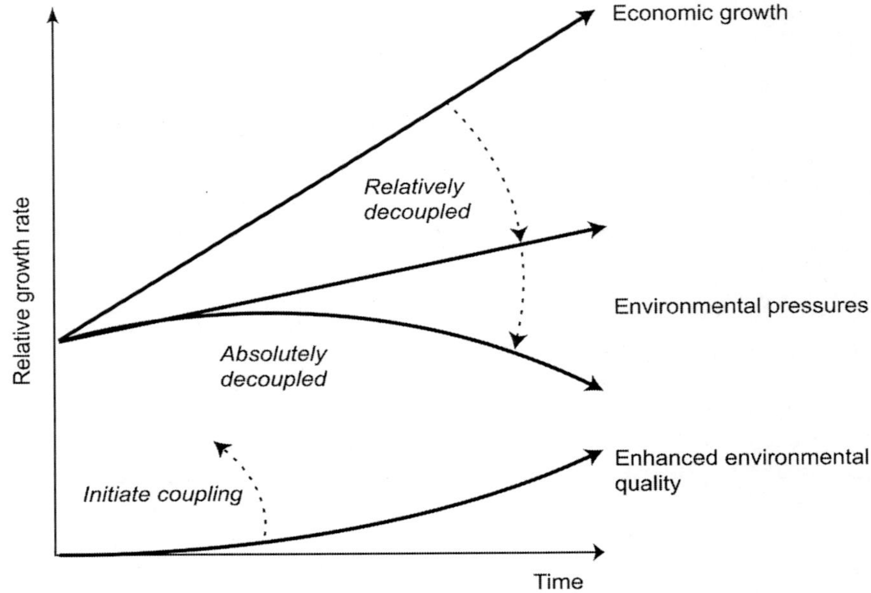

FIGURE 16.2 Peaking and tailing dynamics in decoupling growth from fossil fuel use

Source: Hargroves and Smith (2005: 17).

Ensuring energy supply systems are resilient to challenges and threats, and sustainable into the future, are key considerations in the design of cities. Considering the challenges and impacts on energy systems of peak oil, climate change and population growth collectively, it is clear that multiple innovations and transitions are needed across various aspects of both transportation and stationary energy systems. Innovation and transition will focus on reducing reliance on petroleum for mobility, enabling a transition from petroleum-fueled vehicles to alternative fuels (such as electric, hydrogen and biofuels) and developing capacity within the electricity grid to support increased demand from plug-in electric cars (with potential to benefit from such use to balance demand and supply over short-term fluctuations).

It will be important to build the capacity of the public to reduce their electricity demand and to shift their demand from peak periods. Equally, it will be important to build the capacity within the grid to cater for significant increases in distributed, renewable sources of electricity. Through developing a detailed understanding of electricity demand patterns, strategies can be created for reducing and shifting demand according to supply.

A notable example of an initiative using a holistic approach to transitioning energy systems is the Sustainable Technological Development program in the Netherlands. In 2001, the Dutch government adopted strong measures to transition toward a sustainable energy system in the Fourth Dutch National Environmental Policy Plan (NMP4). The NMP4 called for "system innovation" to achieve these goals, drawing on Dutch experience in the early 1990s with the Sustainable Technological Development (STD) program, in which they demonstrated methods for managing transitions in large, complex systems. The plan emphasized that persistent environmental problems such as climate change cannot be solved by intensifying existing policies, and that a transformation of the major systems and practices would be needed (Kern and Smith 2008).

Transition efforts were based on the observation that sustainability problems are deeply rooted in social structures, and that system innovations and transitions are complex and multidimensional, resulting from a collective set of many societal and technological developments (Vergragt 2005). As drawn from Vergragt (2005) and Rotmans et al. (2001), key aspects of the program approach included a process of back-casting, a "learning by doing" approach and small- to medium-scale experiments. First, based on scientific evidence, the Dutch government worked backward from a pre-determined vision to establish immediate and short-term bipartisan steps that would facilitate the process. Second, they conducted experiments in multi-stakeholder settings, on a small to medium scale, to foster collective learning processes among stakeholders. Third, they used small- to medium-scale experiments to demonstrate and communicate new possibilities, rather than to initially try to transform entire areas or sectors.

In summary, the STD program was a world first in calling for dramatic "leapfrog" technological, cultural and structural changes to address sustainability issues. Through these experiences, the Dutch showed that system innovations and transformations could be managed and developed in a step-by-step process. More recent international examples of systemic attempts to transition cities toward sustainable development include the German "Morgenstadt" (MCFI 2014) and European "Energy Cities" (Energy Cities n.d.) – collaborative initiatives developed to build capacity for a transition to sustainable urban energy systems. In Australia, the authors have been involved in Townsville City Council's journey to transition beyond fossil fuels as the dominant source of energy, as described in the following sections.

A case study of energy driving city planning: Townsville City Council

Townsville is the largest city in northern Queensland, with a current population of more than 200,000 residents. The population is expected to reach 225,000–361,000 by 2036 (Queensland Government 2013). Its economy is diverse, with significant industries including retail trade, health and educational services, government administration and defense, construction, mining, manufacturing, and property and business services (TCC n.d.). In addition to population growth, as in much of North Queensland, the warm, tropical climate is contributing to significant increases in electricity demand for air conditioning. Strong economic growth is also driving increases in electricity demand and, in particular, increases in peak demand (UNQRDO 2010).

Electricity suppled to Townsville from the National Electricity Market (NEM) is primarily through long transmission lines from coal-fired power stations located in Central Queensland, with losses of 10 percent during transmission (AEMO 2014). The cost of supplying electricity with such losses, along with maintaining the transmission lines themselves, is substantial (TCC n.d.b). Wholesale electricity customers in North Queensland currently pay around five times as much in Transmission Use of System charges as central Queensland customers. Modeling suggests that growth in electricity demand is such to require either investment in additional generation capacity in North Queensland, or upgrading transmission lines from Central Queensland (UNQRDOS 2010). Currently, the economic case for developing new coal-fired or gas-fired generators in North Queensland is unfavorable, making such options unlikely (UNQRDOS 2010). In addition, remoteness creates significant difficulties. For instance, during Cyclone Yasi (2011), power supplies to nearly one-third of Ergon Energy's customer base were disrupted, exposing the vulnerability of Townsville's electricity supply (QRA n.d.).

Within this context, over the last decade, the City of Townsville has been transitioning to a resilient, low carbon, alternative energy future by recognizing, and taking advantage of, strategic opportunities for interventions and investments to build capacity in energy efficiency, demand management, and renewable and distributed energy systems. Understanding, awareness and expertise in alternative energy systems (TCC 2013a) have been built through pilot projects, active community and stakeholder engagement, and collaborative whole-of-system approaches to enable a mainstream transition in the future, while retaining current overarching stability and cost-effectiveness provided by the nationally dominant Australian National Electricity Market (NEM) and petroleum-fueled, automobile-based transportation system.

We reference five programs to discuss how Townsville has built capacity to enable a transition toward high levels of renewable energy penetration and electric vehicles. These programs have used sophisticated demand management and energy-efficiency interventions to help stabilize energy demand, facilitating the increased supply of energy through alternative technology systems. As shown in Figure 16.3, such interventions fit within a whole-of-system approach to the city's journey and energy is at the center of strategic considerations. These early initiatives have succeeded through a strong focus on projects that offer immediate economic benefits, in addition to the critical role played by champions within government and the community. In particular, a strong and collaborative relationship between the Townsville City Council and the city's main electricity utility, Ergon Energy, has resulted in several key projects that have produced optimal and synergistic outcomes for Council, Ergon Energy and the community – through reducing electricity demand, delaying investment requirements in electricity infrastructure and reducing household costs.

As reflected in Figure 16.3, an important component of the whole-of-system approach is leading through example. Townsville City Council has taken a leadership role in demonstrating and encouraging renewable energy to build capacity citywide. This includes installing 180 kW of solar photovoltaic (PV) on Townsville City Council buildings, and installing a 6 kW grid-connected vertical axis wind turbine as a demonstration project to educate people about wind power and attract tourists to the city (State of Environment n.d.; CT n.d.). In addition, a small-scale wind project has been developed to assist with water reticulation involving storm water and wetland management (TCC n.d.c).

Council trialed two methane generators, with subsequent purchase and upgrade for use as part of the Carbon Neutral Water Recycling program, capturing methane from both the Vantassel Street Landfill and the wastewater sludge (TQSC n.d.). Also, the council converted a Toyota Prius car into a hybrid electric plug-in vehicle, and established pilot charging stations around Townsville. In summary, Townsville

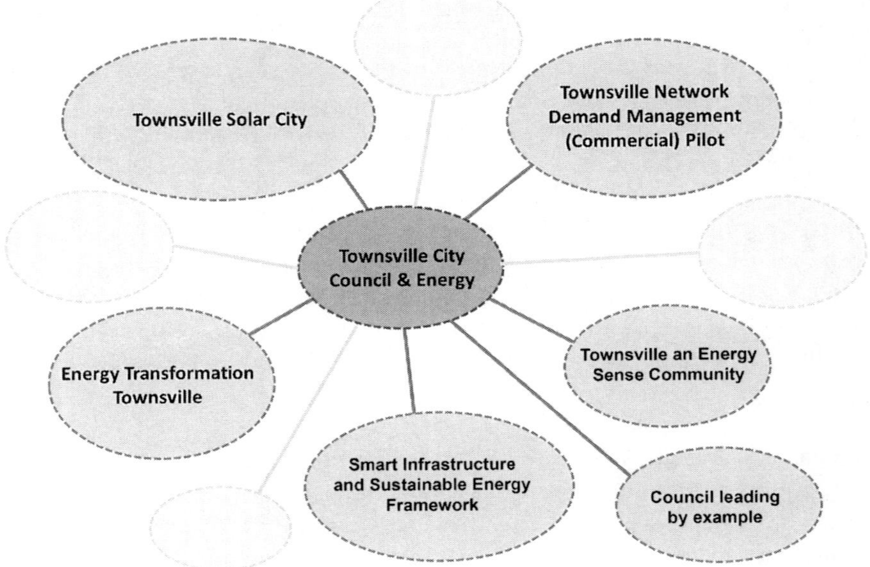

FIGURE 16.3 Mind-map of Townsville City Council's programs: Energy-related initiatives are core to planning

demonstrates the viability of adopting world-leading, holistic approaches to reduce the vulnerability of a mid-sized city to climate change and resource shortages, while also building economic resilience and future-oriented industries.

The Townsville Queensland solar city program

Since it was announced in 2004, the Australian government's A$93.8 million Solar Cities program has funded seven Australian cities, including Townsville, to demonstrate the economic and environmental impacts of cost-reflective pricing, the concentrated uptake of solar, demand management and smart metering technologies, and to identify and address barriers to distributed solar photo-voltaic technology, energy efficiency and electricity demand management for grid connected urban areas. The Townsville Solar Cities Consortium (TQSC 2013) received A$15 million from the Australian government, the Queensland government contributing a further A$5 million, and the consortium members A$11 million toward the A$32 million project. Table 16.1 summarizes the achievements of the Townsville Queensland Solar City project.

There were a number of initiatives implemented through the Townsville Queensland Solar City project, which took advantage of investment thresholds for electricity network upgrades and the potential for synergistic benefits for industry stakeholders and the broader community. Details of four such projects follow.

TABLE 16.1 Summary of Solar City achievements by Townsville City Council

Objective	*Target*	*Achievement*
Reduce maximum demand*	27% improvement on business as usual	46%
Reduce electricity consumption*	25% improvement on business as usual	46%
Defer undersea cable*	Within 6 years	Now at least 8 years
Reduce greenhouse gas emissions	By 50,000 tonnes	53,000 tonnes to June 2012
Increase uptake of renewable energy	Install 1 MW of hosted photovoltaic systems	1.084 MW installed in 212 systems
Reduce costs to consumers*	By at least A$1,000,000	A$1,784,000 to June 2012
Build capability in the community for sustainable living	Develop campaigns in the community, schools and businesses using the city's solar program	Eco-electricity tours, Cool Roofs, Centre of Excellence, Smart Lifestyle Expo
Demonstrate energy efficient housing	Develop continuous improvement in residential and medium density housing	Three efficient apartment complexes completed, energy efficient construction now business as usual
Energy efficient office accommodation	Investigate best practices for energy efficiency office accommodation	Wide research, workshops, report produced
Energy efficient refurbishment of heritage office building	Reduce the energy demand in a 128-year-old office building	25% reduction in demand from the grid

* Specific to Magnetic Island.
Source: Adapted from TQSC (2013).

Magnetic Island: Ergon Energy initiative

In 2003, Ergon Energy realized that in order to satisfy rising demand, a third undersea cable was needed immediately to double the voltage connecting Magnetic Island to the mainland electricity network from 11 to 22 kilovolts. The investment cost was estimated at A$18.6 million at net present value in 2006 (Bruce et al. 2013: iv). Ergon Energy identified this as an opportunity to trial a comprehensive application of demand management, energy efficiency and renewable energy, with potential for knowledge gained to inform future management of the electricity grid and integration of solar PV into the statewide electricity network (Bruce et al. 2013). Consequently, in collaboration with Townsville City Council, a bid for Solar Cities funding was developed and a consortium of industry, community and academic partners activated around their common overarching objective (TQSC 2013).

The project achieved a 46 percent reduction in annual peak demand compared with 2012–13 predictions and a 27 percent improvement on the project's target, translating into collective savings of A$1.784 million for customers over the short life of the project (Anon. 2014; Bruce et al. 2013; TCC 2013b). As a result, Ergon has been able to defer investment in the proposed undersea supply cable for a decade, saving an estimated A$20 million. The high penetration of solar has enabled Ergon Energy (Anon. 2014) to test a range of different technologies and ways of addressing issues with PV technology and high penetration rates, to inform future efforts throughout their electricity network.

Community behaviour change: Cool roofs initiative

A comprehensive study into energy demand reduction opportunities through behavior change was piloted and then widely implemented as a rigorous and repeatable program, in collaboration with the Natural Edge Project, then hosted by Griffith University (Australian Government n.d.). The Townsville Residential Energy Demand (TRED) initiative drew on international behavior change methods and theories, including community-based social marketing (McKenzie-Mohr 2011) and thematic communication (Ham 1992). Using a world-first hybrid approach, a comprehensive investigation of the literature found 280 energy-efficient behaviors applicable in the home (Hargroves et al. 2008), subsequently evaluated for impact and likelihood of implementation, to identify the most critical opportunities to impact the energy demand of households. Council selected six target behaviors for community behavior change, prioritizing the cool roofs initiative for the first pilot program given findings that this could reduce household demand for air conditioning by 10–15 percent (TQSC 2013).

A range of behavior change tools and strategies are currently being tested in this pilot program to encourage households to paint their roofs white, and analysis of results will inform future efforts to encourage other key energy-saving behaviors (Hargroves et al. 2010). Through TRED, Townsville City Council is building capacity to engage and assist the community to understand and change their behavior, a critical component of a systemic transition to alternative energy systems.

Renewable energy and energy efficiency retrofit trials

A number of large-scale renewable energy and energy-efficiency initiatives were implemented in Townsville as part of the Solar Cities project, across various building types and business models. In one such model, Ergon Energy owned the PV infrastructure and all the electricity produced was fed back into the grid, with a combination of residential, commercial and public premises and facilities hosting the panels.

This was an effective approach, although reduced costs of PV and associated federal subsidies to households have adversely affected the model's viability (TQSC 2013). A second example involved a private company (Chester Holdings) joining the Townsville Solar Cities Consortium in 2011 and subsequently retrofitting their 1885 heritage-listed building in central Townsville to reduce network energy demand by 25 percent (TQSC 2013).

Energy efficient residential development

Given strong population growth in Townsville, new residential developments need to be compatible with future energy systems. As part of the Townsville Solar Cities Consortium, Honeycombes Property Group piloted the use of sustainable design features in the first stage of the Itara medium-density residential development in Townsville. These features included energy-efficient fans, high levels of insulation, tinted windows, screens and demand-response enabling devices (DRED) for air conditioning. In addition, the apartments were fitted with energy-efficient lighting, smart meters and a centralized gas hot water system. Subsequent evaluation of the Itara complex has resulted in these features becoming standard inclusions in Honeycombes Property Group developments (TCC 2013b).

Network demand management (commercial) pilot

In partnership with the Queensland State Government Department, Ergon Energy (EE n.d.) initiated a range of commercial pilot projects in Townsville to reduce demand on the electricity network. Their aim was to develop tools and expertise in network demand management to enable Ergon Energy (EE 2013) to delay future investments in network upgrades. Ergon Energy contributed A$9.7 million of funding assistance for projects, anticipating cost savings for transmission, distribution and generation of electricity in the network. Commercial customers participating in the program gained substantial cost savings from reduced electricity consumption and many received a range of co-benefits, such as reduced capital and maintenance costs for equipment, saving space and improving the performance of lighting and HVAC systems (EE n.d.).

Three initiatives demonstrate the potential to develop pilot projects that integrate energy efficiency, demand management and renewable energy innovations. The first initiative was the District Cooling–James Cook University (JCU 2014) demonstration where, rather than upgrade their electricity infrastructure, JCU opted to dramatically reduce their energy demand. In addition to a range of energy efficiency and demand measures, JCU developed the largest central district cooling (CDC) system in the Southern Hemisphere, which was completed in 2009 (JCU 2012, 2014; Queensland Government et al. 2011).

The second initiative was a cooling feasibility study for the central business district (CBD), which proposed to replace existing air conditioning systems in the CBD. This included a "tolling" – a third party would finance, build, own and operate the chilled water production facility and distribution systems, using electricity supplied by Ergon Energy, and then the utility would charge end users for the supply of chilled water. A detailed feasibility study led to a decision to halt the project until future review (EE 2013). This example highlights that, although pilot projects may not be successful, they provide timely and risk-sensitive insights to inform decisions regarding costly full-scale interventions.

The third initiative involved a small not-for-profit organization, Kith and Kin, which operated in a 400 m^2 office space in central Townsville. The Ergon Energy (EE 2011) Network Demand Management Pilot Program awarded Kith and Kin a nominal A$2,300 to implement retrofits in order to reduce demand by 10 KVA, measures that resulted in actual reductions of 11.75 KVA, reducing the organization's energy bills by 45 percent, so that more of their funding could be allocated to client services.

Townsville: An energy sense community

The two power utilities Ergon Energy collaborated with Townsville City Council to explore the potential to drive demand reduction and peak load shifting through a series of pilots with residential communities and with commercial and industrial clients. The Energy Sense Community Program has had three streams designed to collectively work toward an energy-smart "medium-term future" by 2020: the electricity distribution grid, the residential community and the commercial and industrial energy user (IBM 2011). The program aimed to trial the use of smart asset management techniques and technologies to defer planned network investments within the 2010–15 period and to gather Smart Grid–related evidence to support a regulatory reset in the 2015–20 period.

As part of the Energy Sense Program, Ergon Energy identified an opportunity to defer a A$30 million investment in the construction of a substation (at Mt. Saint John) by two years, by reducing coincident peak demand through customer demand management. Under this initiative, Ergon Energy has evaluated commercial and industrial customers and identified a value proposition suitable to their operations to encourage them to make changes to reduce their peak electricity demand. Customer demand management mechanisms include improved building management systems and installing proven energy-efficient technologies, such as lights, air conditioners, equipment and machinery (EE 2013). This cooperative approach with customers has provided cost savings to both Ergon Energy and their customers, and built capacity within the community to manage and reduce energy demand.

Smart infrastructure and sustainable energy framework

The Townsville CBD Smart Infrastructure and Sustainable Energy Framework has built on Townsville's success – in piloting and demonstrating a wide range of demand management, energy efficiency, renewable energy and sustainable design innovations – by leveraging opportunities to create and build implementable projects and processes to achieve high impact, low cost energy efficiency, delivering benefits to property owners, occupants and the broader public (TCC 2013b). A variety of key sub-projects have been identified, each designed to deliver energy demand reductions and cost savings, while improving inner-city living and CBD commerce. The framework was designed to enable integration of such sub-projects, so as to form a whole system of interrelating and interacting projects achieving compounding efficiencies and operational cost reduction once implemented.

Energy transformation Townsville

Townsville City Council was awarded A$813,000 in federal funding through the Community Energy Efficiency Program, part of the Clean Energy Fund. This funding was matched dollar-for-dollar basis by the Townsville City Council to implement city energy conservation measures. The three key elements involved implementing energy conservation measures on council-owned buildings and facilities, introducing smarter instrumentation and data management systems to develop a better understanding of energy demand patterns and council-wide energy management systems, and developing community education and capacity-building programs to assist the community to engage in energy conservation behaviors. The target annual savings of more than 850,000 kWh was expected to save council more than A$100,000 per annum (p.a.) in utility costs (Staff Writers 2012; TCC 2013a; TQSC 2013).

IBM Smarter Cities challenge

Townsville was selected as one of 100 global cities to participate in the IBM Smarter Cities Challenge in 2010 based on their leadership in demonstrating the feasibility and benefits of sustainable, integrated renewable energy supply and demand management, as well as partnership approaches with the electricity utility and the community as a whole (IBM 2011).

A team of experts from different business units and geographies within IBM worked with the Townsville community to develop a strategy to build on their successes to date. A systems approach guided the understanding of the challenges and opportunities for Townsville – including, most critically, those stemming from peak oil, peak network demand, energy conservation, climate change and energy security, water supply and conservation, and changing demographics and health care costs – to create a set of integrated recommendations for Townsville to become a smart and resilient city. The IBM (2011) recommendations included efforts:

- To establish a focused city sustainability hub as a platform that aggregates digital services, information and knowledge from private and public sources to allow government, academics, businesses and citizens to easily connect, learn, organize and collaborate in ways that change individual and group behavior.
- To leverage the federal government's investment in the National Broadband Network and the selection of Townsville as one of five pilot sites for early connection, to underpin digital innovation, commerce and collaboration.
- To establish citywide open data architecture and agenda to facilitate data sharing between the government, universities, business, and broader community to allow for non-linear, collaborative learning and innovation.
- To develop guidelines for open data management, including to protect privacy of individuals and organizations as well as to ensure the quality, reliability and interoperability of datasets.
- To develop a set of sustainability key performance indicators for Townsville to direct the city's actions, and support evaluation of progress toward its goal of being the leading tropical city in sustainability.
- To select a project for piloting open data and open apps with a small group of stakeholders, to provide learning and demonstration of these concepts before broadening efforts to a citywide approach.
- To integrate digital technologies into physical learning environments to expand access to education, and to enhance the way in which people learn.

To implement these recommendations, Townsville City Council partnered and collaborated with IBM Research, James Cook University, Ergon Energy, local environmental products and services businesses, community non-governmental organizations and natural resource management groups, and information and communication technology businesses to research, develop and deliver trial projects that could be scaled up across the city. One of these projects was the Australian Urban Research Infrastructure Network (AURIN n.d.), which involved several Australian universities in establishing the Townsville Data Hub to bring energy and water consumption data and climate data, together with council's land use, geospatial and demographic information, to identify strategies for sustainability and resilience.

Implications for building capacity

This chapter has discussed the need to transition the focus of city planning to managing energy flows. This will require a shift in context and priorities across a variety of planning tools, from strategic planning and policy documents through to development application requirements, incentive structures and fees. Actions to enable this shift include immediate and longer-term tasks that address environmental priorities for energy production with regard to climate change mitigation, and matching the diverse range of building, transportation and other demands for electricity with a range of appropriate supply options.

Through strategic planning and identifying critical opportunities for interventions and investments, capacity can cost-effectively be built within existing energy systems to facilitate whole-of-city transitions to sustainable, post-petroleum energy systems in the future. Such opportunities might include delaying substantial capacity-increasing investments in energy infrastructure through demand management, developing pilot projects to trial and build capacity in the use of alternative energy systems, and building awareness and capacity for energy efficiency and demand management throughout the community.

Part of the process of decoupling cities from fossil fuel based resource consumption involves building capacity for reducing demand as well as implementing alternative energy supply options (Desha and Hargroves 2014). The ability to build capacity for cost-effectively enabling the types of energy infrastructure discussed will play a major role in determining the type of experience residents will have in "life after petroleum." Figure 16.4 indicates that the longer we wait to transition, the steeper the transition curve and the greater the impacts on societies around the planet.

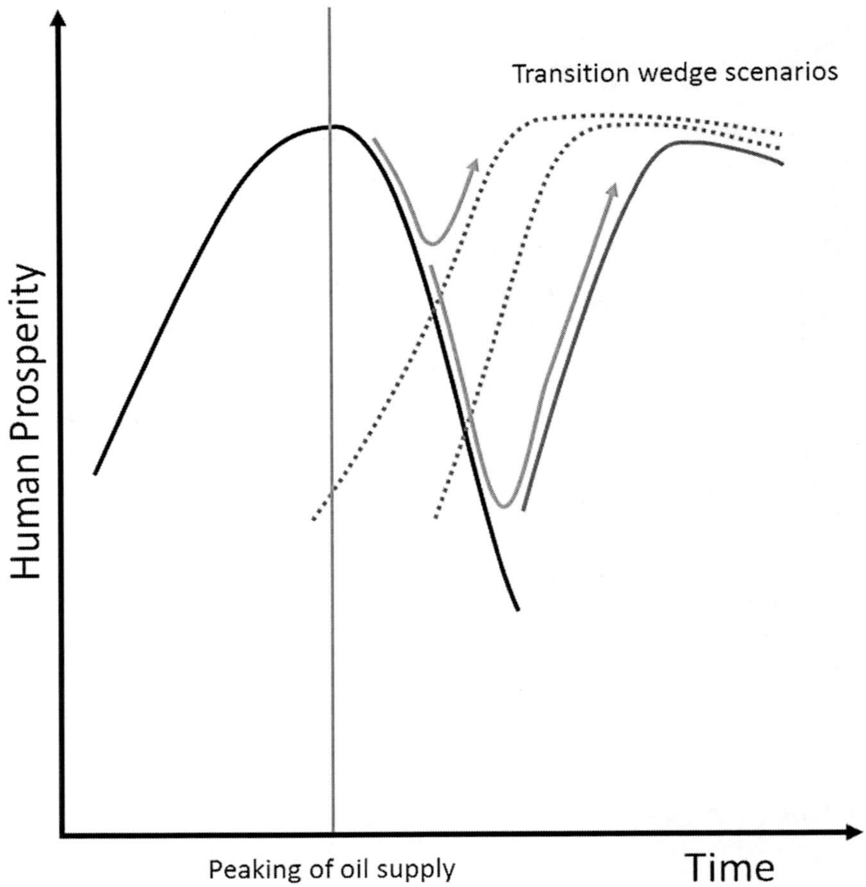

FIGURE 16.4 Transition scenarios and human prosperity beyond "peak oil"

Going forward, a robust strategy for capacity building will comprise temporal and sectoral dimensions, including targeting action-based knowledge and skills training for action with regard to identified short- and long-term priorities. Depending on the priority, such training will have a range of lead times for development and delivery to ensure that goals can be met. All planning will need to address uncertainty in the timing and differential scales of challenges, such as climate change and peak oil, with an imperative for early action to minimize risk.

Knowledge and skills training will need to be delivered to sectors where it will be of optimum use and to include different goals for formal undergraduate education, ongoing professional education and community capacity building for citizens. Strategies will need to identify critical opportunities for small- and large-scale shifts to alternative energy systems, say where existing energy systems are reaching capacity, and where substantial investment is required, or where engaged communities are eager to test and demonstrate new technologies and systems.

Such planning will require attention by a suite of education providers, spanning the vocational education and higher education sectors, primary through to secondary education providers, private training providers and community engagement organizations. Strategic planning will require capacity building of educators and trainers, to ensure that knowledge and skills being developed are appropriate to the intended learning outcomes and associated priorities being addressed. Depending on the priorities and time frames, there may be significant benefit in seeking out partnerships and collaborations between key stakeholders, including various levels of government, electricity utilities, car manufacturers and citizens (among others) to enable integrated, project-oriented capacity building that provides benefits to multiple stakeholders and concurrently achieves outcomes in the energy transition.

Conclusion

In light of the energy complexity discussed in this chapter, there is a clear and urgent need for decision makers to foster locally appropriate, integrated and multifaceted projects that can foster conditions conducive to transitioning away from fossil fuels as quickly as practical, with minimum fuss. In order to address the great challenges ahead, we suggest a suite of action-oriented responses, as follows.

Consider whole-of-system opportunities that provide synergies that may not be possible by considering any aspect of an energy system in isolation. Seek out "win-win" situations where investments in energy efficiency, demand management, renewable energy or alternative technologies can result in cost savings while building capacity for future transitions. Cultivate a culture of "learning by doing" and purposeful experimentation, with intentional evaluation and learning from initiatives to inform future efforts. Develop collaborative approaches to energy management, including multiple levels of government, industry and the community in order to leverage funding opportunities and enable comprehensive and synergistic innovations. Start small with pilot projects focusing on smaller communities or pilot innovations or interventions to gather experience and evidence for widespread and integrated application of interrelated and mutually reinforcing innovations. Harness existing opportunities for action, for example the potential to defer investment in energy infrastructure through demand management, and the availability of government or industry funding packages for innovation and demonstration. Monitor and communicate outcomes of innovations to ensure that they build awareness and support for more widespread application.

With such considerations in mind, and supported by professionals and citizens armed with relevant knowledge and skills, our cities' decision makers face a challenging but actionable agenda for transitioning to a sustainable energy future.

References

AEMO (2014) *List of NEM Regions and Marginal Loss Factors for the 2014–2015 Financial Year*, Sydney: Australian Energy Market Operator.

Anon. (2014) "Magnetic Island Still Reaping Solar City Benefits," *Utility Magazine*, 14 June, accessed 15 September — www.utilitymagazine.com.au/magnetic-island-still-reaping-solar-city-benefits/

AURIN (n.d.) *Townsville Data Hub*, Melbourne: Australian Urban Research Infrastructure Network (University of Melbourne), accessed 15 December 2015 — http://aurin.org.au/projects/data-hubs/townsville-data-hub/

Australian Government (n.d.) *Townsville Queensland Solar City – Townsville Residential Energy Demand Project (2009/10)*, accessed 15 September 2015 — www.townsvillesolarcity.com.au/ProjectOverview/Citysolar CommunityCapacityBuildingProgram/BehaviourChangeProgram/tabid/134/Default.aspx

Beaumont, J. and P. Keys (1982) *Future Cities: Spatial Analysis of Energy Issues*, New York: Research Studies Press.

Bruce, A., Heslop, S., MacGill, I. and M. Watt (2013) *Magnetic Island and Townsville Solar City: A Case Study of Increasing PV Penetration in Electricity Networks*, report by University of New South Wales for Australian PV Association, accessed 15 September — http://apvi.org.au/sites/default/files/documents/APVA%20Reports/Magnetic%20 Island%20High%20PV%20Penetration%20Case%20Study%20Report.pdf

Burby, R. and A. Flemming Bell (eds) (1979) *Energy and the Community*, Cambridge: Ballinger.

CT (n.d.) *The Strand Wind Project Sustainable Townsville*, Townsville: City of Townsville, accessed 15 September 2015 — www.soe-townsville.org/strandwindproject/wind_project_data_sheet.pdf

Desha, C. and K. Hargroves (2014) "A Peaking and Tailing Approach to Education and Curriculum Renewal for Sustainable Development," *Sustainability*, accessed 15 September 2015 — www.mdpi.com/2071–1050/6/7/4181

EE (2013) *Ergon Energy Demand Management Outcomes Report 2012/13*, Townsville: Ergon Energy.

——— (2011) *Network Demand Management Case Study – Kith and Kin*, Townsville: Ergon Energy.

——— (n.d.) "Townsville Commercial Demand Management Pilot," (page) Ergon Energy site (Townsville), accessed 15 September 2015 — www.ergon.com.au/energy-conservation/demand-management/electricity-demand-trials/townsville-commercial-demand-management-pilot

Einstein, A. (1905) "Does the Inertia of a Body Depend Upon its Energy Content?" *Annalen der Physik* 18: 639–41, accessed 15 September 2015 — http://einsteinpapers.press.princeton.edu/vol2-trans/186

Energy Cities (n.d.) *30 Proposals for the Energy Transition of Cities and Towns*, Besançon: Energy Cities, accessed 15 September — www.scribd.com/doc/97202740/30-Proposals-for-the-Energy-Transition-of-cities-and-towns

Gleeson, B. (2008) "Critical Commentary. Waking from the Dream: An Australian Perspective on Urban Resilience," *Urban Studies* 45(13): 2653–68.

Gray, R. and B. Gleeson (2007) "Energy Demands of Urban Living: What Role for Planning?" *Proceedings of 3rd State of Australian Cities Conference*, 28–30 November, Adelaide, Australia.

Gray, R., Gleeson, B. and M. Burke (2010) "Urban Consolidation, Household Greenhouse Emissions and the Role of Planning," *Urban Policy and Research* 28(3): 335–46.

Ham, S. (1992) *Environmental Interpretation, A Practical Guide for People with Big Ideas and Small Budgets*, Golden, CO: Fulcrum.

Hargroves, K., Desha, C. and A. Reeve (2008) *Identification and Assessment of Homeowner Behaviours Related to Reducing Residential Energy Demand: Report to the Townsville CitySolar Community Capacity Building Program*, The Natural Edge Project, Griffith University, Brisbane.

Hargroves, K., Reeve, A. and C. Desha (2010) *Investigation into Tools and Strategies for Behaviour Change Programs on Reducing Residential Energy Demand: Report to the Townsville CitySolar Community Capacity Building Program*, The Natural Edge Project, Griffith University, Brisbane.

Hargroves, K. and M. Smith (eds) (2005) *The Natural Advantage of Nations: Business Opportunities, Innovation and Governance in the 21st Century*, London: Routledge.

IBM (2011) *IBM Smarter Cities Challenge, Townsville Report*, Armonk, NY: IBM Corporate Citizenship and Corporate Affairs, accessed 15 September 2015 — www.townsville.qld.gov.au/community/sustainability/docu ments/ibm%20smarter%20cities%20challenge%20townsville%20report.pdf

JCU (2014) *Campus District Cooling (CDC) System – Averting an Energy Crisis*, Townsville: James Cook University, accessed 01 July 2014 — www.jcu.edu.au/tropeco/sustainatjcu/energy/JCUPRD1_070531.html

——— (2012) *2012 JCU Impact Report, Northern Queensland Operations*, Townsville: James Cook University, accessed 15 September 2015 — www.jcu.edu.au/cppo/public/groups/everyone/documents/annual_report/jcu_127691.pdf

Kern, F. and A. Smith (2008) "Restructuring Energy Systems for Sustainability? Energy Transition Policy in the Netherlands," *Transition towards Sustainable Energy Systems* 36(11): 4093–103.

Lerch, D. (2007) *Post Carbon Cities: Planning for Energy and Climate Uncertainty*, Rutgers: Post Carbon Press.

Lovins, A., Datta, K., Bustnes, O., Koomey, J. and N. Glasgow (2004) *Winning the Oil End-Game: Innovation for Profits, Jobs and Security*, Boulder, CO: Rocky Mountain Institute.

Lovins, A., Odum, M. and J. Rowe (2011) *Reinventing Fire: Bold Business Solutions for the New Energy Era*, New York: Chelsea Green.

Madlener, R. and Y. Sunak (2011) "Impacts of Urbanization on Urban Structures and Energy Demand: What can we Learn for Urban Energy Planning and Urbanization Management?" *Sustainable Cities and Society* 1(1): 45–53.

MCFI (2014) Morgenstadt City of the Future Initiative (site), accessed 20 September 2015 — www.morgenstadt.de/en.html

McKenzie-Mohr, D. (2011) *Fostering Sustainable Behavior: An Introduction to Community-Based Social Marketing* (3rd edn), Gabriola Island, British Columbia: New Society.

Newman, P., Beatley, T. and H. Boyer (2009) *Resilient Cities: Responding to Peak Oil and Climate Change*, New York: Island Press.

NYISO (2003) *Blackout August 14, 2003 – Final Report*, Rensselaer: New York Independent System Operator.

OECD (1995) *Urban Energy Handbook: Good Local Practice, OECD Group on Urban Affairs*, London: Organisation for Economic Co-operation and Development.

Owens, S. (1986) *Energy, Planning and Urban Form*, London: Pion Press.

QRA (n.d.) *Planning for Stronger, More Resilient Electrical Infrastructure*, Brisbane: Queensland Reconstruction Authority, accessed 15 September 2015 — http://qldreconstruction.org.au/u/lib/cms2/improving-resilience-electrical-infrastructure-full.pdf

Queensland Government (2013) *Queensland Government Population Projections* (2013 edn), Brisbane, accessed 15 September 2015 — www.qgso.qld.gov.au/products/publications/qld-govt-pop-proj/qld-govt-pop-proj.pdf

Queensland Government, Ergon Energy and James Cook University (2011) *Network Demand Management Case Study – James Cook University Douglas Campus, Townsville*, April, accessed 15 September 2015 — www.ergon.com.au/__data/assets/pdf_file/0006/72609/Case-Study-James-Cook-University.pdf

Rees, W. (1992) "Ecological Footprints and Appropriated Carrying Capacity: What Urban Economics Leaves Out," *Environment and Urbanization* 4(2): 121–30.

Rotmans, J., Kemp, R. and M. van Asselt (2001) "More Evolution than Revolution: Transition Management in Public Policy," *Journal of Futures Studies, Strategic Thinking and Policy* 3(1): 15–31.

Smith, M., Hargroves, K. and C. Desha (2010) *Cents and Sustainability: Securing Our Common Future by Decoupling Economic Growth from Environmental Pressures*, London: Routledge.

Staff Writers (2012) "Townsville Energy Project Attracts Federal Interest," *Government News*, 19 June, accessed 15 September 2015 — www.governmentnews.com.au/2012/06/townsville-energy-project-attracts-federal-interest/

State of Environment (n.d.) *The Strand Wind Project*, Townsville: Townsville City Council, accessed 15 September — www.soe-townsville.org/strandwindproject/

TCC (2013a) *Briefing Paper on Townsville: Queensland Solar City (A Historical Record)*, Townsville: Townsville City Council.

——— (2013b) *Townsville CBD: Smart Infrastructure and Sustainable Energy Framework*, Townsville: Townsville City Council.

——— (n.d.a) *City Profile and Statistics*, (page) Townsville: Townsville City Council, accessed 15 September 2015 — www.townsville.qld.gov.au/townsville/cityprofile/pages/default.aspx

——— (n.d.b) *Energy*, (page), Townsville: Townsville City Council, accessed 15 September 2015 — www.townsville.qld.gov.au/community/sustainability/climate/pages/default.aspx

——— (n.d.c) *The Lakes Wind Project, Stormwater Quality Management Powered by the Wind*, Townsville: Townsville City Council, accessed 15 September 2015 — www.soe-townsville.org/lakeswindproject/

TQSC (2013) *Townsville Queensland Solar City Final Report 2006–2013*, Townsville: Townsville Queensland Solar City.

——— (n.d.) "Settlement" (page) *State of Environment Townsville* (site) Townsville: Townsville Queensland Solar City, accessed 15 September 2015 — www.soe-townsville.org/humansettlement.html

UNCTAD (2010) *The Future Energy Matrix and Renewable Energy: Implications for Energy and Food Security*, note from the United Nations Conference on Trade and Development Secretariat, TD/B/C.I/MEM.2/8, accessed 15 September 2015 — http://unctad.org/en/docs/cimem2d8_en.pdf

UN Habitat (n.d.) "Climate change" (page) United Nations Habitat (site), accessed 15 September 2015 — http://unhabitat.org/urban-themes/climate-change/

UNQRDO (2010) *Growing Queensland's Renewable Energy Electricity Sector*, submission from United Northern Queensland Regional Development Organisation, accessed 15 September 2015 — www.parliament.qld.gov.au/documents/tableOffice/CommSubs/RenewEnergy/047.pdf

Vergragt, P.J. (2005) "Back-Casting for Environmental Sustainability: From STD and SusHouse Towards Implementation," in M. Weber and J. Hemmelskamp (eds) *Towards Environmental Innovation Systems*, Berlin: Springer, 301–18.

17

THE ROLE OF TELECOMMUNICATION IN POST-PETROLEUM PLANNING

Tooran Alizadeh

The Australian government has been constructing a National Broadband Network (NBN), first announced in 2009, to provide high-speed broadband for Australians by the end of 2020. While scholarship on oil vulnerability by the urban planning community slowly grows, the potential role of publically funded telecommunication technology in the post-petroleum era remains a marginalized topic. While Australia is by no means the only nation where intensive and accelerating telecommunication infrastructure development is taking place, its high vulnerability to oil (Newman and Kenworthy 1999) makes it a fascinating case to understand the potential role of the new technology in the post-petroleum era. Understanding the Australian multi-level planning systems and their responses to the new infrastructure offers valuable insights for cities and urban regions in other nations.

This chapter questions to what extent current planning strategies and policies at federal, state and local levels respond to the urban and regional potential of the NBN in a highly oil dependent nation. To do so, the chapter first offers a review of oil vulnerability literature in which a polycentric urban pattern has been advocated as the preferred development pattern scenario to tackle the possible urban challenges posed by oil depletion. Second, the role that telecommunication could and should play in the implementation of the preferred development pattern scenario is discussed. Finally, the NBN in Australia is scrutinized as a case study to explore the readiness of contemporary planning systems to embrace the new technology-based opportunities as part of the solution to build energy-efficient cities.

Considering the growing interest in energy efficiency and the number of governments worldwide investing in telecommunication infrastructure and broadband, this chapter addresses policy issues that will impact upon future planning across the world. Major themes discussed in the oil vulnerability debates are identified and structure a multi-level policy analysis of the integration of the NBN and planning systems in Australia. The chapter aims to raise awareness and invite closer examination of the potential role of such new technology in energy security debates, especially the role of the NBN under conditions of energy uncertainty. A comprehensive resolution of the problems identified is not sought. Rather, the more modest objective is to throw new light on planning in Australia and further international understanding of the potentials of the new technology for promoting energy-efficient cities.

Planning for the post-petroleum era

In 2008, when the cost of a barrel of oil reached US$140, the world remembered those who had been predicting a global oil crisis since the first oil shocks in the 1970s. Moreover, the rising fuel costs heightened concerns among planners about energy-efficient cities, raising questions about the future of heavily car dependent metropolitan regions in North America and Australia, which rely on petroleum fuel for transportation (Forrest 1996). Amid the uncertainty around the topic of peak oil, one fact seemed certain: The world of cheap oil was over. This raised a serious dilemma for urban and regional planning, for decades based around cheap fuel for transport (Wight and Newman 2010).

If energy efficiency is a global issue, different cities and regions will be impacted differently depending on their level of vulnerability; on the global scale differences in oil consumption between cities are huge. Data collected by Kenworthy and Laube (2001) on eighty-four cities showed that Australian cities are twice as vulnerable as most European cities. Cities such as Barcelona, Singapore and Tokyo – with similar income levels to Australian cities – are four to five times less vulnerable to oil-based transport issues.

Within cities, variations are similar to those between cities. Dodson and Sipe (2005, 2008a) generated a "vulnerability index for petroleum energy rises" (VIPER) based on Australian Census data, and examined Australian major cities to understand how the socio-economic impacts of any oil crisis would be spatially distributed. Their findings demonstrated that high levels of oil vulnerability in Australian cities are remarkably unevenly distributed across metropolitan regions. Dodson and Sipe (2005, 2008a) highlighted a regressive pattern in which the impacts of higher fuel costs fall on those with least capacity. Localities situated in middle and outer suburbs, mostly occupied by lower-income disadvantaged households (Murphy and Watson 1994; Yates 2002), are most vulnerable to the socio-economic impact of oil price rises. This clearly links in with "locational disadvantage," extensively discussed in the early 1990s (Maher 1994; National Housing Strategy 1992).

Maher (1994) has described outer suburbs as disadvantaged areas that lack facilities necessary to enable a "satisfactory life," requiring residents to take long journeys to access basic services. Such debates in the literature introduce contemporary urban development patterns as partial sources of socio-economic disparity across cities and regions particularly under conditions of energy uncertainty. This has encouraged a number of scholars to outline likely urban development scenarios in the post-petroleum era and assess the relative vulnerability different scenarios to understand what urban pattern might provide the greatest opportunity for responding to contemporary challenges (Newman 2007; Wight and Newman 2010).

Not all of the development scenarios posed are reviewed here. Preferred scenarios described in the literature tend to enhance what planners generally refer to as the "polycentric" city. For example, Wight (2010) described the preferred urban pattern as where most future development is directed into a number of mixed-use, medium-density centers distributed along transit routes, featuring green technology and green buildings with significant landscaping. Trubka et al. (2010a, 2010b, 2010c) explored the substantial cost advantages of such polycentric urban development with respect to infrastructure, transport, greenhouse gas reduction and even health care costs. The oil vulnerability literature (Newman 2007; Wight and Newman 2010) also argues for a series of enhancements to refocus challenges imposed by any energy supply on this preferred urban pattern.

The following discussion attempts to reconceptualize some of the enhancements suggested for polycentric urban development pattern in previous studies, reorienting them to embrace novel possibilities introduced by new telecommunication infrastructure. This introduces a new perspective on the oil depletion challenges, bringing telecommunication from the periphery to the center of oil debates in close association with public transport, governance and sustainable urban development patterns.

Telecommunication and post-petroleum planning

Over the last few decades a number of studies concerning metropolitan population and employment redistribution have speculated on the effects of telecommunication technologies on urban form. The discussions mostly permeated two broad urban research traditions of decentralization versus centralization trends (Audirac and Fitzgerald 2003). Moreover, there is an emerging trend (Alizadeh 2009; Clifford 2002; Kotkin 2000) that acknowledges both centrifugal and centripetal forces shape the urban form of the information age metropolitan cities. The resulting urban form in this new trend is described as a polycentric pattern of spatially distributed network of information and communication technology agglomerations interconnected via high-speed transportation and digital networks (Audirac 2005).

Since research on a subject as dynamic and volatile as telecommunication technology is seldom conclusive, the descriptive findings here remain tentative, but a body of heterogeneous research regarding the new technology and urban form could be drawn usefully into the oil vulnerability debates. Research examining the spatial implications of telecommunication indicate that cities take a more polycentric and dynamic form as they integrate more advanced telecommunication systems, for instance concentrating employment in urban sub-centers that rival the central business district (CBD) (Anas et al. 1998; Audirac 2002). Audirac (2005) has suggested that the new emphasis on a polycentric pattern stresses the spatial form of the post-industrial city because it translates into time-sensitive rather than distance-sensitive development patterns – what Castells (1996) has called the dominance of the "space of flows" rather than "space of places."

Since telecommunication technology seems to be pushing for the polycentric pattern of urban development preferred in the oil vulnerability debates, the golden question for urban scholars is how telecommunication can contribute to the implementation of the polycentric urban pattern in the context of oil depletion. Certainly oil vulnerability literature has already noted that the capacity to choose alternative means of travel to work is critical for communities to survive, let alone flourish, once demand for petrol outstrips supply resulting in ongoing price rises (Dodson and Sipe 2005).

The main solution offered by oil vulnerability scholars focuses on upgrades of public transport (Newman and Kenworthy 1999) and minimization of transport costs in a polycentric urban pattern (Lyons et al. 1990; Newman and Kenworthy 1992). Newman and Kenworthy (1999) have advocated for upgrades of public transport, suggesting that increasing residential densities in a polycentric urban pattern is a means of achieving higher public transport use. Mees (2000), by comparison, has argued that high-quality services operating as an integrated network connecting clusters of urban villages is the key to generating public transport patronage. To sum up, Dodson and Sipe (2005) have suggested that the optimal solution is probably a combination of both.

Moreover, from within the transport literature, a growing number of studies seek to appraise and advocate non-motorist modes of transport, including the potential of biking and walking in any oil crisis, as they could play specific roles in supplying local mobility and access to and from mass public transport nodes for longer-distance trips (Burke and Bonham 2010; Murphy 2008). Nevertheless, in reality, transport disadvantage remains a critical issue in many North American and Australian cities (Dodson and Sipe 2005) and, for some time, oil vulnerability scholars have sought travel alternatives based on the interrelationship among four main elements of transportation, labor, land and housing markets (Dodson and Sipe 2008b; Forrest 1996; Gordon and Richardson 1989).

That said, there is great scope in the medium term for urban and regional planners to re-evaluate telecommunication-based options that could reduce the need for travel and offer alternative modes of work based on telecommute rather than daily commute. Oil vulnerability researchers have so far dedicated very little attention to investigating the potential of telecommunication in a peak oil situation, and have only hoped that a combination of new technology and good urban planning can enable us to make better cities in the future (Wight and Newman 2010). For example, Newman (2007) briefly mentioned the value of Internet and video conference facilities in maintaining global interactions in an oil depletion scenario.

Furthermore, a growing trend in the literature attempts to reconceptualize oil vulnerability as a governance deficit within cities (see Steele et al., Chapter 6 in this book). This research identifies a number of gaps in current governmental institutional arrangements in cities, which actually serve to intensify the urban pressures associated with oil risk and vulnerability. There is a call for a polycentric governance system substituting the tendency to restrain civic input with a commitment to civic enlistment and enrichment in decision making (Marshall 2009). The idea is to further resilience by acknowledging the roles and synergies between decentralized and centralized approaches. Steele and Gleeson (2010), for example, have argued that Australia's urban planning challenge in relation to oil vulnerability is to weave a robust and coherent polycentric governance model, and to offer and implement a visionary policy and action agenda, including multi-level energy conservation and efficiency, large-scale conversion to renewable energy, major changes in the transport systems, investment in low emission technology and rapid reform of the tax system.

Yigitcanlar and Han (2010) have rightfully warned that efficient and effective urban management systems for cities require intelligent and integrated management mechanisms in which telecommunication infrastructure plays an important role in monitoring and managing activities and forms the backbone of urban management. Following this line of argument, in order to have a democratically defensible pathway to the polycentric governance model, adoption of a comprehensive and interactive e-governance model using new telecommunication-based opportunities seems inevitable. There is a consensus in the literature that the implementation of e-government increases access, delivers better public services, has internal efficiencies, and supports political accountability and participatory democracy (Bekkers 2003; Irani et al. 2005; Jesuale 2006; Moody 2007). Perhaps, the missing benefit to add to this list relates to the critical role that e-governance – empowered by ubiquitous telecommunication infrastructure – could play in keeping citizens and different sectors of governments connected to one another in an oil risk scenario that results in a less physically mobile world.

The challenge of peak oil seems daunting, but raises a number of opportunities as well as problems (Dunlop 2010). It could offer a real chance for two very different types of infrastructure, that is, telecommunication and public transport, to complement each other and define dominant modes of travel as partly virtual and partly physical. Peak oil could be the final straw that urban governance systems have needed to seriously work with new telecommunication-based opportunities and adopt comprehensive e-governance systems that facilitate e-democracy in a polycentric governance model. However, this would require contemporary planning systems to both acknowledge the serious nature of oil vulnerability debates and embrace the opportunities that the fast growing telecommunication infrastructure offers to address urban and regional planning issues, including threats imposed by peak oil.

Contemporary planning and telecommunication

The link between contemporary planning and infrastructure has been a topic of investigation for numerous studies (Graham and Marvin 2001; Neuman and Smith 2010). More recently, planning literature has shown growing interest in broadband technology as the key telecommunication infrastructure (Alizadeh and Shearer 2015; Alizadeh et al. 2014) and the backbone of the emerging knowledge economy (Eskelinena et al. 2008; Ford and Koutsky 2005). Over the last decade, many national governments, including in the UK (Galloway 2007), Korea (Kelly et al. 2003) and Spain (Gerrand 2006) have been developing policy and implementing telecommunication infrastructure based on assumptions that such novel technology-based infrastructure will increase productivity and innovation, and guarantee their long-term economic competitiveness (Willson et al. 2009).

Close governmental involvement with this new infrastructure is partly founded on literature generated in the first decade or so of the twenty-first century, which has argued that a certain level of government intervention is inevitable in order to achieve social equity, fulfill a universal service obligation and enhance the provision of telecommunications services to non-profitable areas (Faulhaber and Hogendorn 2000; Galloway 2007; Mcmahon and Salant 2001). However, it would be an extremely complex task to establish beyond doubt that infrastructure per se results in improvements in urban development indicators (Ford and Koutsky 2005; Lee et al. 2005; Strategic Networks Group 2003).

At the same time, there is a new line of studies warning against the urban management mentality that sees large-scale urban infrastructure as a panacea for urban problems and as a replacement for strategic urban spatial planning. As Dodson (2009) points out, this "infrastructure turn" is problematic for a number of reasons, not least the way that it narrows the focus of urban planning and spatial strategy-making. Following this line of argument, the Australian NBN provides an interesting case in which a large-scale, centrally funded telecommunication infrastructure could potentially provide a number of opportunities for urban and regional planning. Moreover, the fact that Australia's major cities are among the most car dependent in the Global North, ranking only behind US cities (Newman and Kenworthy 1999) makes a useful case for examining questions about the role of fast-developing telecommunication infrastructure in such oil vulnerable settings. Lessons learned in the extreme case of Australian cities are transferable to other cities around the world, particularly those eager on the so-called Americanization of cities concept (Pizarro 2002) and those that have not appropriately controlled urban sprawl through planning regulations.

Australia's national broadband network

In response to the shortcomings in Australia's telecommunication infrastructure (Barr 2008; Given 2008; Middleton 2009; Middleton and Chang 2008), the NBN initiative was announced in April 2009 – to connect all Australians to broadband, including the vast majority to superfast broadband (NBN Co. 2010a). The NBN is the largest single infrastructure project introduced in Australia with an investment initially estimated at up to A$43 billion over eight years (DBCDE 2010). After a trial phase on the large island state of Tasmania, the rollout of the NBN on mainland Australia started in March 2010 with the announcement of the five first release sites (Minister's Media Release 2009; NBN Co. 2010b). Second-release sites, including fourteen new locations and five sites adjacent to the existing first-release sites, were announced later in July 2010 (NBN Co. 2010c).

Finally, in October 2011, NBN Co. released a twelve-month national rollout schedule plan to cover half a million premises and a total of sixty locations in all Australian states and territories (NBN Co. 2011). Additionally, in March 2012, a three-year large-scale rollout plan was announced to complete (or get underway) NBN fiber in one third of the country, in areas with more than 3.5 million premises in 1,500 communities (NBN Ltd. 2012). However, the 2013 federal election changed the fate of the NBN (NBN Co. 2013) when the newly elected government decided to reassess the scale of the national fiber project, and put the first stage of the large-scale NBN rollout on hold.

As of October 2013, only 300,000 premises listed in the first large-scale rollout phase were guaranteed connection directly to the fiber network (Duke 2013; Turner 2013). For the rest of the nation, the Australian government is now building the NBN using a multi-technology mix incorporating fiber-to-the-node, hybrid fiber-coaxial and fiber-to-the-basement technologies.

Now a highly politicized issue, there has been little space for evidence-based research on the merits or otherwise of the Coalition government's model of the NBN versus the Australian Labor Party's (ALP) model of the NBN, and their respective technical foundations. This chapter does not offer a particular viewpoint of either model, instead investigating the extent to which planning strategies and policies at federal, state and local levels address the urban and regional possibilities of the high-speed broadband in the context of a highly oil dependent nation.

Federal NBN policy and planning

After some decades of relative neglect, the past several years have witnessed a resurgence of the Australian government's interest in urban challenges with infrastructure and environment as particular areas of policy concern. In 2008 the intergovernmental forum of national and state governments, the Council of Australian Governments (COAG), was reinvigorated with a specific city agenda to address policy coordination and development issues. COAG recognized the need to provide for nationally significant economic infrastructure, and tasked Infrastructure Australia (IA) with the mission of identifying infrastructure priorities. Established under the IA Act 2008, IA is a statutory body that reports regularly to COAG through the Federal Minister for Infrastructure and Transport. The first IA document prepared for COAG included "a national broadband network" as one of seven items to meet gaps and deficiencies in national infrastructure (IA 2008a). While the NBN was largely developed as a stand-alone part of the ALP 2007 policy platform, it remained a major component of the Coalition government's policy platform (Alizadeh et al. 2011; DBCDE 2011b).

There have been high expectations about NBN's ability to resolve multiple socio-economic-environmental issues. The main planning matters discussed in Australian government policy and documents are spatial equity in access to advanced services for those living in regional and remote settlements (IA 2008b). The NBN has been noted as a key tool in reducing spatial frictions, managing population distribution and enabling employment suburbanization (DSEWPC 2011: 79). This view follows a line of thinking that had been developed in Australian government policy documents focusing on the community and social justice impacts of broadband systems within the context of a highly uneven continental settlement pattern (DCITA 2005a, 2005b, 2006).

In May 2011, the Australian government announced a *National Digital Economic Strategy* (DBCDE 2011b) that outlined a vision for Australia to become a leading global digital economy by 2020. A number of initiatives were introduced to enable more Australians – particularly those in non-metropolitan areas – to benefit from the e-health, e-education and e-business opportunities of the NBN. The strategy (DBCDE 2011b) described an NBN-empowered digital economy as a means to assist in managing challenges imposed by climate change, by improving efficiency in use of energy infrastructure and transport systems, promoting increased adoption of teleworking and reducing pressure on capital cities (through facilitating businesses operations online in regional Australia). *Advancing Australia as a Digital Economy* (DBCDE 2013) offered an update, laying out the next steps toward delivering the government's 2020 vision. This 2013 update provided an overview of dozens of individual initiatives, outlined a number of new initiatives and reported progress made by the Australian government to embrace the digital future.

The 2011 and 2013 national strategies defined the inevitable transition of Australia to a digital economy as a "market-led phenomenon" without setting clear roles for either state or local governments. There were few instructions to state governments to explicitly incorporate the NBN into state and metropolitan strategic planning (Alizadeh et al. 2011), let alone in the context of reducing oil dependency and improving energy efficiency. Moreover, the Coalition coming to power in the 2013 federal election put the national digital strategies on hold. Advancement in digital economy at the national level does not seem to be a priority. It does not show up in the Coalition's "five pillar economy," which is based on manufacturing innovation, advanced services, agriculture exports, education and research, and mining exports. This shift at the federal level could set back Australia's position on a highly digitally oriented world stage, and hinder any timely comprehensive transformation to the digital economy (DBCDE 2011b).

State NBN policy and planning

Since the turn of the century, state governments across Australia began to prepare metropolitan planning strategies for their capital cities, which accommodated more than two-thirds of the national population, in a distinctively Australian paradigm (Searle and Bunker 2010). They prepared strategies that centered on land-use arrangements and linked to other state-level infrastructure plans to guide their respective metropolitan regions for a period of twenty to twenty-five years (Searle and Bunker 2010). All recognized high car dependency in their regions, yet did not intervene in any radical or explicit way to address the possibility of peak oil and its effects on the projected growth in each region (Newman 2007). Subsequently, metropolitan strategies were prepared across Australia to guide urban and regional development during the NBN rollout and its post-construction phases. Focusing on New South Wales (NSW), this section investigates the metropolitan planning documents produced to guide the rollout and post-construction periods of the NBN in Sydney, the capital city of NSW and Australia's largest metropolis.

In December 2005, the NSW government released a metropolitan strategy for Sydney through to 2031 — *City of Cities: A Plan for Sydney's Future* (Department of Planning 2005). After a wide range of critiques, it was amended and updated and, in December 2010, was superseded by *Metropolitan Strategy Review: Sydney Towards 2036* (Department of Planning 2010). The new metropolitan strategy was released when the NBN had become a prominent feature of federal policy but mentioned little about the upcoming telecommunication infrastructure. The 2010 document was meant to incorporate the principles of the NSW government (2010) *Metropolitan Transport Plan 2010: Connecting the City of Cities* and to integrate transport and land use. However, looking at both 2005 and 2010 metropolitan strategies reveals that infrastructure was simply referred to as "transport and others," and recognized very little integration between metropolitan and infrastructure plans.

Sydney is strongly geographically patterned for oil vulnerability. The most vulnerable areas are beyond 20 km from the CBD, particularly to the west, while the least vulnerable areas are in the inner northern Sydney (from the harbor mouth in the inner northeast to north of the CBD) and the area immediately around and east of the Sydney CBD (Dodson and Sipe 2005, 2008a). Dodson and Sipe (2008b) studied changes in oil vulnerability across Sydney during 2001–2006. They noted that, while the spatial patterns of oil vulnerability remained, Sydney as a whole became much more oil vulnerable. Unfortunately, neither the 2005 nor the 2010 metropolitan strategies explicitly acknowledged any need to reduce Sydney's high dependency on oil. They boasted about Sydney's global position, economic competitiveness in the Asia-Pacific, and high proportion of employment in the financial services at the national level. Yet, both documents assumed a business-as-usual approach and failed to acknowledge the possibility of any interruption in the form of oil depletion. Consequently, metropolitan planning for Sydney fell short of acknowledging the potential economic and social role of telecommunication, under any oil crisis scenario, and had no intent or strategy to work with the NBN during its rollout and post-construction phases (Alizadeh 2011). Furthermore, even the metropolitan strategy in its latest format has not properly integrated national strategies to advance Australia's position in the digital economy (DBCDE 2013).

The metropolitan strategy through to 2031 presented a vision for metropolitan areas to operate as a city of cities, and supported a metropolis made up of five key cities and twenty-two strategic centers. The stated goal was to strengthen all five key metropolitan centers, not just the two harbor cities of the CBD and North Sydney, but also the western Sydney centers of Parramatta, Liverpool and Penrith. Although resembling discussions around a polycentric city as the preferred urban development scenario of the peak oil debates, in this case the devil was in the details. The strategy divided the five key cities into two very different categories: the CBD and North Sydney became "Global Sydney"; the three other cities that were more oil vulnerable were labeled "Regional Cities" and received less attention in terms of employment opportunities and public transport.

Previous studies focusing on the Sydney metropolitan region had argued that, with further development of telecommunication technology, effective spatial dispersion of jobs into the outer suburbs and centers (particularly Parramatta) was possible (Alizadeh 2011). More specifically Forrest (1996) had referred to issues around peak oil, and called for polycentric urban patterns embedded with technology-based opportunities that could change the nature of home-work linkages and reduce daily commutes (Gordon and Richardson 1989).

If this failure to define any explicit role for digital infrastructure to address some of the urban development issues in Sydney is generalized, it poses the risk that subsequent rounds of metropolitan plans and strategies prepared across other states might also overlook the technology-based opportunities of improving resilience in a highly oil dependent nation. However, there are other planning actors, notably local governments, to which we now turn our attention.

Local NBN policy and planning

From the perspective of Australia being a highly oil dependent nation, what was the role defined for and played by Australian local governments during the NBN rollout process? To answer this question, it needs to be noted that oil depletion scenarios suggest substantial declines in mobility and increased localization (Stone and Mees 2010) with correspondingly greater roles for local planning and local governments. As discussed earlier, oil vulnerability literature has already established disparities in impacts on different localities within metropolitan regions, which once again emphasizes the need to design and implement locally based policies and action plans to address challenges posed by peak oil.

As discussed, a comprehensive, interactive multi-level e-government approach could facilitate the implementation of a polycentric governance model in the context of oil risk and security (Steele and Gleeson 2010). Indeed, a Digital Local Government program was outlined as an initiative in both the National Digital Economy Strategy (DBCDE 2011b) and Advancing Australia as a Digital Economy (DBCDE 2013) to facilitate the adoption of e-governance at the local government level. The Digital Local Government program aimed to enhance or replace the existing council service with an online service delivered to the local community via NBN, using cloud-based applications. The program provided a platform to host online video-based community forums on topical issues such as major development proposals (DBCDE 2011a). An overview has shown that it has helped local governments to deliver better public service and to increase their internal efficiency (Alizadeh 2015). It has enabled a wider cross section of interested residents to participate in local government consultative processes through video-based community forums (Bekkers 2003; Irani et al. 2005), which have the potential to address some concerns raised in the oil vulnerability literature about the need to establish truly democratic governance systems (Marshall 2009).

Nevertheless, recent studies (Alizadeh and Shearer 2015) report a sense of uncertainty across local governments, as the majority of federally funded initiatives fail to be renewed under the new Coalition government's policies, and local governments do not have the capacity to keep programs running without such funding, at least until they become financially viable. Furthermore, since its introduction, a growing number of local governments have shown interest in the socio-economic opportunities provided by the NBN, and local digital economic strategies are emerging to increase the possibilities for local communities. Some have achieved international recognition for their attempts, such as Melbourne and Whittlesea (Victoria), the Gold Coast and Ipswich (Queensland) and Prospect (South Australia), which were nominated Australian Intelligent Communities – that understand the enormous challenges of the broadband economy, and have taken conscious steps to create an economy capable of prospering in it – by the Intelligent Communities Forum (ICF 2012), which named Ipswich in the global Top 7 Intelligent Communities for 2015. Ipswich is the first Australian city to make it into the top seven.

However, the majority of Australian local governments still have not incorporated broadband opportunities into current economic plans, which could hinder their benefiting from the socio-economic potential of such infrastructure in numerous areas, including telework (DBCDE 2011b). These discrepancies in local planning and policies across local governments in Australia is working as an enormous regulatory barrier for Australian businesses.

Conclusion

Australian cities currently face high levels of oil vulnerability that demand an integrated, visionary policy and action agenda. Several years ago Newman et al. (2009) warned that new urban policies and planning practices were needed to avoid, remedy or mitigate the impacts of oil depletion. Investigating possible effects of peak oil on urban living shows the very real risks that need to be assessed but also reveals opportunities to address a number of structural issues, such as reducing car dependency, applying sustainable technologies for energy, water and waste that work best in the polycentric urban form, and embracing new telecommunication-based life- and work-styles that are wise, both economically and environmentally. Therefore, a positive response to challenges posed by oil depletion would help to effectively achieve a range of other planning objectives and yield a more sustainable urban outcome.

This chapter set out to reconceptualize the role of major telecommunication infrastructure in an oil dependent context and to examine the NBN in Australia as a case study to explore the readiness of the contemporary planning systems to embrace such new technology-based opportunities as part of the solution to build energy-efficient cities. A more comprehensive understanding of links between telecommunication-based opportunities and oil vulnerability debates was attempted, by discussing how fast-growing and novel telecommunication infrastructure could enhance the polycentric urban form as the preferred development pattern scenario in the context of oil depletion. The chapter opened discussions on the role that the new technology could play in offering effective telecommute options and in setting the stage for a polycentric governance model through comprehensive e-governance options. Multilevel discussions have been initiated on the implications of telecommunication infrastructure at federal, state and local planning levels by reference to the hefty ongoing investment in the NBN, here understood as both a challenge and an opportunity facing planners of Australia's future. This examination revealed that adequate and consistent strategies and policies at different levels of government are required to deliver the full potential of the NBN.

Through appropriate strategies and policies, telecommunication could help tackle some of the major planning issues in Australia, including those posed by peak oil and climate change. However, this analysis of the patchwork of federal, state and local strategies, policies and initiatives – developed in documents written to guide the NBN rollout and the post-construction phases – failed to explicitly refer to oil peak and the challenges it could impose on planning in Australia. Documents produced at the state and local levels, in particular, had difficulty making any clear link between telecommunication infrastructures and planning challenges. The two national strategies explicitly promoting a digital economy seem to be on hold as a result of federal policy changes, and the uncertainties of decision making at the federal level are sending the wrong (or unclear) signals to state and local governments. All this has hindered any concerted attempt to link digital opportunities to mainstream planning systems in the context of an oil depletion scenario.

In summary, just as the NBN rollout is ongoing, this chapter is only a beginning in the exploration of the multi-level implications of the telecommunication infrastructure for planning systems in the context of a highly oil dependent nation. The discussions offered here seek to raise awareness of the potential role of the new technology in energy security debates, and to invite closer examination of planning strategies and policies and stances toward fast-growing telecommunication infrastructure and its role under conditions of energy uncertainty. There is a need to conduct further evidence-based research in this area. We need to develop a more comprehensive understanding of the role defined for, and played by, planning systems through time and especially in relation to NBN dynamics in Australia. Oil debates must move from the periphery to the center of planning discussions in close concert with policy discourses in relation to public transport, governance and climate change.

Lessons learned in the severely oil dependent case of Australian cities are transferable to similar contexts, such the US and Canada, and others that follow the Americanization of cities. Furthermore, these discussions contribute to awareness among planners and policy makers all around the world about the importance of energy security, and further our understanding of the potential of telecommunication for sustainable urban living.

References

Alizadeh, T. (2015) "Local Government Planning and High-Speed Broadband," *Journal of Urban Technology* 22(4): 23–43.

——— (2011) "Urban Implications of Telework: Policy Gap in Sydney Metropolitan Planning?," paper presented at 3rd World Planning Schools Congress, 4–8 July, Perth.

——— (2009) "Urban Design in the Digital Age: A Literature Review of Telework and Wired Communities," *Journal of Urbanism* 2(3): 195–213.

Alizadeh, T. and H. Shearer (2015) "A Snapshot of High-Speed Broadband Responses at Local Government Level in Australia: A Marriage Between Federally Funded Initiatives and Locally Driven Innovations?" *Australian Planner* 52(1): 42–50.

Alizadeh, T., Sipe, N. and J. Dodson (2014) "Spatial Planning and High Speed Broadband: Australia's National Broadband Network and Metropolitan Planning," *International Planning Studies* 19(3–4): 359–78.

——— (2011) "Metropolitan Planning and NBN: A Comparative Policy Analysis, Sydney vs. Brisbane," paper presented at the State of Australian Cities National Conference, 29 November–2 December, Melbourne, accessed 9 September 2015 — www98.griffith.edu.au/dspace/bitstream/handle/10072/43549/74563_1.pdf?sequence=1

Anas, A., Arnott, R. and K.A. Small (1998) "Urban Spatial Structure," *Journal of Economic Literature* 36(3): 1426–64.

Audirac, I. (2005) "Information Technology and Urban Form: Challenges to Smart Growth," *International Regional Science Review* 28(2): 119–45.

——— (2002) "Information Technology and Urban Form," *Journal of Planning Literature* 17(2): 212–26.

Audirac, I. and J. Fitzgerald (2003) "Information Technology (IT) and Urban Form an Annotated Bibliography of the Urban Deconcentration and Economic Restructuring Literatures," *Journal of Planning Literature* 17(4): 480–511.

Barr, T. (2008) "Broadband Bottleneck: History Revisited," *Media International Australia* 129(November): 129–39.

Bekkers, V. (2003) "e-Government and the Emergence of Virtual Organizations in the Public Sector," *Information Polity* 8: 89–101.

Burke, M. and J. Bonham (2010) "Rethinking Oil Depletion: What Role Can Cycling Really Play in Dispersed Cities?" *Australian Planner* 47(4): 272–83.

Castells, M. (1996) "Megacities and the End of Urban Civilization," *New Perspectives Quarterly* 13: 12–15.

Clifford, J.S. (2002) *Transcending Locality-Driven Lifestyle: The Potential of the Internet to Redefine Our Neighborhood Patterns.* Unpublished Thesis for Doctorate of Design, Harvard Graduate School of Design, Harvard University, Cambridge, MA.

DBCDE (2013) *Advancing Australia as a Digital Economy*, Canberra: Department of Broadband, Communications and the Digital Economy.

——— (2011a) *Digital Local Program Guidelines-Round Two*, Canberra: Department of Broadband, Communications and the Digital Economy.

——— (2011b) *National Digital Economy Strategy*, Canberra: Department of Broadband Communications Digital Economy.

——— (2010) *National Broadband Network, Overview*, Canberra: Department of Broadband, Communications and the Digital Economy.

DCITA (2006) *Broadband Blueprint*, Canberra: Australian Government, Department of Communications, Information Technology and the Arts.

——— (2005a) *Networking the Nation: Evaluation of Outcomes and Impacts*, Canberra: Department of Communications, Information Technology and the Arts, Communications Research Unit.

——— (2005b) *The Role of ICT in Building Communities and Social Capital*, (Discussion Paper), Canberra: Department of Communications, Information Technology and the Arts.

Department of Planning (2010) *Metropolitan Strategy Review: Sydney Towards 2036*, Sydney: NSW Government.

——— (2005) *City of Cities: A Plan for Sydney's Future*, Sydney: NSW Government.

Dodson, J. (2009) "The 'Infrastructure Turn' in Australian Metropolitan Spatial Planning," *International Planning Studies* 14(2): 109–23.

Dodson, J. and N. Sipe (2008a) "Shocking the Suburbs: Urban Location, Homeownership and Oil Vulnerability in the Australian City," *Housing Studies* 23(3): 377–401.

——— (2008b) *Unsettling Suburbia: The New Landscape of Oil and Mortgage Vulnerability in Australian Cities*, Research Paper 17, Brisbane: Urban Research Program.

——— (2005) *Oil Vulnerability in the Australian City*, Research Paper 6, Brisbane: Urban Research Program, Griffith University.

DSEWPC (2011) *Sustainable Australia – Sustainable Communities: A Sustainable Population Strategy for Australia*, Canberra: Department of Sustainability, Environment, Water, Population and Communities.

Duke, J. (2013) "900,000 Premises May Need to Foot Bill Under NBN Revision," *Property Observer*, 14 November, accessed 15 September 2015 — www.propertyobserver.com.au/news/900000-premises-may-need-to-foot-bill-under-nbn-revision-archers/2013111466370

Dunlop, I. (2010) "Peak Oil – Catalyst for a Resilient, Sustainable Society," in S. Cork (ed.) *Resilience and Transformation: Preparing Australia for Uncertain Futures*, Collingwood: CSIRO, 75–80.

Eskelinen, H., Frank, L. and T. Hirvonen (2008) "Does Strategy Matter? A Comparison of Broadband Rollout Policies in Finland and Sweden," *Telecommunications Policy* 32(6): 412–21.

Faulhaber, G.R. and C. Hogendorn (2000) "The Market Structure of Broadband Telecommunications," *Journal of Industrial Economics* 48(3): 305–29.

Ford, G.S. and T.M. Koutsky (2005) "Broadband and Economic Development: A Municipal Case Study from Florida," *Review of Urban & Regional Development Studies* 17(3): 216–29.

Forrest, J. (1996) "Place of Residence, Place of Work: The Spatial Clustering of the Journey to Work in Sydney," *Australian Planner* 33(3): 132–35.

Galloway, L. (2007) "Can Broadband Access Rescue the Rural Economy?" *Journal of Small Business and Enterprise Development* 14(4): 641–53.

Gerrand, P. (2006) "Accelerating Broadband Rollout – Initiatives in Regional Spain," *Telecommunications Journal of Australia* 56(3/4): 84–89.

Given, J. (2008) "Australia's Broadband: How Big Is the Problem?" *Media International Australia* 127(May): 6–10.

Gordon, P. and H. Richardson (1989) "Gasoline Consumption and Cities: A Reply," *American Planning Association Journal* 55: 342–46.

Graham, S. and S. Marvin (2001) *Splintering Urbanism: Networked Infrastructures, Technological Mobilities and the Urban Condition*, London: Routledge.

IA (2008a) *Outline of Infrastructure Australia's Prioritisation Methodology*, Sydney: Infrastructure Australia.

——— (2008b) *A Report to the Council of Australian Government*, Canberra: Australian Government.

ICF (2012) "Awards Program" (page), *Intelligent Community Forum* (site) New York: Intelligent Communities Forum, accessed 15 September 2015 — www.intelligentcommunity.org/index.php?src

Irani, Z., Love, P.E.D., Elliman, T., Jones, S. and M. Themistocleous (2005) "Evaluating e-Government: Learning from the Experiences of Two UK Local Authorities," *Information Systems Journal* 15(1): 61–82.

Jesuale, N. (2006) "Spectrum Policy Issues for State and Local Government," *International Journal of Network Management* 16(2): 89–101.

Kelly, T., Gray, V. and M. Minges (2003) *Broadband Korea: Internet Case Study*, Seoul: International Telecommunications Union.

Kenworthy, J. and F. Laube (2001) *The Millennium Cities Database for Transport*, Brussels: International Union of Public Transport.

Kotkin, J. (2000) *The New Geography: How the Digital Revolution is Reshaping the American Landscape*, New York: Random House.

Lee, H., Oh, S. and Y. Shim (2005) "Do We Need Broadband? Impacts of Broadband in Korea," *Info* 7(4): 47–56.

Lyons, T., Kenworthy, J. and P. Newman. (1990) "Urban Structure and Air Pollution," *Atmospheric Environment* 24(1): 43–48.

Maher, C. (1994) "Residential Mobility, Locational Disadvantage and Spatial Inequality in Australian Cities," *Urban Policy and Research* 12(3): 185–91.

Marshall, G. (2009) "Governance for Sustaining Trust in a Complex World," in S. Cork (ed.) *Brighter Prospects, Enhancing the Resilience of Australia*, Weston, ACT: Shaping the Future: Australia, 25–29, 21.

McMahon, K. and P. Salant (2001) "Strategic Planning for Telecommunications in Rural Communities," *Rural Development Perspectives* 14(3): 2–7.

Mees, P. (2000) *A Very Public Solution: Transport in the Dispersed City*, Melbourne: Melbourne University Press.

Middleton, C. (2009) "Can Broadband Support Environmental Sustainability?" *Telecommunication Journal of Australia* 59(1): 10.11–14.

Middleton, C. and S. Chang (2008) "The Adoption of Broadband Internet in Australia and Canada," in Y.K. Dwivedi, A. Papazafeiropoulou and J. Choudrie (eds) *Handbook of Research on Global Diffusion of Broadband Data Transmission*, Harrisburg, PA: IGI Global, 820–42.

Minister's Media Release (2009) "Tasmania NBN Co limited Established," Canberra: Senator the Hon. Stephen Conroy, Minister for Broadband, Communications and the Digital Economy, accessed 8 February 2012 — www.minister.dbcde.gov.au/media/media_releases/2009/075

Moody, R. (2007) "Assessing the Role of GIS in e-Government: A Tale of e-Participation in Two Cities," *Electronic Government: Lecture Notes in Computer Science* 4656: 354–65.

Murphy, P. (2008) *Plan C: Community Survival Strategies for Peak Oil & Climate Change*, Philadelphia, PA: New Society.

Murphy, P. and S. Watson (1994) "Social Polarization and Australian Cities," *International Journal of Urban and Regional Research* 18(4): 573–90.

National Housing Strategy (1992) *Housing Location and Access to Services*, Canberra: National Housing Strategy.

NBN Co. (2013) "Detailed Data and Updated Website Maps Open Up Australia's Largest Infrastructure Project to More Transparency," 30 October, Melbourne: NBN Co. Ltd, accessed 19 November 2013 — www.nbnco.com.au/corporate-information/media-centre/media-releases/weekly-rollout-metrics.html

——— (2012) "Three Year Rollout Plan For NBN Announced," 29 March, Melbourne: NBN Co. Ltd, accessed 12 December 2012 — www.nbnco.com.au/corporate-information/media-centre/media-releases/nbn-co-announces-three-year-rollout-plan.html

——— (2011) "NBN Co Releases 12-Month National Rollout Plan," Melbourne: NBN Co. Ltd, accessed 12 February 2012 — www.nbnco.com.au/corporate-information/media-centre/media-releases/nbn-co-releases-12-month-national-rollout-plan.html

——— (2010a) *Corporate Plan 2011–2013*, Canberra: NBN Co. Ltd.

——— (2010b) "NBN Co Announces 'First Release' Sites For High-Speed Network," 2 March, Melbourne: NBN Co. Ltd, accessed 15 September 2915 — www.nbnco.com.au/corporate-information/media-centre/media-releases/nbn-co-announces-first-release-sites-for-high-speed-network.html

——— (2010c) "NBN Co Announces Next Rollout Locations," 8 July, Melbourne: NBN Co. Ltd, accessed 12 September 2015 — www.nbnco.com.au/corporate-information/media-centre/media-releases/nbn-co-announces-next-rollout-locations.html

Neuman, M. and S. Smith (2010) "City Planning and Infrastructure: Once and Future Partners," *Journal of Planning History* 9(21): 21–42.

Newman, P. (2007) "Beyond Peak Oil: Will Our Cities Collapse?" *Journal of Urban Technology* 14(2): 15–30.

Newman, P., Beatley, T. and H. Boyer (2009) *Resilient Cities: Responding to Peak Oil and Climate Change*, Washington, DC: Island Press.

Newman, P. and J. Kenworthy (1999) *Sustainability and Cities: Overcoming Automobile Dependence*, Washington, DC: Island Press.

——— (1992) "Transit Oriented Urban Village: Design Solution for the 90s," *Urban Futures Journal* 2(1): 50–58.

NSW Government (2010) *Metropolitan Transport Plan 2010: Connecting the City of Cities*, Sydney: NSW Transport and Infrastructure.

Pizarro, R. (2002) "Exporting the Dream: Hollywood Cinema and Latin American Suburbia," in E. Heikkilaand and R. Pizarro (eds) *Southern California and the World*, New York: Praeger, 179–94.

Searle, G. and R. Bunker (2010) "Metropolitan Strategic Planning: An Australian Paradigm?" *Planning Theory* 9(3): 163–80.

Steele, W.E. and B. Gleeson (2010) "Mind the Governance Gap: Oil Vulnerability and Urban Resilience in Australian Cities," *Australian Planner* 47(4): 302–10.

Stone, J. and P. Mees (2010) "Planning Public Transport Networks in the Post-Petroleum Era," *Australian Planner* 47(4): 263–71.

Strategic Networks Group (2003) *Economic Impact Study of the South Dundas Township Fibre Network*, London: Department of Trade and Industry.

Trubka, R., Newman, P. and B. Darren (2010a) "The Costs of Urban Sprawl (1): Infrastructure and Transportation," *Environment Design Guide* 83: 1–6.

——— (2010b) "The Costs of Urban Sprawl (2): Predicting Greenhouse Gases from Urban Form Parameters," *Environment Design Guide* 84: 1–16.

——— (2010c) "The Costs of Urban Sprawl (3): Physical Activity Links to Healthcare Costs and Productivity," *Environment Design Guide* 85: 1–13.

Turner, A. (2013) "NBN Hookups Put on Hold," Sydney: Sydney Morning Herald, accessed 15 September 2015 — www.smh.com.au/digital-life/digital-life-news/nbn-hookups-put-on-hold-20131106–2wzz4.html

Wight, W. and P. Newman (2010) "Petroleum Depletion Scenarios for Australian Cities," *Australian Planner* 47(4): 232–42.

Willson, P., Marshall P. and J. McCann (2009) "Evaluating the Economic and Social Impact of NBN," paper presented at the 20th Australasian Conference on Information Systems, 2–4 December, Melbourne.

Yates, J. (2002) *Housing Implications of Spatial Social and Structural Change*, Melbourne: Australian Housing and Urban Research Institute.

Yigitcanlar, T. and J.H. Han (2010) "Ubiquitous Eco Cities: Telecommunication Infrastructure, Technology Convergence and Urban Management," *International Journal of Advanced Pervasive and Ubiquitous Computing* 2(1): 1–17.

18

PEAK OIL

Challenges and changes for the air transport industry

Douglas Baker, Nicholas Stevens and Md. Kamruzzaman

The direct, indirect and induced global economic impact of air transport generated US$539 billion to world gross domestic product in 2012. However, when catalytic tourism impacts are included, the industry is estimated to contribute some US$2.4 trillion and employ 58.1 million people globally (ATAG 2014). From a passenger traffic perspective alone, air transport has grown on average 5.7 percent per year between 2002 and 2012 (ICAO 2013b). Industry forecasts anticipate continued growth, and the passenger traffic forecast in revenue passenger kilometers (RPK)[1] "is expected to grow from 5 billion in 2010 to more than 13 billion in 2030, at an average growth rate of 4.9%" (ICAO 2013a: 18). In that "most likely" scenario, international RPKs would grow at 5.1 percent per annum (p.a.) and domestic RPK at 4.4 percent p.a. Even when considering the "pessimistic" scenario, global passenger traffic is forecast to grow at 3.5 percent year on year while, in the same period, actual scheduled passenger traffic is expected to rise from 2.6 billion to 6.1 billion movements worldwide (ICAO 2013b).

Thus, the future seems bright for aviation – globally the outlook for the airline industry is one that has a general upward trend. However, it is an industry of tight margins and is not an easy place to make money or do business. While global net profit for 2013 was approximately US$12.9 billion, with an expected net profit of US$19.7 billion in 2014, this only represents a net profit margin of 2.6 percent (ICAO 2013b). Leaving the relatively flat cargo business aside, from a passenger perspective airlines will only make an average of US$5.94 net profit per passenger. Profit and margins are very fine, and the prospect of peak oil and fossil fuel constrained futures presents an additional dilemma to the viability of aviation and to the cost structure.

Over the last four years the increasing use of air travel across the globe and the steady growth in passenger movements have occurred in a relatively stable energy environment with little change to fuel prices (at approximately US$124 per barrel between 2011 and 2014 – see Figure 18.1). More recently, we have seen a drop in the oil prices as a response to ongoing global financial volatility. While this may be recognized as a positive for the air transport industry, in reality the fluctuations are a sign of increasing uncertainty.

This chapter provides an overview of the challenges, and the level of responsiveness, from the air transport industry to both peak oil and climate change scenarios, using the Australian context where appropriate. The chapter recognizes the broader environmental impacts associated with the aviation industry, including global climate change, carbon emissions, aircraft life cycle and noise. However, peak

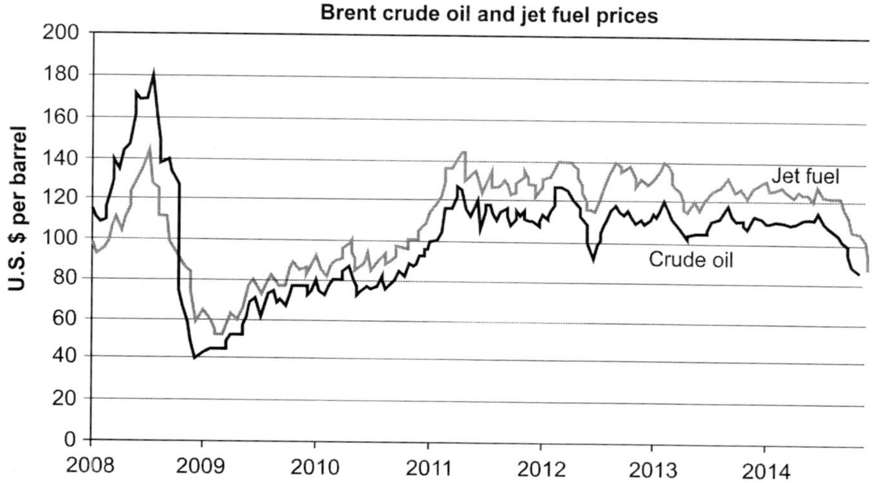

FIGURE 18.1 Jet fuel and crude oil price relationship

Source: IATA (2015).

oil presents a different kind of threat to the industry. Within this context we acknowledge the complexity of the larger air transport environment and focus on the efficiency of operations and design; alternative fuels, with a focus on biofuels; and highlighting future trends for the industry.

The context for air transport

While overall the outlook for the aviation industry is positive, with passenger service expansion in the 5 percent range, the competition is fierce. The International Air Transport Association (IATA) expects that further efficiencies in airline businesses will be required through mergers and joint ventures. IATA forecasts six drivers for the financial outlook of airlines: the economic demand, passenger demand, ancillary revenues, improved industry structure, cargo demand and fuel (IATA 2013).

First, with respect to the economic cycle: world economic output slipped from 2012 to 2013, with advanced economies of the Global North losing 0.1–3.0 percent growth, and emerging and developing economies cooling 0.2–4.7 percent growth. The International Monetary Fund projection for 2015 was for increases in growth in both economies: 3.3 percent, revised down from 3.9 percent, and 4.2 percent, revised down from 5.4 percent (IMF 2014, 2015).

Second, global passenger demand is very strong, with 3.1 billion in 2013, and 6 percent growth in 2014 to 3.3 billion passengers. Third, ancillary revenues – beyond ticket sales – are a key driver for the profitability of airlines already returning an estimated US$13 per passenger, more than double the return on ticketing. Fourth, the consolidation and rationalization of airline operations and individual efficiency initiatives can only help to boost profitability, but with the tight margins of the industry further joint ventures will be necessary.

Fifth, during the global financial crisis (GFC), cargo demand dropped away and was forecast to remain stagnant in the medium term. Indeed, global revenues of around US$60 billion (2007 levels) had been forecast for 2014 and 2015. Nevertheless, cargo demand can be filled through increased belly capacity as passenger services continue to grow.

In line with the international growth of aviation, in Australia there has been a corresponding increase in the need for, and sales of, aviation turbine fuel (ATF) or avtur. ATF is aviation fuel designed for use in gas-powered turbines typically found on common commercial aircraft (Blakey et al. 2011). It is a mixture of hydrocarbons and the most commonly used ATF are Jet A and Jet A-1. (The principal difference between these fuels is the lower freezing point of A-1.) Avgas is used in piston-powered aircraft engines more commonly associated with, but not limited to, private aircraft and general aviation. ATF sales in Australia have doubled from 1,000 megaliters in the September 2003 quarter to 2,000 megaliters in the March 2015 quarter (see Figure 18.2). As such, demand for ATF is forecast to increase. Current aviation fuels are almost exclusively derived from fossil fuels and, when considering fossil fuel constrained peak oil scenarios, the air transport industry recognizes significant challenges. Within the air transport industry, the airline sector is very sensitive to both the cost and supply of fuel. Indeed, fuel is considered to be the principal driver in how and when aircraft are operated (EurActiv 2013).

When considering fuel cost and consumption, the air transport industry recognizes an imperative to build efficiencies in their operations to serve two principal outcomes. The first is that the price of jet fuel has been very closely tied to the crude oil price, and the trend for the crude oil price is currently downward. Second, the industry seeks to build efficiencies in line with global responsibilities to climate change and the management of carbon emissions.

While there has been significant continued growth in international passenger traffic beyond the bottom of the GFC, freight volumes, which are closely related to business confidence, have yet to significantly increase (CAPA 2013). Cargo is one of the major worries for the airline industry, still coming off the downturn of the GFC. Those airlines with dedicated freight businesses are struggling, and this is dragging down their overall profitability, mainly in Asia. Asia-Pacific carriers have nearly 40 percent of the global freight market, and have seen volumes drop and capacity rise (Jaipragas 2014). Much of the freight movement and demand is being managed through the cargo holds of national and international regular passenger transport services.

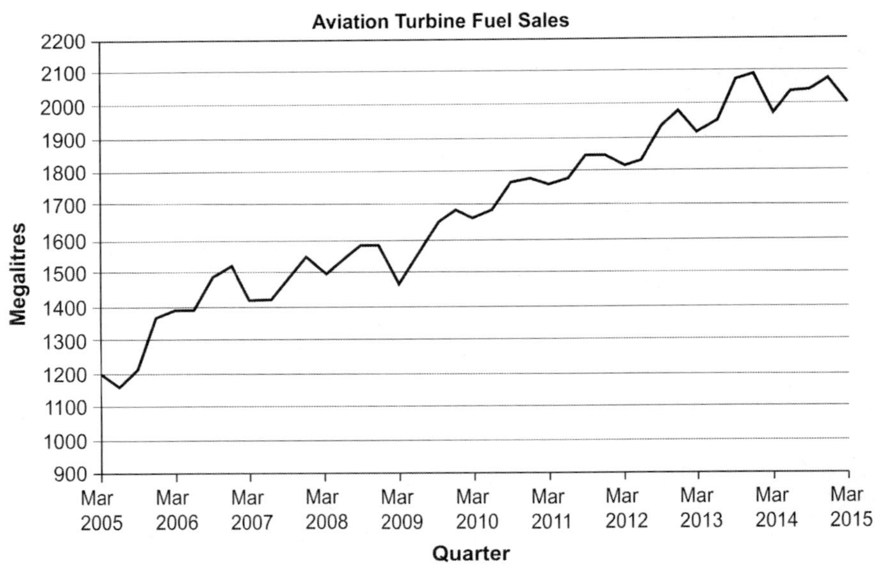

FIGURE 18.2 Aviation turbine fuel sales

Source: BITRE (2015).

Global air transport climate change obligations

The aviation industry has a dedicated United Nations agency, the International Civil Aviation Organization (ICAO), and Australia is a member state of the assembly. Under the provision adopted under the Kyoto Protocol, ICAO has the responsibility of negotiating an international alignment agreement for the aviation industry. The 38th ICAO Assembly held in Montreal (September–October 2013) was, in essence, the last opportunity for ICAO to establish a position for the aviation industry to offer action to deal with climate change. If they had not, the aviation sector's approach to climate change would have been managed under the auspices of the United Nations Framework Convention on Climate Change (UNFCCC) – a move not favored by the air transport industry.

It was at this assembly that significant resolutions were established in committing the aviation industry to tackling climate change. In particular, the commitment to the development and introduction of a global market-based system for climate levies was significant. The Resolution A38–18, *Consolidated Statement of Continuing ICAO Policies and Practices Related to Environmental Protection – Climate Change*, was a significant achievement. This resolution seeks to finalize a global market-based mechanism (MBM) scheme to address emissions from international aviation, for implementation from 2020. Paragraph 19 of the resolution requests ICAO, with the support of member states, to "finalize the work on the technical aspects, environmental and economic impacts and modalities of the possible options for a global MBM scheme, including on its feasibility and practicability, taking into account the need for development of international aviation" (ICAO 2013b: 95).

The form and detail of such MBM will be an issue of ongoing discussion, however ICAO has previously stated that the development of MBMs may include: emissions-related charges and taxes, emissions trading, carbon offsets or a clean development mechanism (ICAO 2008). The benefits of MBM over traditional top-down regulatory measures are that they may "contribute to the achievement of specific environmental goals, at a lower cost, and in a more flexible manner" (ICAO 2013a). ICAO has been investigating the MBM options since 1998, however the preference of ICAO (2001: 3) is the development of an open multilateral emissions trading system for global aviation,

> a system whereby the total amount of emissions is capped and allowances, in the form of permits to emit CO_2, can be bought and sold to meet emission reduction objectives. Such a system could serve as a cost-effective measure to limit or reduce CO_2 emitted by civil aviation in the long term, provided that it is open to all economic sectors.

In 2004, the thirty-fifth session of the ICAO (2008: 17) Assembly endorsed "the further development of an open emissions trading system for international aviation." In that context, *Guidance on the Use of Emissions Trading for Aviation* (Doc 9885) was requested and delivered in 2008. This living document explored the operational boundaries, regulatory considerations, trading units, trading system elements, and the key administrative procedures (such as monitoring, reporting, verification and enforcement) for emissions trading systems (ICAO 2008).

While the aviation sector is estimated to contribute around 2 percent of the world's total human-induced CO_2 emissions (ATAG 2013; Lee et al. 2010), it one of the few industries that is seeking to take a global response to emission reductions. The air transport industry, represented by the Air Transport Action Group (ATAG), is finding consensus in response to tackling climate change and presenting a united front on environmental issues. ATAG is a not-for-profit association that represents a global air transport industry collective. Members include airports, airlines, aviation business councils, aerospace industries and manufacturers, civil aviation regulators, air traffic management, and ground transportation and communications providers.

The global agreements established under ATAG will require international consensus on the MBMs for the management of aviation greenhouse gas emissions. This is important because the air transport industry would prefer not to deal with unilateral measures such as state-based emissions trading schemes. Such schemes have been argued to undermine the "safe, orderly and efficient function of today's international air transport system" (ATAG 2012: 4), whereas a global and multilateral approach will minimize the number of measures that airlines have to deal with. A global infrastructure such as air transport requires global approaches to climate change.

The ATAG has outlined a four-pillar strategy, declared in 2008, for prioritizing efficiency as a means to limit environment impact. However, it is recognized that these changes also make good economic sense to the air transport industry. A summary of the four pillars follows.

1. *Invest in New Technology*
   This pillar represents the ongoing advances in airframe, engine, and lightweight components design and developments in alternative jet fuels.
2. *Fly Using More Efficient Operations*
   Here efficiency gains are delivered by reducing weight on board; for example, better matching fuel loads to actual weight than anticipated weight, utilizing alternate or on-ground power units where possible and establishing more direct landing procedures.
3. *Build and Use Efficient Infrastructure*
   The implementation of more efficient air traffic management systems and the reduction of infrastructure duplication, which results in inefficient routing and flight pattern. Airport infrastructure also has a role to play in enabling the most efficient use of aviation assets.
4. *Use Effective, Global, Market-Based Measures*
   The use of these economic MBMs is anticipated to have one of the largest impacts for the management of aviation emissions.

Aircraft design for fuel efficiency

Increasing fuel efficiency through aircraft design has been a continuous theme in the aviation industry over the past fifty years. The range of design improvements has included the supercritical wing, which reduced shock waves and drag, improved cruising speeds, and increased fuel efficiency. More recently, the winglet or wing tip addition by Boeing and Aviation Partners provides a more efficient wing by reducing kinetic energy. According to Thomas et al. (2008), this innovation alone on 1,500 aircraft in 2008 saved almost 150 million gallons of fuel per year (or 17,000 gallons per hour). The winglet is now a standard fit on the Boeing 737 series.

Fuel efficiency gains are also made through the design of modern aircraft. Currently the Boeing 787 Dreamliner, a long-range, midsized jetliner is considered one of the most fuel-efficient aircraft in the world. This efficiency is achieved largely through the use of composite materials and lightweight internal fittings and twin engines. The use of lithium ion batteries for some avionics functions has also saved weight, although not without incident (Williard et al. 2013). The larger four-engine Airbus A380 is close behind on efficiency, able to move 200 more passengers than the 787 on each flight. The newly released A350 is seen as a more equal comparison to the 787 as a twin-engine, carbon fiber–reinforced, long-range jetliner.

In November 2013, Boeing released its intentions for the 777X, building on the reputation of the 777, but expecting to be the largest and most fuel-efficient twinjet in the world (Robarts 2013). It features folding wing tips, giving it a larger wingspan for efficiency, but allowing the use of current airport infrastructures. It is expected to be operating from 2020.

Increasing engine efficiencies

Advances in engine technologies are also expected to deliver better fuel efficiencies. Engine-maker Pratt & Whitney is developing a composite material fan-drive gear system engine, which is expected to be lighter and cut fuel consumption by 16 percent (Marsh 2014). The PurePower line of engines already has 3,500 orders from five aircraft manufacturers, including Airbus.

Drop-in technologies

Considering the time that it takes for technology transition in the aviation industry, the best outcome for a move to a cleaner aviation industry are drop-in technologies. These innovations allow the current generation of aircraft and airport infrastructures to remain unchanged. The use of alternate propulsion methods, such as liquid nitrogen or electricity (fuel cells, solar and hybrid turbine-electric propulsion) will require significant redesign of air frames and aircraft more broadly (Kivits et al. 2010). Further, when considering the anticipated growth in annual global passenger traffic, there is a very limited time frame to plan and develop the necessary airport and aviation infrastructures to support completely reconfigured design innovations. While engine and aeronautical technologies have made modern aircraft 70 percent more fuel-efficient than first-generation jets, aircraft will continue to need high-energy liquid fuels (ATAG 2011).

Biofuels

Alternative fuels can be categorized into two groups, depending on whether the product increases security of supply (i.e. enables continuity of supply in the face of peak oil scenarios) or provides a reduced environmental footprint (Blakey et al. 2011). The aviation industry is concerned with both groups but, in the context of this work, the principal concern is with the latter group.

For every ton of traditional aviation fuel saved, an equivalent amount of 3.16 tons of CO_2 is avoided (ICAO 2013b). It is expected that as biofuels become available and blended in greater quantities they may cut aviation CO_2 emissions by up to 80 percent. However, the current unavailability of biofuels is a major hurdle (ICAO 2013b). Regardless, it has been estimated that – even if commercial aviation replaced 6 percent of fossil-based jet fuels with fuels derived from biomass sources, such as algae – the overall carbon footprint could be reduced by 5 percent (ATAG 2011).

Advances in technology have created biofuels with a chemical structure very similar to Jet A-1. This has significant advantages for the conversion potential away from fossil fuels, because these next generation biofuels can be added to Jet A-1 supplies without issue. Significantly, it means also that there is no requirement for new engines or changes in aircraft technologies. While aviation is generally considered a late adopter of biofuel technologies, this has allowed the industry to learn from the substitution missteps of the automobile industry. An additional advantage for the transition away from fossil-based fuels is that the aviation sector has a relatively condensed fuel distribution system (airports), making uptake more efficient. Biofuels realistically present the only viable commercial alternate fuel option, however there are constraints in the supply chains of biofuel manufacture and feedstock availabilities.

Currently the principal problem – despite the evidence, approvals and known necessity for the use of biofuels – is cost. Jet A-1 fuel accounts for over 30 percent of airlines' operating costs, and biofuels cost three to five times more than regular jet fuel. They are expensive because aviation biofuel is manufactured in single quantities for particular purposes, so it is still a specialist product. As production increases, the price is expected to come down. However, the industry requires continued research, and public and private sector support, to stimulate and expedite these processes.

It is important to note that not all biofuels are equal when it comes to sourcing and production. The development of aviation biofuel should not be at the expense of biodiversity and food security elsewhere. There is little to be gained by resolving current challenges in ways that establish additional environmental problems – palm oil plantations and industries being an important case in point.

Over the past decade the application of biofuels has moved from obscurity almost to the mainstream, with much of the application of biofuel for road and maritime use. The daily use of aviation biofuels continues to be elusive. On a global scale, Gegg et al. (2014) outline the drivers and constraints for the market development of aviation biofuels, arguing that the single most important factor for the development of aviation biofuels is the need to reduce carbon emissions. Other longer-term drivers include energy security, volatile oil prices, legislation requiring change (such as the EU Emissions Trading System), new business opportunities and an overall lack of alternative technologies. Identified constraints to the uptake of biofuels are all closely linked, and include high production costs, lack of investment (within the context of the global economic downturn), lack of sustainable feedstock supplies, lack of legislation to support aviation biofuels, strict environmental controls for biofuel production and the lack of supply chain certification.

Australia and biofuels

In considering some of the supply-based issues, Australia and New Zealand are seeking to establish a "local commercially viable supply chain" to support ambitious ongoing blended aviation fuel targets (CSIRO 2011). Australia and New Zealand are looking to achieve a 5 percent bio-derived aviation fuel share by 2020, expanding to 40 percent by 2050. This would result in savings of over A$2 billion per year on jet fuel imports and a 17 percent reduction of greenhouse gas emissions p.a. relative to petroleum-based aviation fuel (CSIRO 2011: 7).

The benefits of establishing a sustainable aviation fuels industry in Australia and New Zealand are anticipated to flow directly into employment and rural and regional development, as well as supporting the Australasian tourism industry. The Sustainable Aviation Fuel Road Map is explicitly about establishing energy security for the aviation sector in the face of uncertain oil prices (CSIRO 2011: 17). The CSIRO (2011: 19) has estimated that in the coming decade the Australia-New Zealand region "has the potential to supply almost half of the local aviation sector's fuel needs from biomass and supply all its needs over the long term." To this end, the Road Map acknowledges and outlines key challenges and opportunities for the establishment of sustainable aviation fuels industry.

Adapted from CSIRO (2011: 6–7), such challenges and opportunities arise from the following factors and predictions: sustainable aviation fuels derived from biomass are a feasible option; there is sufficient existing sustainable biomass to support a local bio-derived jet fuel industry; the Australian region is strongly positioned to produce sustainable aviation fuels; there will be challenges in the scale-up of economically viable feedstock production; there will be high demand among industries for biofuels; investment by the refining sector will be impacted by uncertainty; aviation fuel distribution infrastructure will not require significant modification, however access arrangements for bio-derived jet fuel suppliers will need to be established; and local production of sustainable aviation fuels will bring significant economic, social and environmental benefits.

Biofuel uncertainty

A significant uncertainty for biofuel production is that there are currently no commercial-scale plants for bio-derived jet fuel anywhere in the world – it is a specialist product. Additionally, there is still uncertainty around the sourcing and pricing of appropriate feedstock for any large-scale aviation biofuel production. A recent study undertook a techno-economic analysis of renewable aviation fuel from three sources with current proven technologies. Results showed that biofuels from feedstock sources would be competitive with a standard 42-gallon barrel of crude oil for equivalent volumes at US$1,343 (microalgae), US$374 (Pongamia seeds) and US$301 (sugarcane). However, when the research considered expected technological and market developments for the conversion processes and the use of by products, it was estimated that the comparative prices may be reduced to US$385, US$255 and US$168 per barrel, respectively (Klein-Marcuschamer et al. 2013).

Despite the apparent ease of drop-in, there remains the need for further testing of the effects of biofuels on aircraft materials and, in particular, elastomers present in aviation fuel systems. These elastomers are the "elastic polymers" used as sealants, O-rings, gaskets, hosing and tubing. Biofuels have different impacts on these materials than regular aviation fuel, which requires ongoing research on materials and standard-setting (Austin et al. 2013; Ricklick et al. 2012).

Ongoing uncertainty in the fossil-based fuel commodity markets could also lead to delays in investment in biomass production and refining infrastructure. There is a history of fluctuations in the crude price, removing the incentives for investment and advancement of alternative fuel technologies, particularly within the automobile industry. If fossil fuel remains affordable then it will continue to be the dominant fuel despite environmental motivation.

The pricing of crude oil is crucial to the viability of biofuels as realistic substitutes for aviation fuel. In the short to medium term the cost of production of fossil-based liquid fuels still seems likely to be below the competitive price for aviation biofuels. On price alone, putting CO_2 emissions to one side, it will remain cheaper to extract and produce fossil-based aviation fuels from shale and oil sands than to produce biofuels for aviation. Indeed, the extraction of "tight oil" or "shale oil" (held within the rock below more easily extracted liquid deposits) has become significantly more economically viable. US oil production expanded from 305 million tons in 2008 to an estimated 499 million tons in 2013, with most of this attributable to shale oil (Aguilera and Radetzki 2013).

Operational and infrastructure efficiency

A significant use of aviation fuel occurs outside of direct route flying. Fuel is burned through taxiing, runway queuing and circling for landing clearance. Auxiliary power unit usage while the aircraft is docked provides energy savings and tugs may be used to position aircraft – all measures designed to reduce the fuel burned prior to takeoff (Williams 2008). Outside of these ground operations are the routes that airlines operate. For example, "hubbing" is the practice of routing passengers and goods through specific "central" airports while they are still in transit. It is used by airlines to maximize their potential connections at the hub airport, while also allowing for the consolidation of passengers and goods to larger aircraft for transit on higher frequency routes (Loo et al. 2014; Nero and Black 1998). However, the decision to hub impacts on the airlines overall fuel burn from the additional landing and takeoff.

The flight path is another important factor for energy savings. How the aircraft glides into the destination impacts on the extent of fuel burn. In 1999, the Intergovernmental Panel on Climate Change estimated that air traffic management and operational improvements could affect up to 18 percent improvement on fuel burn.

Flight delays also provide a significant burden to the economy. In an FAA-funded study, Ball et al. (2010) estimated that the cost of all US air transport delays in 2007 included costs of US$8.3 billion in losses to airlines for crew, fuel and maintenance; US$16.7 billion in passenger impact due to time lost from schedule buffer, delayed flights, flight cancellations and missed connections; and US$3.9 billion from lost demand (passengers who avoid air travel because of delays).

The present state of air traffic management provides a significant opportunity to improve flight efficiencies. ATAG (2013) argue that current state-based navigation systems need improving to allow cross-border arrangements (especially significant in Europe and Asia where borders are crossed frequently in single flight paths), to improve civil and military cooperation to allow for more flexible use of airspace (large tracts of airspace are reserved for military use and this is often vacant), and to establish effective regulatory practice to improve the present air traffic management operations.

ATAG argues for a seamless navigation system to allow efficient flight paths, coordinated approaches and landings, and efficient taxing and takeoffs. Predetermined air traffic routes that are controlled by obsolete infrastructure, consisting of radio-based navigation supported by ground-based beacons, often do not offer the best and most efficient flight paths. Required navigational performance (RNP) provides a much more efficient approach to streamlining flight paths to improve on-time performance and reduce delays and congestion. RNP benefits from new technology, such as satellite-based systems to improve the flight paths and accuracy, without relying on ground-based navigation. Australia, through Qantas and Air Services Australia, has been using this approach to improve flight paths – starting in 2007 with the Brisbane Green Project, the world's first RNP-integrated project (ATAG 2013).

From the Australian perspective, multilateral partnerships between air navigation service providers and airlines to improve operational efficiency include ASPIRE (Asia-Pacific) and INSPIRE (Indian Ocean). The focus of these partnerships is to lessen the environmental impact of aviation across the Pacific and to focus on efficiencies on fuel and emission reductions. Trials have been run to improve flight efficiency by testing methods such as the following:

User-preferred routes: These are custom-designed routes that take into account the winds, type of aircraft and aircraft weight to estimate the best flight path.

Flexible track: Over the ocean the application of user preferred routes is not feasible, but the flexible track, published daily, provides a means for air traffic to divert to more efficient routes and take advantage of winds and weather conditions.

Dynamic airborne reroute procedures: These procedures allow the flight to adapt to wind and weather conditions while in route – to change the flight pattern to suit climatic conditions to improve fuel efficiency. ASPIRE (2011) provides more detail on the metrics that have been applied in the Pacific region and the savings that have been incurred with tests.

Reduced vertical separation minima: Modern aircraft are fitted with guidance systems that allow for reduction of vertical separation between aircraft. This provides for more efficient use of fuel-efficient altitudes and less overall climb.

Continuous descent arrival and tailored arrival: The optimization of descent procedures allows the most efficient glide paths into airports, when engines can be reduced to provide minimal thrust and trajectories are established for the most energy efficient arrival, resulting in lower noise. According to ATAG (2013) aircraft that use the continuous descent arrival approach achieve much more efficient fuel burn during the descent and arrival phases.

The reduction of hubbing has made significant savings in energy, both through the elimination of landings and takeoffs and by the elevation maintained by the aircraft. Aircraft fuel consumption is high for short trips, primarily because of the takeoffs, landings and frequent delays in hubbing. Fuel burning falls with longer trips, but this efficiency can be impacted due to the weight of fuel required for long trips (Gilbert and Perl 2008). According to Thomas et al. (2008), one of the most significant ways for the aviation sector to reduce fuel use is with the reduction of unnecessary stops. Direct flights can be increased by producing new generation aircraft, such as the 787 or A350 models, that can carry 250–350 passengers and achieve ranges of 14,800–15,750 km (Thomas et al. 2008). However, much of the airport infrastructure presently used is based on hubbing. In the short term, energy efficiency may fall victim to politics and airports needing to maintain hubbing status.

Future trends in air transport

The notion of urbanization and aviation megacities will be increasingly important. By 2030 more than 60 percent of the world's population, forecast to be 8.4 billion, will live in urban centers, up from 50 percent in 2012 (Seto et al. 2012; UN 2012). This represents a world urban population of more than five billion people in 2030.

Using a Delphi process, Linz (2012) developed forty projections for the future of the aviation industry in the short-term horizon of 2025. Using a pool of fifty-seven experts, who all took part in at least two Delphi rounds, the projections were identified as having achieved "consensus" or "high impact estimations" when achieving a consensus of 60 percent and 55 percent, respectively. As such, Linz (2012) outlines that, in accordance with the experts, the future of aviation in 2025 will see passenger and business aviation customers demand "easy" air transportation without wasting time. Self-service is widely accepted in both passenger and cargo services, however passengers will seek further unbundling of services while cargo customers will seek more integration (door-to-door and one-stop shops). Virtual meeting and video conferencing has substantially increased in line with better information and communications technology (ICT) integration, with implications for business and conference travel.

The travel demands to, from and within the emerging markets of newly industrialized economies, such as Brazil, India and China, will continue to be the major driver of global aviation growth in 2025. Overall growth in long-haul national and international air transport has been more rapid than regional and short-haul services. A significant impediment for the aviation industry is anticipated to be capacity, that is, both airport infrastructure and airspace capacity. However, international satellite-based air traffic control systems are expected to be available by 2025, allowing for better allocation of airspace.

When considering aviation business models, the growth of low-cost carriers will outstrip full-service carriers, and low-cost carriers will increase their market share of business travelers. Mergers and joint ventures will continue to provide efficiency dividends for airlines and more efficient connections for consumers (IATA 2013).

Conclusion

There are significant challenges for air transportation in Australia and internationally when facing fossil fuel constrained futures. There is a rising volatility in global fuel prices, set against high levels of competition, increasing costs and marginal profits. Despite this context, forecasts indicate that the air transport industry is resilient, flexible and can adapt quickly to change. The demand for air services, and indeed global reliance on these services, is becoming further entrenched. The ability of the aviation sector to adapt to peak oil is predicated on drivers and constraints. Changing technology is certainly a driver for the development of more efficient aircraft both in the development of more fuel-efficient engines and the design of lighter, more streamlined bodies. In addition, technology also offers a change in the way aircraft are managed on the ground and in the air, with routing, takeoff and landing technologies.

Biofuels are coming to the fore at a rapid rate, offering the opportunity to provide more energy security and business opportunities in Australia and overseas. However, it is also apparent that mainstreaming biofuels continues to be limited by their specialist nature and the increasing affordability of "tight" oil. A single certainty is that the air transport industry will continue to be powered by high-energy liquid fuels, but their blend will be an economic decision.

It is clear that the air transport industry will continue to innovate and build efficiencies to ensure both economic viability and meet commitments to reduce greenhouse gas emissions. The industry recognizes the value of a global response to change and has led in the establishment of international agreements. In many regards the air transport industry provides a global standard for addressing institutional barriers to navigation and data cooperation, and the organization of international agreements on MBM for the management of environmental impact. The challenges that the air transport industry faces are not significantly different from other transport sectors that require transformation. Institutional barriers continue to impact how change is made, while economic considerations are always key in determining the rate at which that change occurs.

Note

1 The term "revenue passenger kilometers" is a measure of traffic and is derived by multiplying the number of revenue-paying passengers by the distance traveled. RPK is considered to be a more accurate volume estimator than simply passengers carried because it combines measures of distances traveled with passengers carried (ABS 2014).

References

ABS (2014) *Australian Systems of National Accounts – Concepts, Sources and Methods*, Australian Bureau of Statistics, Cat. No. 5216.0, Sydney: Commonwealth of Australia.

Aguilera, R.F. and M. Radetzki (2013) "Shale Gas and Oil: Fundamentally Changing Global Energy Markets," *Oil & Gas Journal* 111(12): 54.

ASPIRE (2011) *Annual Report for 2011*, Metric Performance Appendix, Asia and Pacific Initiative to Reduce Emissions/ATO Communications, accessed 5 September 2015 — www.aspire-green.com/mediapublications/docs/metricsappendix.pdf

ATAG (2014) *Aviation Benefits Beyond Borders*, April, Geneva: Air Transport Action Group.

——— (2013) Air Transport Action Group, *Revolutionizing Air Traffic Management*, May, Geneva: Air Transport Action Group, accessed 5 September 2015 — www.atag.org/our-publications/latest.html

——— (2012) *A Sustainable Flight Path Towards Reducing Emissions*, (position paper presented by the aviation industry to UNFCCC delegates at COP18 in Doha, November), accessed 5 September 2015 — www.atag.org/our-publications/latest.html

——— (2011) *Beginner's Guide to Aviation Biofuels* (2nd edn), Geneva: Air Transport Action Group, accessed 5 September 2015 — www.atag.org/our-publications/latest.html

Austin, W. Stanley, D. and M. Thom (2013) *Testing the Effects of a Bio-Derived Alternative Aviation Gasoline on Aircraft Materials in Comparison to 100 Low Lead Aviation Gasoline*, Aviation Technology Graduate Student Publications, Paper 29, West Lafayette: Purdue University, accessed 7 September 2015 — http://docs.lib.purdue.edu/atgrads/29

Ball, M., Barnhart, C., Dresner, M., Hansen, M., Neels, K., Odoni, A., Peterson, E., Sherry, L., Trani, A. and B. Zou (2010) *Total Delay Impact Study: A Comprehensive Assessment of the Costs and Impacts of Flight Delay in the United States*, Berkeley: University of California, National Centre of Excellence for Aviation Operations Research.

BITRE (2015) *Aviation Turbine Fuel Sales*, Canberra: Bureau of Infrastructure, Transport and Regional Economics, accessed 16 August 2015 — https://bitre.gov.au/statistics/aviation/av_fuel_sales.aspx#anc_avtur

Blakey, S., Rye, L. and C.W. Wilson (2011) "Aviation Gas Turbine Alternative Fuels: A Review," *Proceedings of the Combustion Institute* 33(2): 2863–85.

CAPA (2013) "IATA Raises Profit Forecasts – The World's Airlines Can Now Upgrade from an Espresso to a Sandwich," *Aviation Analysis*, Sydney: CAPA Centre for Aviation, accessed 2 September 2015 — http://centreforaviation.com/analysis/iata-raises-profit-forecasts – the-worlds-airlines-can-now-upgrade-from-an-espresso-to-a-sandwich-113244

CSIRO (2011) *Flight Path to Sustainable Aviation – Towards Establishing a Sustainable Aviation Fuels Industry in Australia and New Zealand*, Sustainable Aviation Fuel Road Map, Dickson, ACT: Commonwealth Scientific and Industrial Research Organization (Australian Government).

EurActiv (2013) *Fuel Efficiency Seen Driving Demand for New Aircraft in EU, Globally*, 4 July, accessed 7 September 2015 — www.euractiv.com/transport/fuel-efficiency-seen-driving-dem-news-529067

Gegg, P., Budd L. and S. Ison (2014) "The Market Development of Aviation Biofuel: Drivers and Constraints," *Journal of Air Transport Management* 39: 34–40.

Gilbert, R. and A. Perl (2008) *Transport Revolutions: Moving People and Freight without Oil*, London: Earthscan.

IATA (2015) "Jet Fuel Price Development," Market Development Outlook: Price Development, International Air Transport Association, accessed 16 August 2015 — www.iata.org/publications/economics/fuel-monitor/Pages/price-development.aspx

——— (2013) "Airline Financial Outlook Strengthens," Press Release No. 69, 12 December, International Air Transport Association, accessed 12 January 2014 — www.iata.org/pressroom/pr/Pages/2013–12–12–01.aspx

ICAO (2013a) *Environmental Report: Destination Green, Aviation and Climate Change*, Environment Branch (ICAO) Montreal: International Civil Aviation Organization, accessed 7 September 2015 — http://cfapp.icao.int/Environmental-Report-2013/

——— (2013b) *Global Air Transport Outlook to 2030*, ICAO Circular, 333.AT/190, Montreal: International Civil Aviation Organization.

——— (2008) *Guidance on the Use of Emissions Trading for Aviation*, Doc. 9885, Montreal: International Civil Aviation Organization, accessed 7 September 2015 – http://ec.europa.eu/clima/policies/transport/aviation/docs/icao_guidance_2008_en.pdf

——— (2001) *Statement from the International Civil Aviation Organization (ICAO) to the Fifteenth Session of the UNFCCC Subsidiary Body for Scientific and Technological Advice (SBSTA)*, Marrakesh: 30 October–6 November 2001, accessed 7 September 2015 — www.icao.int/environmental-protection/Documents/STATEMENTS/sbsta-15.PDF

IMF (2015) *Slower Growth in Emerging Markets, a Gradual Pickup in Advanced Economies*, International Monetary Fund, World Economic and Financial Surveys, World Economic Outlook (WEO) Update, July, accessed 7 September 2015 — www.imf.org/external/pubs/ft/weo/2015/update/02/

——— (2014) "Is the Tide Rising?" *World Economic Outlook (WEO) Update*, International Monetary Fund (World Economic and Financial Surveys), January, accessed 7 September 2015 — www.imf.org/external/pubs/ft/weo/2014/update/01/

Jaipragas, B. (2014) "Struggling Cargo Business 'Dragging Down Asian Airlines'," *Digital Journal*, accessed 15 February 2014 – accessed 7 September 2015 — www.digitaljournal.com/biz/business/struggling-cargo-business-dragging-down-asian-airlines/article/369684

Kivits, R., Charles, M.B. and N. Ryan (2010) "A Post-Carbon Aviation Future: Airports and the Transition to a Cleaner Aviation Sector," *Futures* 42(3): 199–211.

Klein-Marcuschamer, D., Turner, C., Allen, M., Gray, P., Dietzgen, R.G., Gresshoff, P.M. and L.K. Nielsen (2013) "Technoeconomic Analysis of Renewable Aviation Fuel from Microalgae, Pongamia Pinnata, and Sugarcane," *Biofuels, Bioproducts and Biorefining* 7(4): 416–28.

Lee, D.S., Pitari, G., Grewe, V., Gierens, K., Penner, J.E., Petzold, A., Prather, M.J., Schumann, U., Bais, A., Berntsen, T., Iachetti, D., Lim, L.L. and R. Sausen (2010) "Transport Impacts on Atmosphere and Climate: Aviation," *Atmospheric Environment* 44(37): 4678–734.

Linz, M. (2012) "Scenarios for the Aviation Industry: A Delphi-Based Analysis for 2025," *Journal of Air Transport Management* 22: 28–35.

Loo, B.P., Li, L., Psaraki, V. and I. Pagoni (2014) "CO_2 Emissions Associated with Hubbing Activities in Air Transport: An International Comparison," *Journal of Transport Geography* 34: 185–93.

Marsh, G. (2014) "Composites Flying High," *Reinforced Plastics* 58(3): 14–18.

Nero, G. and J.A. Black (1998) "Hub-and-Spoke Networks and the Inclusion of Environmental Costs on Airport Pricing. *Transportation Research Part D: Transport and Environment* 3(5): 275–96.

Ricklick, M., Quintero, S.A. and Kapat, J. (2012) Synthetic Jet Fuels and their Impact in Aircraft Performance and Elastomer Materials," paper presented at the *48th AIAA/ASME/SAE/ASEE Joint Propulsion Conference & Exhibit*, 29 July–1 August, Atlanta.

Robarts, S. (2013) "Boeing's Fuel-Efficient 777x Features Flooding Wings," *Gizmag*, 18 November accessed 6 January 2014 — www.gizmag.com/boeing-launches-777x/29818/

Seto, K.C., Güneralp, B. and L. R. Hutyra (2012) "Global Forecasts of Urban Expansion to 2030 and Direct Impacts on Biodiversity and Carbon Pools," *Proceedings of the National Academy of Sciences* 109(40): 16083–88.

Thomas, G., Morris, G., Forbes Smith, C., Creedy, S. and R. Pepper (2008) *Plane Simple Truth: Clearing the Air on Aviation's Environmental Impact*, Perth: Aerospace Technical Publications International.

United Nations (2012) *World Population Prospects: The 2012 Revision* File POP/1–1: Total Population (both sexes combined) by major area, region and country, annually for 1950–2100 (thousands), United Nations Population Division, Department of Economic and Social Affairs, accessed 7 September 2015 — http://esa.un.org/wpp/Excel-Data/population.htm

Williams, V. (2008) "Impacts of the Aviation Sector on the Environment in Environmentally Conscious Transportation," in M.W. Kutz (ed) *Environmentally Conscious Transportation*, New Jersey: Wiley, 301–29.

Williard, N., He, W., Hendricks, C. and M. Pecht (2013) "Lessons Learned from the 787 Dreamliner Issue on Lithium-Ion Battery Reliability," *Energies* 6(9): 4682–95.

Conclusion

19

PLANNING AND PETROLEUM FUTURES

Research directions

Neil Sipe, Jago Dodson and Anitra Nelson

This book has explored the problem of urban oil vulnerability, specifically the imperative to plan cities so that they are prepared for an era of petroleum constraint. This final chapter discusses some key major findings and conclusions from the preceding chapters along with certain "planning after petroleum" issues that were not given chapter-length coverage. The authors identify some significant future research directions that should be undertaken over the next several years to enable cities to better plan for petroleum constraint.

Key insights

The findings and conclusions from chapters in this introductory collection can be distilled into four broad themes that not only offer key insights but also either imply or suggestively point in a variety of directions for future research. Brief discussions of these themes follow but a lateral connection within, and implication of, these themes is identified first: the breadth and depth of issues involved with addressing the future "age beyond oil" challenges the discipline and professional practice of planners to be more interdisciplinary and engaged with citizens. Rather than relying on prescriptive, traditional solutions, these issues require planners to be creative, experimental and bold. This will alter the entire training of planners and their lifelong professional development.

Uncertainty and volatility

A major theme running through the chapters involves the persistent problem of uncertainty and volatility in petroleum markets, how this intersects with the material consumption practices of urban life and the types of planning strategies needed to deal with such instability and complexity. Chapter 2 by Jago Dodson and Chapter 3 by Samuel Alexander focused on this theme in particular. They showed that there is little consensus among experts as to what the future holds. There is, however, general agreement that there will be continued volatility and uncertainty in terms of petroleum prices, supply and demand.

There can be little debate that this pattern has been the main oil market feature over the past decade, with near-record highs and lows of oil prices that no one had predicted. Perhaps the most significant feature of the unpredictable nature of this particular market is the current and future impact of recent developments: (1) global and grassroots initiatives to avoid climate change; (2) outcomes of the twenty-first session of the Conference of Parties (COP21) of the United Nations Framework Convention on Climate Change (December 2015), which heralded an end to fossil fuel use by the end of this century; and (3) major divestment campaigns in the Global North that seek to reduce fossil fuel investment.

Planning barriers

In addition to uncertainty, cities face many barriers to planning more proactively to respond to petroleum constraint and oil vulnerability. Underlying problems with moving away from a reliance on petroleum as a liquid fuel were identified in several chapters, particularly by Tony Matthews and Jago Dodson in Chapter 4 and by Douglas Baker, Nicholas Stevens and Md. Kamruzzaman in Chapter 18. The latter contribution identified a series of ways that fuel sources and management of air transport equipment and services can be changed to make the air transport sector more fuel-efficient. While this cluster of problems tends to be technical, the other set mainly involves institutions and governance.

A consistent theme across many of the chapters is that institutional barriers remain a challenging impediment to constructive and adaptive change in cities exposed to petroleum risks. All the chapters identify and explore such barriers, and suggest ways for overcoming them. In Chapter 16, Cheryl Desha and Angela Reeve propose – and show in their case study – that experiential learning by doing can be an effective means of achieving change. Similarly, in Chapter 14, Brendan F.D. Barrett and Ralph Horne argue that to achieve prospective change, multi-stakeholder action is required at the local level of city or municipal governance, as observed in the energy transitions activities that they report.

Deliberative planning

Another major theme across the chapters is the deliberative act of planning, central to *Planning After Petroleum: Preparing Cities for the Age Beyond Oil*. For instance, in Chapter 6, Wendy Steele, Lisa de Kleyn and Katelyn Samson have argued that planners must prioritize the plight of those who are most vulnerable to the negative impacts of petroleum constraint and climate mitigation. This is a matter not only of social justice and environmental justice but also of appropriate governance in a new era for humanity. Given that addressing climate change demands alterations in everyday practices, citizens need to be more directly involved in, and more widely benefit from, deliberative planning. This is because city planning provides a context for, and therefore facilitates changes in, mundane practices.

This key mobilizing potential for planning in addressing petroleum constraint is also made clear by John Whitelegg in Chapter 7. Whitelegg calls for a re-engineered planning profession that supports the movement for "cities of short distances," where walking becomes a much more significant mode than it is currently. Whitelegg points out that this change would reinstate walking to the place it has had in human history and introduce us to new ways of being and doing in a more complex future. There are many planning approaches to making cities more pedestrian-friendly that coincide with reforms to support greater bicycle use. In Chapter 8, Jennifer Bonham and Matthew Burke advocate for restructuring that dovetails into public transport efficiencies suggested by John Stone and the late Paul Mees in Chapter 10. Walking and cycling are natural and significant complements to a public transport system with central nodes; planners of paths for pedestrians and cyclists need to take account of their potential for facilitating use of public transport.

In Chapter 9, Scott Sharpe and Paul Tranter show that conscientiously planning for improvements to choices of mobility in child-friendly cities creates benefits not only for physical health but also sociality and communities of place. That all such improvements require dedicated long-term planning is a fact assessed with salutary conclusions by Jago Dodson and Neil Sipe in Chapter 5. However, in Chapter 11 they describe and operationalize their "vulnerability assessment for mortgage, petroleum, and inflation risks and expenditure" (VAMPIRE) index, an advance that points in the direction of adverse circumstances creating opportunities. Deliberative planning requires such tools for balancing evidence within a bigger picture.

Challenge as opportunity: Innovation

Another major theme that emerged in this collection was the identification of certain unexpected benefits – already provided by volatility in oil prices – to spur innovation. One example is advancing applications that depend on high-speed Internet technology, which Tooran Alizadeh explored in Chapter 17. Another beneficiary, discussed by Baker et al. in Chapter 18, is the airline industry, which has been driven to seek efficiencies in design and operations due to oil price concerns.

In Chapter 13, Jason Byrne suggested various ways to remake cities to be more inclusive of nature, such as converting disused road and parking spaces into greenspace and vegetating the tops of buildings and small public lots of land, and populating tiny parks and recreational spaces with potted plants.

Our contributors conclude that, rather than perceive constraints as barriers, planners are best advised to adopt open approaches of opportunity, creativity, experimentation and innovation.

Opportunities for future research

While contributors to this collection considered many questions for planners of post-petroleum cities, it was not possible to cover many major problems, let alone every conceivable angle. Some notable gaps deserve discussion, including the following.

Most of the chapters focused on transport, land use and urban systems. However, petroleum is used by, and affects, many other sectors of society not directly considered by our contributors. Environmental impacts include air pollution and greenhouse gas (GHG) emissions from motor vehicles, as well as electric generation plants and manufacturing facilities. The impacts are direct, immediate and localized (as with many air pollutants) through to globalized, long-term influences (as with GHGs). Meanwhile, such pollution enters the water cycle to damage soils, landscapes, bays and oceans. With each damaging pollutant, city planners confront issues at a range of scales and time frames, and various approaches and strategies, including prevention, mitigation, management and compensation.

In addition, contemporary urban utilities – power generation stations, water supply and distribution, and wastewater collection and treatment – have relied to a high degree on petroleum fuel for their operation. Infrastructure disruptions due to petroleum constraint cannot be dismissed lightly. Similarly, the wider economic impacts of high oil prices and, in turn, the need to move beyond petroleum as a fuel could damage gross productivity, employment and income, and put pressure on raising inflation and debt. Obviously, these effects have an urban dimension but, because they tend to be consequential and secondary rather than direct, we decided not to cover them. Nevertheless, this does not diminish their significance in terms of future research, which offers fields for numerous yet-to-be-written books.

Even though a significant part of the book focused on the transport sector, we could not cover all transport issues. One key omission was freight movement across land, sea and air. The movement of goods for consumers, and materials and equipment for producers, is critical for the functioning of cities, but we decided to focus on the movement of people and passengers. Even so, we did not exhaust all angles of the latter. For instance, we did not examine disruptive transport technologies that are gaining much media attention as we write, such as autonomous vehicles, driverless cars and aerial drones. While a truly driverless car might still be years away, drones are ready to use and could, for example, markedly change how small parcels are delivered, with implications for the freight sector. A number of companies, including Australia Post, Singapore Post and Amazon, already have (or are trialing) the use of drones for small parcel delivery and are awaiting government approval (Chang 2015). A shift to electric drones might reduce the need for petroleum fuel last-mile delivery vehicles, just as electric buses – currently capable of traveling several hundreds of kilometers on a single charge – promise to transform public transport (Platt 2016).

Of course, it would have been useful to have included a chapter on electric vehicles (EVs) or other alternative fuel vehicles. However, when this book was first conceptualized, the uptake of these types of vehicles was negligible, and they were insignificant in post-petroleum discussions. Even now, early in 2016, there is a very limited evidence base to inform an academic assessment of the likely introduction of EVs at more than experimental scales. Nor is there a developed literature on the planning implications of EVs as opposed to their engineering. The uptake of EVs is a clear area for further social and scientific investigation.

Finally, we did not include any chapter-length discussion of "peak demand" for energy, a topic that has gained research and scholarly interest over 2015, particularly given declining petroleum prices. Given that this debate about peak demand is so important, the next section sketches the debate, specifically regarding further research.

Emerging post-petroleum issues

The uncertainty about petroleum futures, the range of potentially disruptive transport and petroleum technologies and debates about peak demand mean that there are plenty of topics for researchers to examine. Here we tease out some directions in a few of these topics.

Peak demand

The question of peak energy demand is a topic of scientific and economistic interest. Some believe that peak demand for petroleum may have been reached and that this is one of the reasons for the drop in oil prices. There are several reasons given for an apparent peak in energy demand. One major reason is the growth of alternative fuel sources for electricity generation, for example, the US Energy Information Agency (EIA 2012) has reported that, in 2010, only 1 percent of electric power was generated from petroleum – a significant drop from 1978, when 22 percent of electricity was generated from oil. Another major reason is due to more fuel-efficient vehicles, EVs and alternatively powered vehicles.

However, there is no consensus that peak demand has occurred, or even will occur, in the future. Rapier (2015) does not believe that peak demand will occur because there are no scalable replacements for oil such as there are for coal. As highlighted on many occasions throughout this book, most transport modes – cars, trucks, airplanes and ships – are dependent on petroleum as a fuel source. To date, sales of petroleum-based vehicles have increased much faster than any of the alternatives, be they vehicles using biofuel or EVs. According to the Electric Drive Transportation Association (EDTA 2016), EVs comprised only a 2.87 percent share of total vehicle sales in the US, in 2015, with almost 500,000 sold compared to all vehicle sales of 17.4 million. This represents a fall from a high of 3.84 percent in 2013 and 3.47 percent in 2014. Similar trends exist in other countries.

Other experts (Jaffe 2015) believe that global demand for petroleum will peak sometime over the next twenty years. Jaffe (2015) challenges conventional wisdom suggesting that a growing global middle class will ensure high petroleum demand and prices, instead believing that

> exponential gains in industrial productivity, software-assisted logistics, rapid urbanization, increased political turmoil in key regions of the developing world, and large bets on renewable energy are among the many factors that will combine to slow the previous breakneck growth for oil.

The influences on demand, and demand-supply dynamics that affect price, will continue to be topics of great interest for economists and for petroleum and other industry and scholarly experts monitoring the activities of producers and consumers. With respect to planners, questions abound. If peak demand does occur, what impacts would that have on our cities? What would the implications be for land-use planning? How might travel, in all its forms, be affected?

Electric vehicles

One area of research would involve developing a better understanding of how EVs might affect travel behavior and land-use planning. How would increasing vehicle fuel efficiency and reducing GHGs affect vehicle use and distances traveled? The potential exists for EVs to increase vehicle use, lower use of public transit and active travel modes, and lead to more traffic congestion and dispersed land uses. How would increased uptake of EVs impact on transport disadvantage? A related topic deals with determining what impact the electrification of the motor vehicle fleet would have. Would there be sufficient electricity to power a growing number of vehicles?

One country with numerous reasons to increase EV use is China. The Chinese government is providing incentives to encourage EV uptake and, if incentives continue, will make China the leader in EV sales. In 2014, EV sales in China totaled 38,163, but increased dramatically in the first nine months of 2015, with more than 136,733 sold (Young 2015). Bloomberg News (2015) reports that China's goal is to have five million EVs on the road by 2020. This goal is driven by government regulations to increase fuel efficiency and to reduce air pollution. The 2020 standard for fuel efficiency is 5 L/100km, down from 6.9 L/100km in 2015. If China can achieve these EV goals, then it will significantly reduce the country's petroleum use.

Mobility services

A further area for future research is the question of mobility services. The analysis in Chapter 15 by Tiebei Li, Neil Sipe and Jago Dodson suggests that alternatively fueled vehicles, or EVs, will not reduce transport disadvantage because such vehicles will be out of reach for many lower-income residences. Nevertheless, the idea of mobility as a service has the potential to have an impact across all travelers regardless of their income. The first example of a mobility service is Ubigo in Sweden, which was trialed in November 2015. Ubigo integrates a range of travel modes under one application. The vision for the company is to provide urban households with access to reliable and flexible mobility rather than car ownership. Their product is similar to a mobile phone service with customized packages, including components such as text, local phone calls, international phone calls and data. With Ubigo it would involve a customized package of car rentals, public transport, commuter transport, bike rental, taxi or limo hire and videoconferencing. Customized to an individual's travel needs, the mobility service model has the potential to overcome transport disadvantage.

A significant area of research involves determining the benefits and costs of a mobility services approach to transport. How would this model impact on travel behavior and land use? Would it help or hinder the transport disadvantaged? Would mobility-oriented development change how land-use planning is done? Would this make the transport-oriented development approach obsolete?

Recommendations for policy research

While the post-petroleum future is uncertain with respect to cities, there are numerous measures that could be undertaken now that would have a positive impact, regardless of what the future holds in terms of petroleum demand, supply and use. Planners are beginning to anticipate future trends, such as post-petroleum transport, and to consider how this might change their approach to urban land-use planning. One initiative is to improve planning effectiveness by increasing mobility between jobs, services, shops and housing. Another initiative rich in research detail is to focus on improving network planning for public transit (the focus of Chapter 10 by Stone and Mees) to make better use of existing public transport infrastructure and to help reduce transport disadvantage and vulnerability.

Perhaps most significantly, it is clear that researchers can only investigate, present evidence and analysis, and offer recommendations and arguments for public debate. Governments not only make decisions but also fund research. Government at all levels should strive to be more supportive of innovation in transport and urban planning. There are too many recent examples of government attempts to stifle innovation and prevent new disruptive transport technologies from getting established. For instance, while some state and local governments have embraced new approaches to mobility, such as peer-to-peer ride sharing services Uber and Lyft, or micro-transport technologies (e.g. so-called hoverboards), many others are trying to make such services illegal or to regulate them out of existence. Many governments will need to radically rethink approaches to mobility under petroleum constraint. Indeed, governments – not just planners and researchers – need to be thinking much more systematically and creatively about planning for the age after petroleum.

Conclusion

Beyond the terrain of planning and wider policy sits an array of grand questions about the process by which cities pass into their post-petroleum future. This book, by virtue of its primary focus on planning questions, scarcely ventured into large-scale societal or urban transformation arising from the collective journey beyond oil. But such change is in prospect as human society moves beyond oil, whether on the local, metropolitan, national or global scale. Some of the theorizing of these transitions has entered the present text. Contributors such as Dodson (2014) have written more in that vein elsewhere.

There is much more to be understood about the process of transition and reformation of our cities post-oil. How will urban transport and planning institutions, many of which owe their existence to the fossil fueled systems that they oversee, respond to the shift in their importance and concerns as city managers? How will politicians adapt their platforms to an era when the "realities" of cheap petroleum, and the urban industrial complex that oil has sustained, begins to fracture and fragment? What cultural responses will attend the post-petroleum shift? Many more questions on this grand theme might be posed. For reasons of space, we have not been able to dedicate extensive attention to these wider concerns. Their delineation and resolution will require much further scholarly effort by this set of authors, and the wider research community, as we advance our knowledge of planning beyond petroleum.

References

Bloomberg News (2015) "China Seen Laying Down $15 Billion Bet on Electric Vehicles," *Bloomberg Business*, 16 December, accessed 4 February 2016 — www.bloomberg.com/news/articles/2015–12–15/china-seen-laying-15-billion-bet-future-of-autos-is-electric

Chang, O. (2015) "Australia Post Could Soon Be Delivering Packages with Drones," *Business Insider Australia*, 31 October, accessed 5 February 2016 — www.businessinsider.com.au/australia-post-could-soon-be-delivering-packages-with-drones-2015–10

Dodson, J. (2014) "Suburbia Under an Energy Transition: A Socio-Technical Perspective," *Urban Studies* 51(7): 1487–505.

EDTA (2016) "Electric Drive Sales Dashboard," *Electric Drive Transportation Association*, accessed 10 February 2016 — http://electricdrive.org/index.php?ht=d/sp/i/20952/pid/20952

EIA (2012) *Fuel Competition in Power Generation and Elasticities of Substitution*, Energy Information Agency Independent Statistics and Analysis, Washington, DC: US Department of Energy.

Jaffe, A.M. (2015, May 5) "Why the World's Appetite for Oil Will Peak Soon," *Wall Street Journal*, 5 May, accessed 1 February 2016 — www.wsj.com/articles/why-the-worlds-appetite-for-oil-will-peak-soon-1430881507

Platt, N. (2016) "Buses Face an Electric Future," *RoyalAuto* (February), accessed 10 February 2016 — www.racv.com.au/wps/wcm/connect/royalauto/home/motoring/public-transport/buses-face-an-electric-future

Rapier, R. (2015) "The Fallacy of Peak Oil Demand," *Forbes*, Energy (page), 22 December, accessed 2 February 2016 — www.forbes.com/sites/rrapier/2015/12/22/the-fallacy-of-peak-oil-demand/#65200bd719cc

Young, A. (2015) "Chinese Consumers Bought Nearly 300% More Electric Cars This Year Compared to 2014," *International Business Times*, 13 December, accessed 5 February 2016 — www.ibtimes.com/chinese-consumers-bought-nearly-300-more-electric-cars-year-compared-2014-2222591

INDEX

Note: Page numbers in *italics* indicate figures and tables.